T.N. Narasimhan
Earth Sciences Division
X5655

DEEP COAL MINING—WASTE DISPOSAL TECHNOLOGY

DEEP COAL MINING
Waste Disposal Technology

William S. Doyle

NOYES DATA CORPORATION
Park Ridge, New Jersey London, England
1976

Library of Congress Catalog Card Number: 76-2323
ISBN: 0-8155-0619-8
Printed in the United States

Published in the United States of America by
Noyes Data Corporation
Noyes Building, Park Ridge, New Jersey 07656

FOREWORD

This review contains carefully excerpted and collated data from diverse and difficult-to-locate sources. As is the case with its companion volume on the strip mining of coal, the information to be found here—technological, geographic and environmental—is based on studies carried out by research teams, very often under the auspices of various governmental agencies. Private industrial research studies are reflected in the recent patents whose text has been excerpted.

A complete list of the reports cited throughout the book is to be found in the bibliography at the end of the book.

Advanced composition and production methods developed by Noyes Data are employed to bring new durably bound books to the reader in a minimum of time. Special techniques are used to close the gap between "manuscript" and "completed book." Industrial technology is progressing so rapidly that time-honored, conventional typesetting, binding and shipping methods are no longer suitable. Delays in the conventional book publishing cycle have been bypassed to provide the user with an effective and convenient means of reviewing up-to-date information in depth.

The Table of Contents is organized in such a way as to serve as a subject index and provides easy access to the information contained in this book.

CONTENTS AND SUBJECT INDEX

INTRODUCTION

The detrimental environmental effects resulting from the deep mining of coal have long been recognized. Drainage from both active and abandoned underground mines has polluted and continues to pollute rivers and streams. Coal refuse banks are also a source of acid mine drainage and of silt.

In addition these unsightly refuse banks may contribute to air pollution as they sometimes catch fire and burn for long periods of time. With renewed interest in coal as an energy source, interest in the abatement of pollution associated with the underground mining of coal has also increased.

This book, based on nineteen government reports and seven patents, describes methods of preventing and controlling this pollution. The various processes for neutralization of acid mine drainage, disposal of the resultant sludges, the role of iron oxidation and methods of oxidizing iron are described. Treatment of acid mine drainage with soil, by ion exchange and reverse osmosis is detailed.

One chapter describes the combined treatment of acid mine drainage and municipal wastewater. The underground disposal of mine wastes and control of drainage from refuse banks are also discussed. Materials and methods for the sealing of refuse banks and the conversion of refuse to usable products by sealing are included. Reclamation of refuse banks by vegetation, as well as a method for extinguishing and removing burning refuse banks is described.

PREVENTIVE POLLUTION CONTROL TECHNIQUES

The material in this chapter is excerpted from PB 241 402 and EPA-430/9-73-011 (see Bibliography, p 390).

CONVENTIONAL UNDERGROUND MINING

Conventional underground mining methods are considered to be those in which underground entry is made by a drift, slope, or shaft (Figure 1.1a, 1.1b and 1.1c), the rock broken by drilling and blasting or with mining machines, and the broken rock removed from the mine for processing. Underground mining methods are usually discussed by the type of extraction process and the means of support of the roof and walls. Underground openings created by the extraction of ore are called stopes. Stoping methods are classified as those in which stopes are open or naturally supported, those in which stopes are artificially supported, those in which stopes are allowed or encouraged to cave, and combinations of supported and caved stoping methods.

FIGURE 1.1: UNDERGROUND BITUMINOUS COAL MINES

a.

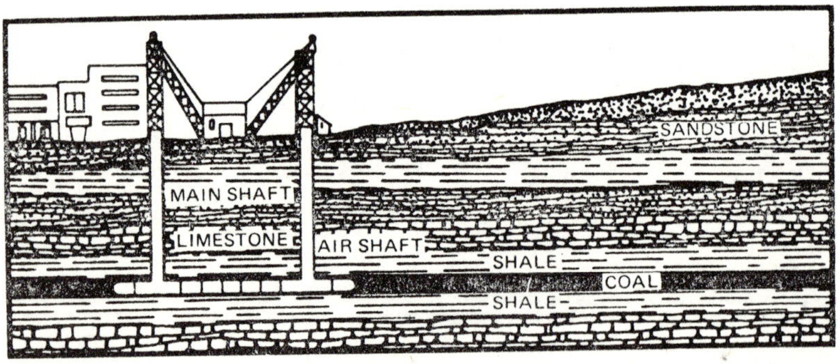

Shaft Mine

(continued)

FIGURE 1.1: (continued)

b.

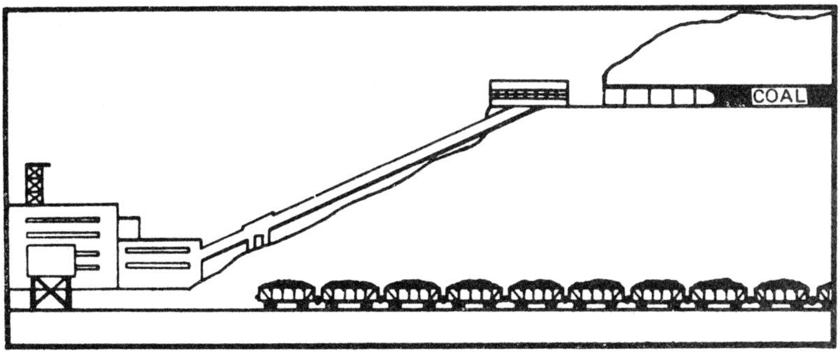

Drift Mine

c.

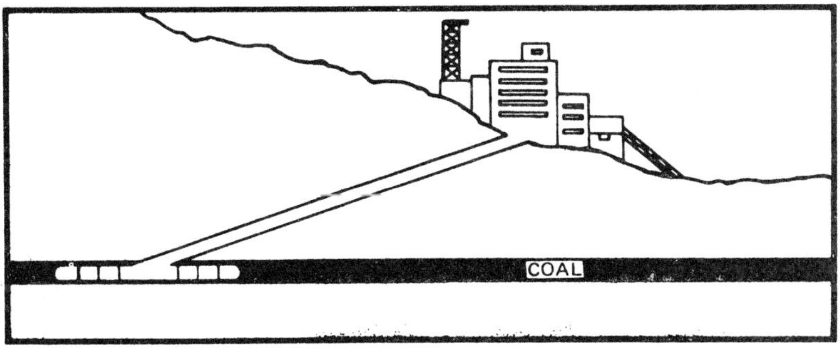

Slope Mine

d.

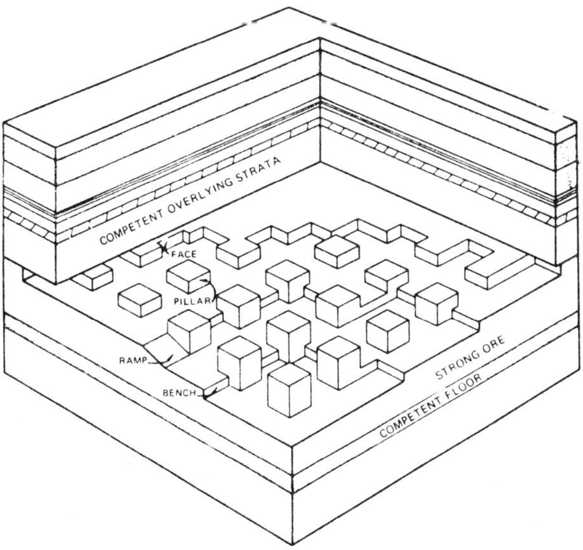

Room and Pillar Mining

Source: PB 241 402

An example of the open or naturally supported stoping method is room and pillar mining of coal (Figure 1.1d). Such methods are used where the openings are sufficiently stable so that little or no artificial support is needed during mining. In coal mining, however, it is usually planned to ultimately remove as many pillars as possible and collapse of the mine roof will then occur after mining is completed.

MINING HYDROLOGY AND GROUNDWATER POLLUTION

Most underground mines will in some way measurably interrupt the exising hydrologic system at the location where they are developed. In the case of an underground mine that reaches below the water table of an unconfined aquifer or intersects a confined aquifer, groundwater will have to be pumped to allow the mine to be worked, and pumping will have to be continued as long as the mine is being operated. During this time, the mine will be a sink, toward which groundwater will flow, and groundwater levels will be lowered in the surrounding area (Figure 1.2).

FIGURE 1.2: CONE OF DEPRESSION FROM MINE DEWATERING

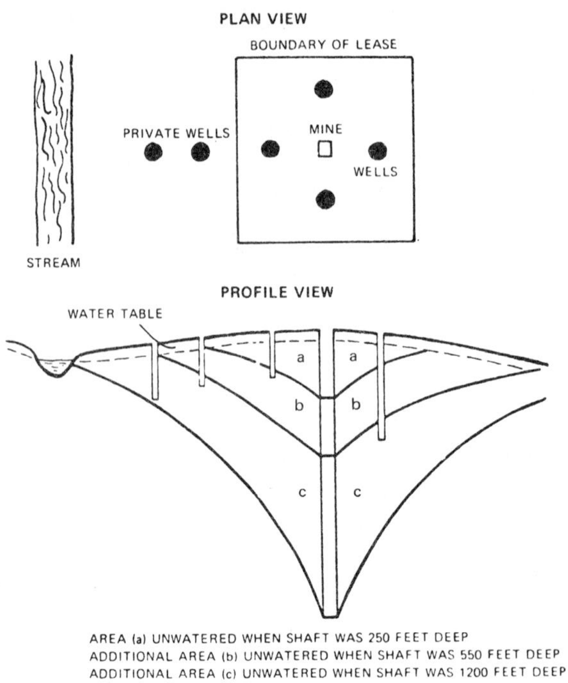

Source: PB 241 402

Dewatering of underground mines may also affect the quality of groundwater. The most serious water quality problem associated with past and present mining in the United States is the formation of acid mine water. Acid mine water is formed when pyrite (FeS_2) and perhaps other sulfide minerals are exposed to the atmosphere as a result of the mining operations.

Although the exact reaction process is still not fully understood, the formation of acid mine water from pyrite is generally illustrated by the equations shown below. The initial reaction that occurs when iron sulfide minerals are exposed to air and water produced ferrous sulfate and sulfuric acid.

$$2FeS_2 + 7O_2 + 2H_2O \longrightarrow 2FeSO_4 + 2H_2SO_4$$

Subsequent oxidation of ferrous sulfate produces ferric sulfate.

$$4FeSO_4 + 2H_2SO_4 + O_2 \longrightarrow 2Fe_2(SO_4)_3 + 2H_2O$$

Depending on physical and chemical conditions, the reaction may then proceed to form ferric hydroxide or basic ferric sulfate.

$$Fe_2(SO_4)_3 + 6H_2O \longrightarrow 2Fe(OH)_3 + 3H_2SO_4$$

and/or

$$Fe_2(SO_4)_3 + 2H_2O \longrightarrow 2Fe(OH)(SO_4) + H_2SO_4$$

Pyrite can also be oxidized by ferric iron as shown below.

$$FeS_2 + 14Fe^{+++} + 8H_2O \longrightarrow 15Fe^{++} + 2SO_4^{-2} + 16H^+$$

Other constituents found in mine drainage are produced by secondary reactions of sulfuric acid with minerals and organic compounds in the mine and along the stream valleys. Such secondary reactions commonly produce concentrations of aluminum, manganese, calcium, and sodium in the drainage waters from coal mining areas.

Many of the underground coal mines in Appalachia are above-drainage drift mines (Figure 1.1b). In many if not all cases, the coal seam and the overlying beds are saturated prior to mining. As a result of mining, these strata are dewatered, either by gravity flow or pumping. After the groundwater table has been lowered, the pyrite that was previously below the water table is exposed to oxygen in the air and oxidation of the pyrite occurs.

The sulfate and iron become soluble as a result of the oxidation and are taken into solution by water percolating through the partially saturated soil and rock above the mine. The acid water then enters the mine workings and flows out by gravity, is pumped out, or percolates through the mine floor and enters the groundwater system (Figures 1.3 and 1.4).

In the condition shown in Figure 1.4b a greater proportion of mine drainage is diverted to the regional groundwater flow system. The acid drainage that directly enters the groundwater system through the mine floor, will, of course, act as a pollutant. Water that flows or is pumped from mines first becomes part of the surface water, but may enter the groundwater system by infiltration at some point as shown in Figure 1.5.

FIGURE 1.3: DRIFT MINING OF COAL

a.

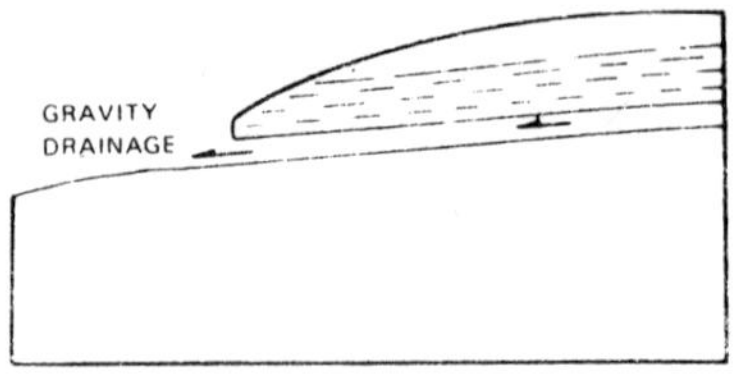

Free Gravity Drainage During Mining

b.

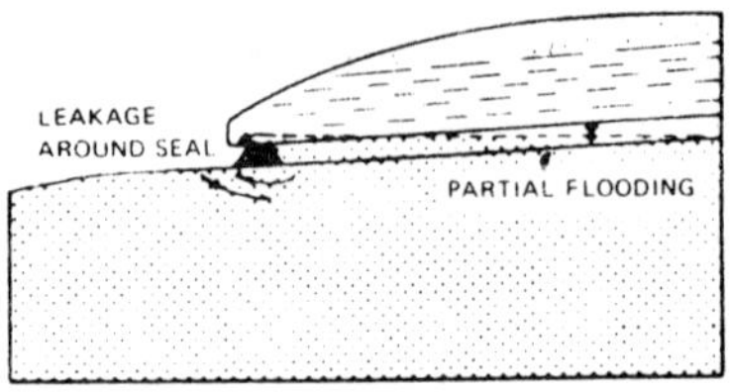

Partial Flooding of Mine After Sealing

c.

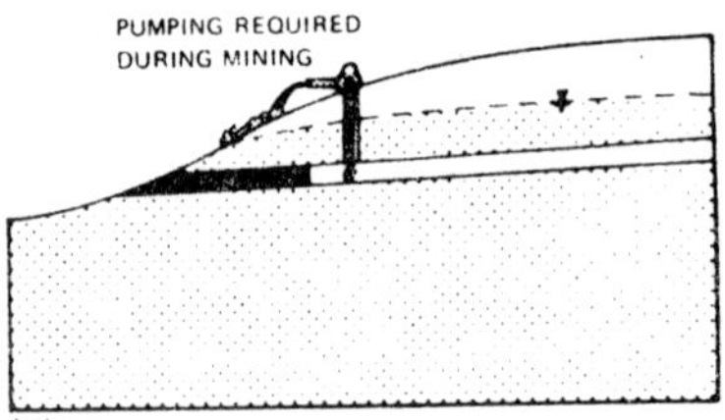

Coal Mined Downdip Requires Pumping During Mining

d.

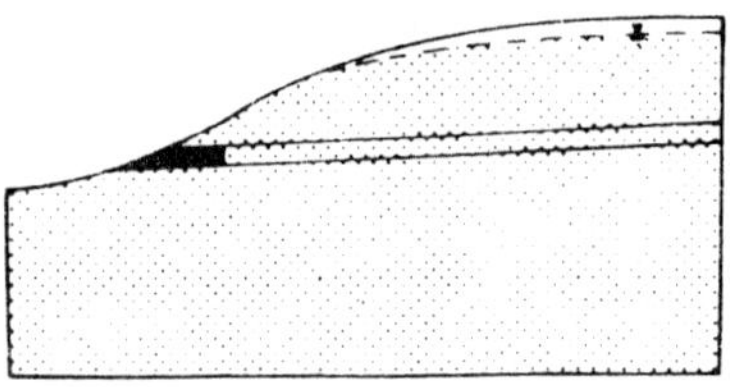

Mine Floods Completely When Abandoned

Source: PB 241 402

FIGURE 1.4: RELATIONSHIP OF UNDERGROUND COAL MINES TO GROUNDWATER FLOW SYSTEMS

a.

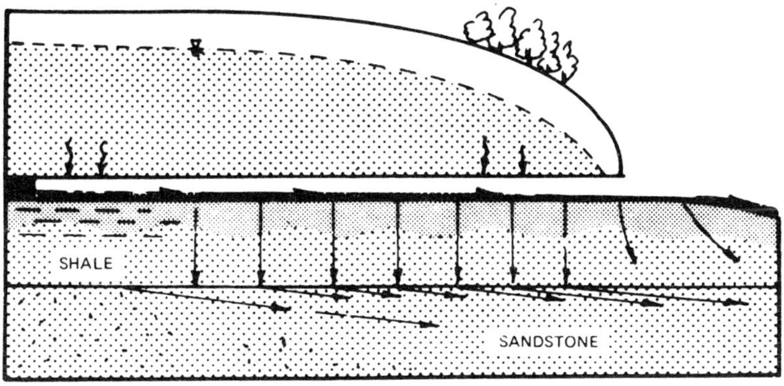

Before Mine Sealing

b.

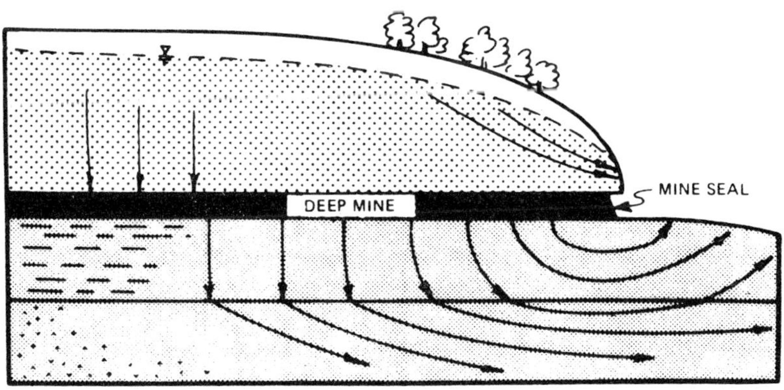

After Sealing and Flooding

Source: PB 241 402

FIGURE 1.5: FLOW OF CONTAMINATED WATER FROM SURFACE SOURCE TO WELL

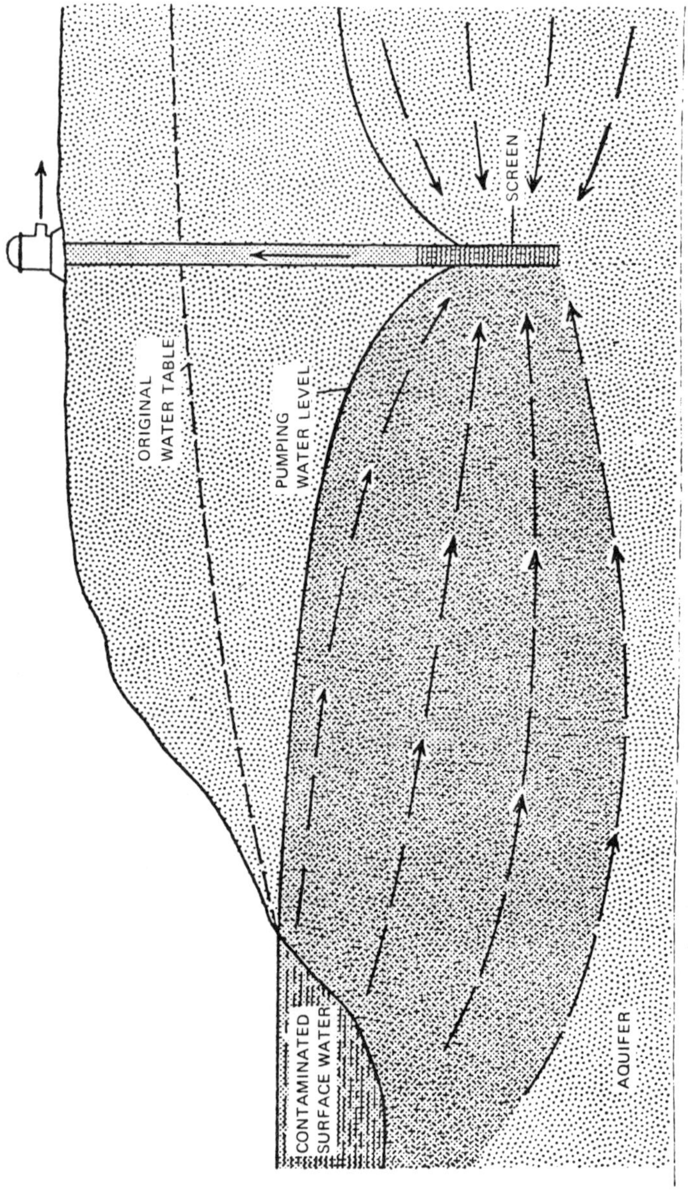

Source: PB 241 402

One method of controlling the production of acid water from abandoned above-drainage drift mines is sealing off the mine in an attempt to flood the mine and thus stop the oxidation of pyrite. Figure 1.4 shows how such sealing may increase the amount of acid water entering the groundwater system.

Acid water is also formed in underground mines that are below the level of the local drainage system. Mines of this type must be continuously pumped during their operation to allow them to be worked. The mine thus becomes the center of a cone of depression in the groundwater table (Figure 1.2). Pyrite in the rocks that are drained during the operation of the mine is thus subject to oxidation and the water being pumped from the mine may be acid.

As long as the mine is being operated and the water discharged at the surface, groundwater pollution would occur principally from infiltration of the pumped water back into the groundwater system. However, when the mine is abandoned, the workings will fill with water, which will take the already oxidized minerals into solution. This polluted water will then enter the flowing groundwater system.

In addition to the types of groundwater pollution mentioned above, the pumping of underground mines may result in the upward flow of saline groundwater in the vicinity of the mine, thus inducing saline water intrusion, as shown in Figure 1.6. The hazard of this happening may be particularly great in some areas of Appalachia where saline water occurs at depths of only 100 to 300 feet.

FIGURE 1.6: CONTAMINATION OF AQUIFER BY SALINE WATER FROM UNDERLYING ROCKS

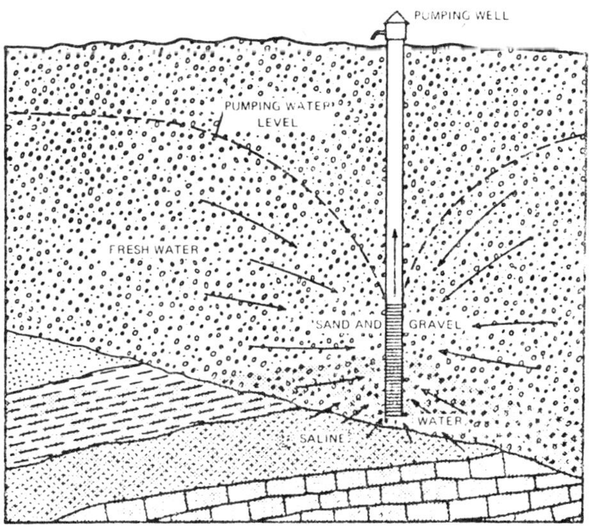

Source: PB 241 402

CONTROLLED MINING PROCEDURES

Many water pollution problems can be avoided by use of water control preplanning for future mining operations. Water producing zones such as faults, fracture zones and aquifers should be identified by photointerpretation, field geology, and core borings. Special provisions can be included in the initial mining plan to avoid these zones. A mine opening can be sited so that there is a good area for deposit of refuse piles, and for location of treatment facilities.

Progression of mining can be planned to allow for backfilling of waste materials. Mine openings can be situated so that there will either be complete inundation or zero discharge on completion of mining. Planned flooding of the mine will indicate the location of the points of highest hydrostatic pressure that will occur. Mineral barriers of sufficient size to withstand this head of water can be allowed to remain in place to permit flooding.

The techniques described below are all control methods that can be incorporated with active underground mines. Most of the techniques are aimed at preventing water pollution after mining is completed. The water pollution generated during mining can be treated to present effluent standards. All of these techniques increase the cost of mining.

Daylighting of underground mines is also presented in this section. It is not a controlled mining method, but it is a means of controlling water pollution from underground mines, and as such is included in this section.

Preplanned Flooding: The principal cause of polluted mine water is the oxidation of sulfide minerals exposed during mining. Flooding of a mine upon completion of mining will greatly reduce further oxidation. The most effective method of achieving flooding is by mining downdip and leaving a barrier at the outcrop, so that flooding will occur naturally (Figure 1.7). Although flooding will control further oxidation of sulfide minerals, it may contribute to groundwater pollution in some cases by dissolving previously oxidized minerals and by increasing the rate of flow into groundwater aquifers (Figure 1.4).

FIGURE 1.7: PREPLANNED FLOODING OF UNDERGROUND DRIFT MINES

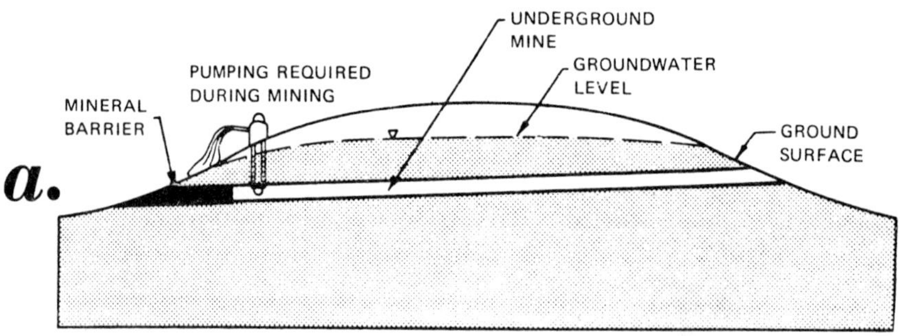

Downdip Mine—During Mining

(continued)

FIGURE 1.7: (continued)

b.

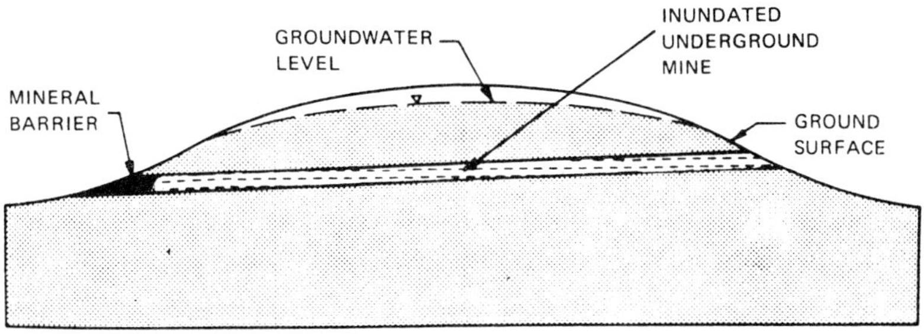

Downdip Mine—After Mining

Source: PB 241 402

Roof Fracture Control: Most of the water entering many underground mines passes vertically through the mine roof from overlying strata. The original source of this water is infiltrating rainfall. Collapse of a mine roof is sometimes responsible for increased vertical flow, particularly in coal mines. Horizontal permeability is characteristically much greater than vertical permeability in the rock units overlying coal mines. These rock units have well developed joint systems which cause vertical flow, except for intercalated beds of shale and clay that inhibit it.

Roof collapse causes widespread fracturing in the strata around a mine roof, and subsequent joint separation far above the roof. These opened joints can tap overlying perched aquifers and provide flow paths to the mine. Roof collapse in shallow mines will often cause surface subsidence. Subsidence fissures collect and then funnel surface runoff directly to the mine. Roof collapse is directly responsible for a large portion of the drainage from many underground mines. This source of water can be substantially reduced by using mining procedures that ameliorate the severity of roof collapse.

Fracturing of overlying strata can be reduced by employing one or a combination of the following: pillars, roof support, limiting the width of openings in which caving will occur, and backfilling of voids with materials.

Pillar mining is accomplished by partial extraction of the mineral resource, leaving the remaining mineral to support the overburden. Also, the geometry or shape of an opening can increase stability of a mine roof. Circular voids reduce stress concentrations that occur at corners of rectangular voids where shear failures usually develop. Timbers and roof bolts add additional support. By limiting the width of openings, the vertical extent of roof rock fracturing can be controlled. This should reduce the vertical extent of joint opening, and therefore reduce the vertical extent of aquifer interception.

Backfilling consists of filling mine voids with waste rock and other materials to aid in supporting overlying strata. It is a common procedure in countries with more limited coal resources. Mine voids can be filled by solid stowing or by hydraulic stowing. Hydraulic backfilling has been used in areas with inadequate pillar support in the anthracite region of eastern Pennsylvania. Backfilling is discussed further under Underground Disposal of Coal Mine Wastes.

Controlled Mineral Extraction: The central theme of this method is to shift emphasis from mining in areas which have a high probability for causing pollution, to areas where pollution is unlikely to occur. Again, it should be understood that pollution potential usually varies greatly, even within small areas. Use of this method assumes that advances in technology will provide for future removal of minerals from high pollution potential areas, with less environmental harm than would occur today. Effective use of this method would require water quality surveys, core boring analyses, and/or review of existing data to define the high pollution potential areas.

Controlled Atmosphere Mining: This technique is similar to mine inundation in that free air oxygen is eliminated from an underground mine. Pollution production is reduced through the reduction of oxygen. This is a very complex mining method that involves replacement of normal mine atmosphere with an oxygen-free, noncombustible gas. The mine workers must wear complex life support and communications systems. This system is reported to be feasible.

Water pollution control continues only as long as the oxygen-free atmosphere is maintained. The mine will revert to normal pollution production conditions when mining is completed, unless the oxygen-free atmosphere is maintained. This is unlikely. More conventional means, such as sealing, will probably be utilized to control pollution after abandonment. The beneficial effects occur only during mining activities. Therefore, the criterion for use of this technique for purposes of water pollution control is economic, based on the cost differential between using this technique or creating the mine water discharge.

Daylighting: Daylighting is performed by completely stripping out underground mines. A feasibility study conducted for the EPA indicates that it is feasible. Daylighting is carried out similar to strip mining, and all surface mining pollution control techniques apply. This technique substitutes a regraded strip mine for an underground mine. Care must be exercised to ensure that the strip mine does not create more pollution than the old underground mine.

There are two general prerequisites necessary to make this technique feasible. There should be sufficient marketable mineral to offset some of the cost of overburden removal, and the underground mine should be a documented pollution source.

WATER INFILTRATION CONTROL

These techniques are designed to reduce the amount of water entering underground mines, and subsequently reducing the amount of drainage exiting the mine. These techniques can often be advantageously employed by active miners to decrease the volume of water that needs to be handled and treated.

Use of these techniques for abandoned mine water pollution control is based on the premise that a decrease in the amount of flow exiting the mine will result in a decrease in the total pollution load. The pollution load is the actual weight of specific pollutants passing a point within a specified time period. The load is calculated by multiplying pollution concentration in the the water by the amount of flow, using appropriate conversion factors. Loads are commonly expressed in the unit kilograms per day (pounds per day).

In order for this technique to be useful in pollution control, the resultant decrease in flow must not be accompanied by a proportionate increase in pollution concentration. If such a trade-off should occur, the pollution load could remain essentially the same. This trade-off is not an entirely unlikely possibility.

Coal mine drainage pollutants result from the oxidation of pyrite. Oxygen and water are required for this oxidation reaction in a nonflooded mine. The relative humidity in an underground coal mine is usually at or near saturation (100% relative humidity). Mine walls are normally damp. Water required for the pollution forming reaction is almost always available. Flushing of the oxidation sites is not even required. Salts resulting from oxidation are hygroscopic and will weep downward exposing the reaction sites to further oxidation.

The point of this discussion is that the availability of oxygen is the oxidation rate controlling factor, and the amount of water flowing through the mine does not control the oxidation rate. Pollution production may be constant within the mine regardless of the flow of water through the mine. Decreasing flow may result in increased pollution concentrations.

Therefore it is possible that decreasing mine drainage could have little or no effect in controlling pollution. Decreased flow may result in decreased water pollution if the amount of drainage is reduced sufficiently to prevent pollution transportation from the mine. In this case, decreases in water pollution coming out to the surface could also result in increases in groundwater pollution.

The techniques discussed in this section can be used to decrease the amount of water flowing through underground mines. Choice of technique and extent of its use will depend on hydrologic conditions in the area and cost effectiveness of the technique. These techniques are not universally applicable. Some mines are already receiving minimal infiltration, and further decreases may be difficult to obtain.

Infiltration generally occurs as a result of rainfall recharge to the ground water reservoir. Water can enter from below, or laterally through the mineral or adjacent rock units. Rock fracture zones and faults have strong influence on groundwater flow patterns. They often collect and convey large quantities of water. Infiltration can usually be reduced by avoiding these zones during mining.

Increasing Surface Runoff: Water infiltration can be decreased by increasing surface water runoff. This technique involves elimination of depressions and grading the surface to increase water velocities. Subsidence depressions often collect and convey large quantities of surface water to underground mines. The amount of water collected depends on size of drainage area tributary to the depression, and annual rainfall and runoff rate. Subsidence holes in stream channels can cause entire streams to enter underlying underground mines. Uneven surfaces caused by agricultural, logging or other surface activities can cause increased infiltration.

Surface runoff can be increased by grading an overlying area to a smoother, better draining configuration. Surface depressions can be filled in and leveled with clay. Stream channels can be flumed, reconstructed with impermeable liners, or diverted around water loss areas. Channel stability under increased flows must be assured.

As with roof fracture control, the overall effect of increased surface runoff should be appraised to insure that the objective of decreased pollution loads is achieved and that groundwater pollution is not accelerated.

Regrading Surface Mines: Surface mines are often responsible for collecting and conveying large quantities of surface water to adjacent or underlying underground mines. Nonregraded surface mines often collect water in an open pit where no surface exit point is available. Many abandoned underground mine outcrop areas have been contour stripped. These surface mines often intercepted underground mine workings, providing a direct hydrologic connection. The surface mine does not have to intercept underground mine workings in order to increase infiltration.

Surface mines on the updip side of underground mines collect water and allow it to enter a permeable coal seam. It then flows along the coal seam to underground mines. Overlying surface mines that collect and entrap water can also be significant sources of infiltration. These surface mines facilitate entry of surface runoff to the groundwater system, which eventually works its way into an underground mine.

Sealing Boreholes and Fracture Zones: Boreholes and fracture zones act as water conduits to underground mines. They are usually vertical, or near vertical, and tap overlying aquifers. They collect and transport groundwater.

Boreholes are commonly present around underground mines and usually remain from earlier mineral exploration efforts. These boreholes can be located and plugged to prevent passage of water. Concrete can be inserted hydraulically to form a seal. Boreholes can be easily sealed from below in an active underground mine. Difficulty can be encountered if sealing has to be performed from the surface. Abandoned holes are often difficult to locate on the surface, and many times they will be blocked by debris.

Fracture zones are often major conduits of water. They increase vertical movement of water and can cause large lateral movements. Fracture zones are usually vertically oriented planar type features. Their location can be plotted by experienced personnel using aerial photography. Permeability of these zones can be reduced by drilling and grouting. Holes are drilled into zone and grout is inserted hydraulically. Care must be taken to ensure that the boreholes are located in the fracture zone at the point of grouting. There are various types of grout available; however, concrete is commonly used.

Interception of Aquifers: This technique takes advantage of the natural geologic and hydrologic systems surrounding a mined area. It involves use of boreholes, casing, and pumps to transfer water from one point to another in order to reduce water flow into an underground mine. The techniques are theoretical and will require development and demonstration to establish feasibility.

A complete hydrogeologic site evaluation of a mined area to determine aquifer characteristics and water flow systems is required prior to implementation. Most underground mines receive water from overlying aquifers. Several techniques can be employed to tap these aquifers and reduce the amount of water entering a mine. Overlying aquifers can be drilled and the water pumped to the surface (Figure 1.8). Boreholes can also be drilled through aquifers, passing through an underground mine and into underlying aquifers (Figure 1.9a). The boreholes must be cased through the mined zone (it collects water from the overlying aquifer, passes it through the mine zone for discharge to an underlying aquifer). The underlying aquifer must be capable of accepting the anticipated flow.

A variation on this technique is to drill holes into the underground mine, casing and grouting the borehole through the zone forming the roof. The boreholes are then connected by pipes and the water carried outside the mine. The uncased or perforated-casing portion of the borehole collects water from overlying aquifers and passes it into the piping system for conveyance out of the mine, never contacting pollution-forming materials (Figure 1.9b).

Boreholes, pumps and piping systems can also be used to convey acid mine drainage to a nearby alkaline aquifer, or alkaline underground mine, to encourage mixing, neutralization and settling of precipitates. Mine sealing, another preventive pollution control technique, will be discussed in the following chapter.

FIGURE 1.8: INTERCEPTION OF MINE WATER BY PUMPING OF OVERLYING AQUIFERS

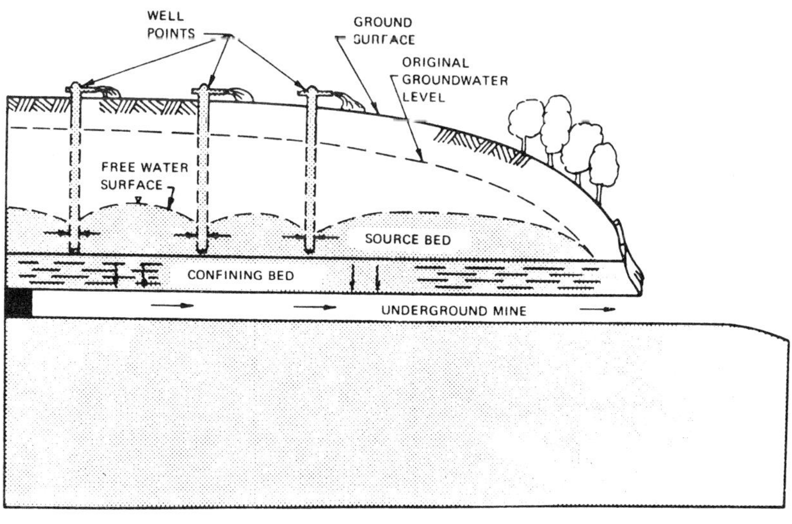

Source: PB 241 402

FIGURE 1.9: INTERCEPTION OF MINE WATER BY GRAVITY WELLS

a.

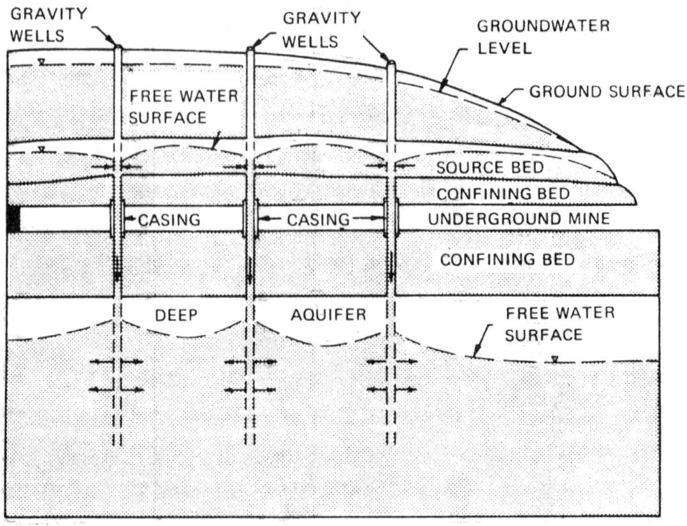

b.

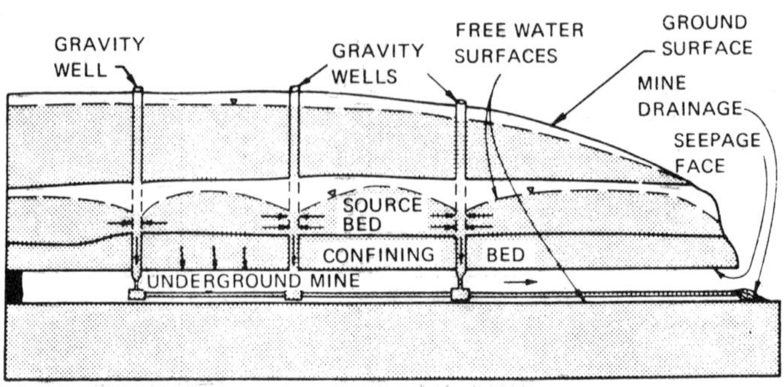

INTERCEPTION OF AQUIFERS

Source: PB 241 402

MINE SEALING

The material in this chapter is excerpted from EPA-430/9-73-011, EPA-670/2-73-081 and EPA-670/2-73-092 (see Bibliography, p 390).

Mine sealing is usually employed to promote inundation of underground mine workings to reduce oxidation of pyritic materials. Seals have also been used to prevent the entrance of air or water to the underground mine.

Mine sealing for purposes of inundation involves construction of a physical barrier in a mine opening to prevent passage of water. A barrier must be designed to withstand the maximum expected pressure (head) of water that will be exerted against it. Sealing underground mines is somewhat analogous to creating a surface water impoundment: a major portion of the dam structure would already be in place and the seal merely closes the opening.

Engineering considerations are also similar to these for surface impoundment design. The entire dam structure must be capable of withstanding exerted pressure, and leakage rates must be determined. Underground mine seals have seldom been successful due to lack of consideration of leakage rates and weak points. Seals can be designed to withstand a large amount of pressure, but the seal is only a small part of the impoundment structure. The perimeter of the mine forms most of the impoundment, and often it is not capable of withstanding any significant amount of pressure. The first step in mine sealing is to obtain and analyze all pertinent available site data including geology, mine maps, locations of sink holes, hydrologic data, rock hydraulic characteristics, borehole logs, location of strip mines, mineral structure contours and aerial photogrammetric mapping.

Sealing feasibility and practical limits of inundation are then determined. Limits of the expected mine pool are plotted on a mine map. Areas where water pressure will be exerted are then delineated, including the expected pressure to be exerted for each area. Hydraulic evaluation of all pressure areas is required to determine if existing barriers are capable of withstanding the applied pressure. Areas deemed incapable of withstanding the anticipated applied pressure must be evaluated further. This evaluation is to determine whether additional measures, such as grouting, would be successful in rendering the areas capable of withstanding expected water pressure. If the required work is technologically or economically not practical, the desired mine pool level will have to be lowered, or the sealing program abandoned.

As previously stated, mine seals can be designed to hold reasonable heads of water. Mine sealing problems generally occur from natural weak spots such as the outcrop, fractures and subsidence. The mineral and natural rock systems around underground mined areas usually become more permeable due to the nature of the rock and the disturbance caused by mining.

The outcrop area normally is the weakest link in an underground impoundment. The mineral outcrop is generally of nonuniform thickness. Mining approached very close to the outcrop in some areas, resulting in very little material remaining to withstand any water pressure. Surface mined crop areas are seldom capable of withstanding significant water pressure. Mine roof collapse in a flooded zone will provide highly permeable zones, allowing water to escape, thus preventing extensive flooding. Physical failure of outcrop areas will sometimes occur, but more often the increased pressure results in seepage through the permeable zones, preventing significant amounts of water level increase. Water can also be lost through the mine floor.

The second step in mine sealing is to determine the ability of the natural system to withstand water pressure. The ability of a natural barrier to withstand pressure can be increased by use of grouting and sealing subsidence holes. The practicality of implementing these procedures will have to be evaluated.

Unfortunately, precise data is seldom available for evaluation of mine conditions. The engineer designing the system may have inadequate knowledge of how the natural system will react. The hydrologic characteristics of mined areas are highly complex and variable from mine to mine. The natural response of the area to changes in hydrostatic pressure is even more complex, and is difficult to approximate without voluminous amounts of data and computer analysis. Mine sealing decisions and design are therefore judgmental projects that should involve personnel that are expert in mine and groundwater hydraulics.

A mine seal can be constructed in many ways, using many different types of material. Any material capable of withstanding water pressure has theoretical application. A mine seal must have internal strength capable of withstanding the water pressure, and it must be tied into the floor, roof, and sides of a mine opening. Sufficient internal strength is easily obtained. Physically anchoring the seals to the mine opening is much more difficult. Many mine seals leak around their edges due to poor anchoring.

The natural rock and mineral surrounding the seal area is usually fractured, fissured, uneven and unstable. Leakage occurs through this permeable zone because of the inability of most seals to penetrate surrounding materials. Most mine seals are incapable of providing effective sealing between the top of the seal and the roof rock. Extensive grouting directly around the seal area can help to tie the seal into the surrounding rock and reduce perimeter leakage. Grouting should occur very close (within 1 meter) to the sides of the seal and directly into the overlying roof to insure effective grout penetration. Curtain grouting extending outward along the outcrop from the seal is usually employed with most mine seals where appreciable heads are expected. This serves to decrease permeability and reduce the amount of leakage that is bypassing the seal. Seal failure has also occurred due to its being constructed in moving water. Water flow should be stopped prior to placement of a seal, especially when using liquid or alkaline sealants such as grout, concrete or gel. Flow can be stopped by pumping an existing mine pool (if one is present) or by construction of an impounding dike and piping system behind the seal area.

Mine sealing can be a very dangerous operation. The ultimate water level behind the seal is seldom controlled, and excessive water pressures can build up, resulting in a mine seal or outcrop failure. Sudden release of large quantities of water can have devastating downstream effects. Subsequent flooding can and has caused loss of life and massive property damages. Release of large quantities of polluted water can also result in far-reaching downstream fish kills.

Excessive increase in water pressure can be prevented through use of boreholes drilled into a mine pool area from above. Boreholes are drilled from a surface elevation equal to the maximum permissible mine pool elevation. Boreholes should be cased and protected to insure that they remain open. When the mine pool reaches the maximum elevation, the boreholes begin to flow and thus prevent further water level increase. Boreholes must be of sufficient size and distribution to be capable of transmitting the greatest expected inflow to the mine pool without allowing further water level rises. This is a natural, gravity operating system that does not require supervision.

A mine pool drawdown system should also be installed during mine sealing. This usually consists of a pipe constructed through the mine seal which has a manually operated valve. When the valve is opened, water can flow through the seal, lowering the pool elevation to its original presealing level. This is mainly a safety device, but it could be used to drain a mine pool if future mineral extraction should be desired in a flooded area without destroying the seal.

Special legal considerations are involved in mine sealing. Adjacent mineral extraction is often difficult or impossible after sealing. Mineral rights owners may have valid damage claims. The extent of damage is often difficult to establish and could lead to prolonged legal disputes.

Many of the sealing techniques described in the sections following have not had notable success in mine water control. In most cases the lack of success is not due to failure of the seal itself. The lack of success is more often attributable to the manner of placement and lack of proper consideration for the natural system.

SEALING TECHNIQUES

Double Bulkhead Seals

The technique involves placement of two retaining bulkheads in a mine opening followed by placement of a seal in the space between the bulkheads (Figure 2.1). Bulkheads can be placed from a mine portal, if it is open and accessible, or through vertical boreholes from above. Grout or concrete is then placed between the bulkheads via pipes through the front bulkhead, if accessible, or from vertical boreholes. Two types of double bulkhead mine seals have been successfully demonstrated. In inaccessible mine entryways a grouted seal has been used, and for accessible mines quick setting concrete seals have proven effective.

FIGURE 2.1: DOUBLE BULKHEAD SEAL

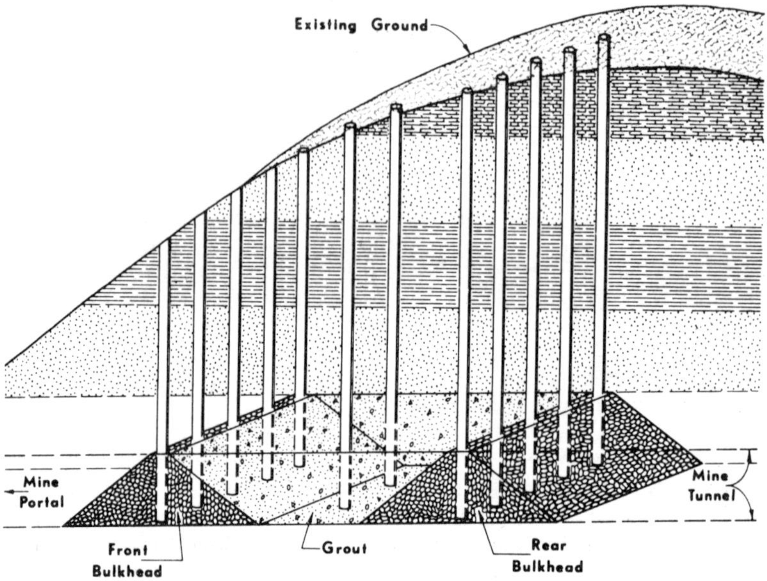

Source: EPA-430/9-73-011

Grouted double bulkhead seals have been recently constructed at Moraine State Park, Pennsylvania, under a state reclamation program. This method utilized dry, coarse aggregate for front and rear bulkheads placed through drill holes. The bulkheads were then grouted to form solid front and rear seals. Water was pumped out of the center cavity between the two bulkheads by newly placed drill holes. Concrete was poured into the space between the two bulkheads. These same mine seals have also been successfully installed without grouting the retaining bulkheads.

Use of double bulkhead seals for accessible mine entries has been attempted only a few times, primarily by the Halliburton Company under contract to the EPA. A quick-setting slurry consisting of water, cement, bentonite and sodium silicate was used to construct the two bulkheads. The void between the bulkheads was filled with a special light concrete composed of Portland cement, fly ash, bentonite and water, pumped through a grout pipe. In another case, this void was filled with pneumatically pumped limestone aggregate, which was then grouted with light concrete.

Double bulkhead seals have been successfully demonstrated and appear capable of withstanding relatively large amounts of water pressure. The maximum pressure exerted has been limited to 10.7 meters (35 feet) of head. However, these seals should be capable of greater pressures as installation procedures improve.

Grout curtains are required for total effectiveness. Seal leakages generally occur through the bottom and around the sides of a seal. It is difficult to get a good seal at the mine roof because of slumping. The perimeter of a seal should be well grouted. Special grouting procedures are explained under "Grout Curtains." Double bulkhead seals without curtain grouting range in cost from $7,500 to $15,000 per seal. These seals have cost as much as $50,000 in certain instances. Quick-setting double bulkhead seals in accessible entry mines average $9,500 per seal without curtain grouting.

Gunite Seals

This technique involves use of gunite, a pneumatically-placed low slump concrete, to rapidly and effectively seal mine openings. Gunite is projected by an air jet directly into place and, by proper adjustment of the mix and nozzle, will stand vertically eliminating the need for forms. An area is selected for the seal within sound or reasonably sound zones and the roof, sides and floor are shaped so that the space will be a form for a tapered gunite "plug" and to provide clean surfaces against which to construct the seal.

A wood bulkhead is constructed at the inner limit of the seal to support the initial placement of gunite. The seal is then constructed by placing successive thin layers of gunite (Figure 2.2). Care must be taken to remove "rebound" from the floor as construction progresses to avoid a permeable zone in that area.

A seal constructed by this process completely fills the opening in which it is placed and should provide an effective seal, particularly if an expansive type cement is used. Since the seal is pneumatically placed against the perimeter of the opening (which has been first shaped and cleaned) and since the shape of the seal is such that it becomes tighter as the pressure against it increases without relying on flexural characteristics, it should be particularly effective in sealing against higher hydraulic heads.

This technique shows promise of being an excellent underground mine seal. The seal forms a good bond with the mine opening. However, grout curtains are still deemed necessary.

Costs will generally average $260 per cubic meter ($200 per cubic yard) of gunite installation. Excavation, cleaning, and shoring in the mine opening will entail additional expense. A complete gunite seal in a standard mine opening is estimated to cost $13,000.

Single Bulkhead Seals

Single bulkhead mine seals are generally composed of a grouted aggregate bulkhead. They can be constructed of other materials, such as masonry block. The seals are placed remotely by using boreholes from above (Figure 2.3), or constructed directly in the opening, if accessible (Figure 2.4). Exploratory and observation boreholes are used for remote installation to determine the size and condition of the opening, and to locate the best sealing location.

FIGURE 2.2: TYPICAL GUNITE SEAL

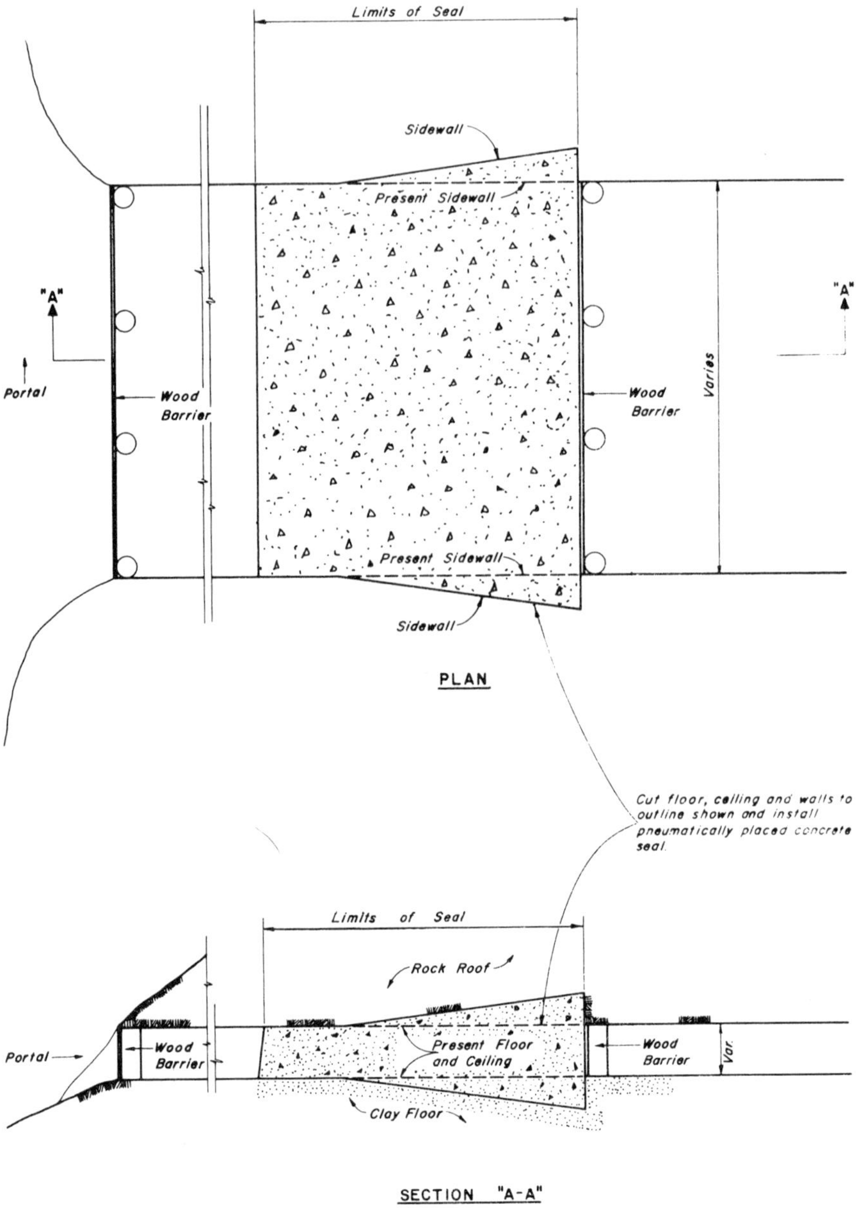

Source: EPA-430/9-73-011

FIGURE 2.3: INACESSIBLE ENTRY SINGLE BULKHEAD SEAL

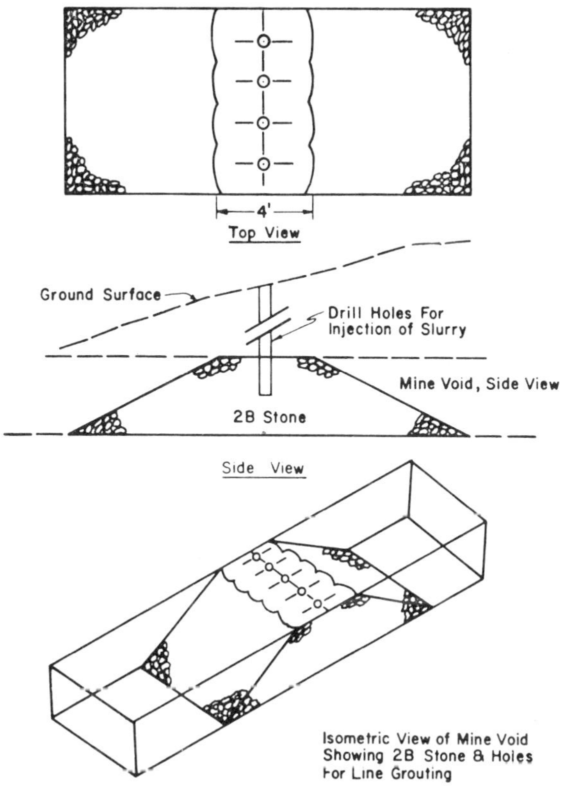

Source: EPA-430/9-73-011

An aggregate bulkhead is placed either through vertical boreholes or directly from the mine opening. The aggregate is then grouted, using a quick-setting slurry composed of water, cement, bentonite and sodium silicate.

The slurry is introduced either through pipes inserted into the mine opening or vertically through boreholes from above. The aggregate usually slumps away from the mine roof during grouting, and additional grout must be added. Curtain grouting is usually needed to tie the seal to the surrounding rock and mineral.

Single bulkhead seals utilizing a concrete block wall have also been used. These walls are highly susceptible to damage and could not be used where high water pressures are expected. The grouted aggregate single bulkhead seal has been successfully demonstrated. Some of these seals were unsuccessful, but this was possibly due to poor anchoring with the existing rock and mineral or incomplete grouting of the aggregate. These seals should be used where the expected water pressure will be low. The double bulkhead seal is better for high pressure.

FIGURE 2.4: ACCESSIBLE ENTRY SINGLE BULKHEAD SEAL

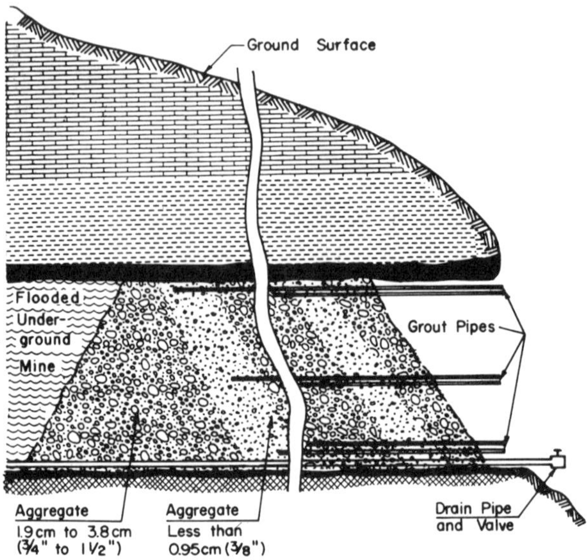

Source: EPA-430/9-73-011

The conrete block seal has only limited usefulness, under low head conditions. Long term effectiveness of a concrete block wall is questionable. A concrete block wall seal can be strengthened and protected from weathering if earth is backfilled against it.

Accessible entry aggregate seals cost approximately $3,500 each. Remotely placed aggregate seals cost about $2,100 each. Concrete block wall seals cost in the range of $1,400 to $6,000 each.

Grout Curtains

Grout curtains are used in conjunction with other types of mine sealing to reduce permeability around a mine seal and other seepage areas. Grouting is only applicable where void areas are small.

Grout is commonly used around, and extending away from, mine seals. It tends to fill voids between a mine seal and the mine entryway, and to decrease permeability in adjacent rock. This will reduce seepage bypassing a seal area. The grout is generally pressure injected from boreholes placed on 3 meter centers. Pressure forces the liquid grout from the borehole into permeable zones of the rock units. The grout sets or solidifies in small voids and greatly decreases permeability. Effectiveness of grouting is difficult to determine during the operation. There is no way of knowing where it is going and where it is collecting. The three meter center spacing is somewhat arbitrary: the spacing should be sufficient to ensure that the entire space between holes receives grout. A grout curtain in conjunction with a double bulkhead seal is shown in Figure 2.5.

FIGURE 2.5: GROUT CURTAIN WITH DOUBLE BULKHEAD SEAL

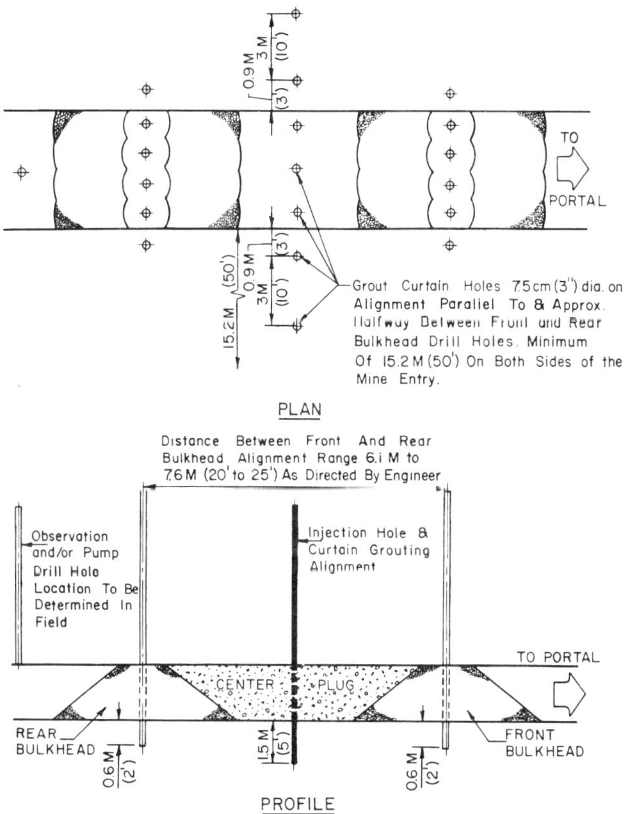

Source: EPA-430/9-73-011

Grout curtains can also be placed in areas of permeable or weak outcrops during mine sealing. This serves to decrease leakage rates and strengthens the outcrop to decrease failure possibilities. Grout curtains have been successfully utilized many times. The success of grout curtains in leak areas has not been documented by before-and-after flow recording, but it is said to be effective. Efficiency of grout curtains is dependent on the manner of injection. Effectiveness may be increased by use of packers to grout individual zones, instead of attempting to grout the entire hole at one time. Grout effectiveness can be further enhanced by varying the viscosity at the mixer. The viscosity is increased in permeable zones and areas with larger voids. The machine operator can tell when viscosity changes are required by observing flow rates. Grouting is more effective if aggregate is placed in large voids encountered during drilling.

Grouting will also be more effective if the first grout holes are placed within 1 meter (3 feet) of the seal. Grout holes should be drilled into the seal from above. Contact of a mine seal with the roof and sides of a mine opening is a weak spot where leakage commonly occurs. Extensive grouting in these areas will improve effectiveness of a seal.

If there is a possibility of a leak in the bottom of a mine seal, the holes drilled into the top of the seal should continue on through for grouting the bottom of the seal. The bottom of a concrete seal is likely to leak if it is placed in moving acid water. Flowing acid water can dissolve cement, leaving behind porous sand and aggregate. Remotely placed double bulkhead seals may leak at the bottom.

Slumping of the front and rear retaining piles of aggregate may introduce aggregate into the bottom of the center void. The concrete center plug material may be unable to flow into this aggregate, leaving a permeable zone.

Grouting curtains generally cost in the range of $80 to $260 per meter ($25 to $80 per linear foot) of curtain.

Clay Seals

Clay can be used to form a hydraulic underground mine seal where low water pressure is expected. The mine opening is first cleaned of debris and loose rock. Clay is then compacted into the opening to form a seal. A good quality plastic clay should be used to ensure impermeability, and to enable the clay to flow into cracks and voids along the walls and roof of the seal area.

Hand placement and compaction would probably yield a better anchor between the plug and the seal area. Earth should be backfilled and compacted around the mine opening and over the seal to hold it in place and prevent the clay from flowing under pressure. A cross section of an underground mine sealed with clay in conjunction with backfilling of a surface mine is shown in Figure 2.6.

This should prove to be an effective and inexpensive mine sealing technique for low water pressure installations. The mine opening has to be accessible in order to use a clay plug.

Effectiveness of the seal will depend on quality of the clay, manner of placement, and physical condition of the seal area. Clay seals may be capable of withstanding 10 meters (approximately 30 feet) of head of water under ideal conditions.

Costs will depend on availability of suitable clay, its acquisition, and transportation costs. Costs will also vary in response to difficulty in preparing the seal area, and the installation. Costs will generally range from $1,200 to $4,000 per seal.

FIGURE 2.6: CLAY SEAL

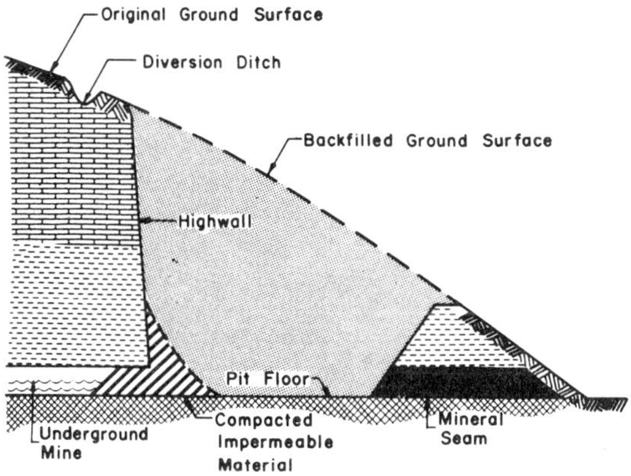

Source: EPA-430/9-73-011

Permeable Aggregate Seal

Permeable aggregate underground mine sealing involves use of ungrouted alkaline aggregate material that will neutralize acid water passing through it. This causes formation of precipitates, which progressively clog the pores in the aggregate. Theoretically, the precipitate continues to form and clog all of the pores in the aggregate, until the permeable aggregate seal actually becomes a solid, single bulkhead seal of aggregate and precipitate material.

An example of this technique is use of limestone aggregate to seal underground coal mines that discharge acid mine drainage (Figure 2.7). Here the acid water is neutralized during its flow through a limestone aggregate plug, causing iron hydroxide and calcium sulfate to precipitate, filling the pores of the aggregate and producing a solid bulkhead type seal. Limestone aggregate seals were demonstrated by the EPA in West Virginia.

The seals helped attain various degrees of mine inundation. A 1.8 meter (6 feet) head of water was reported behind one seal. The seals continued to leak, meaning that precipitates have not completely clogged the pores, or the precipitates are unable to withstand the water pressure.

Increases in pH and alkalinity, and decreases in acidity of the discharge, showed the neutralizing ability of the seal. The neutralization is only temporary and is expected to decrease as the limestone aggregate becomes coated with precipitate. The long term value of this type of seal would be due to its capabilities to withstand water pressure and cause inundation. Its long term capabilities in this respect have not yet been demonstrated.

Slumping of the aggregate, causing an opening at the mine roof, has been a problem. Grouting of the opening may help solve this problem. An average cost estimate of $7,500 per seal has been suggested.

FIGURE 2.7: PERMEABLE LIMESTONE AGGREGATE SEAL

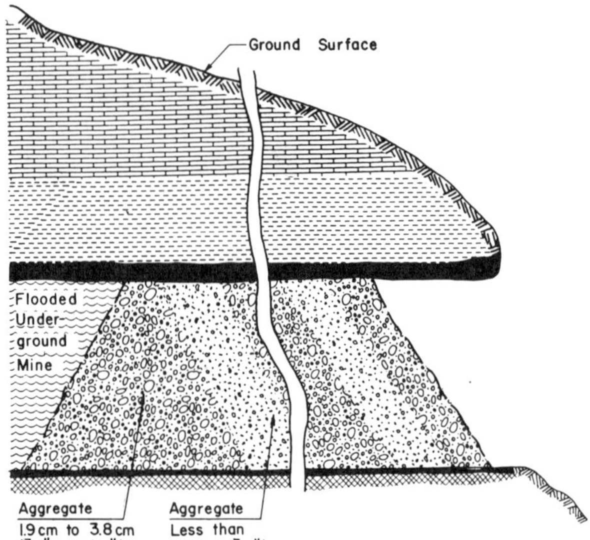

Source: EPA-430/9-73-011

Grout Bag Seals

This method of underground mine sealing involves placement of expendable grout containers to seal accessible mine openings. Seals are constructed by a vertical series of cement grout slurry filled nylon or cotton bags, which decrease in length from bottom to top of the entryway.

The container placed on the floor of the entryway is 6.1 meters long, with other dimensions matching those of the mine itself. The container is positioned in the mine and filled with slurry, which causes it to conform to the shape of the mine opening.

As each container hardens, a new, shorter one is placed above it and filled with slurry. This process is repeated until the top container, measuring about 3.1 meters in length, has been positioned and filled. See Figure 2.8.

A grout bag seal was unsuccessfully demonstrated near Coalton, West Virginia. Leakage occurred around the contact between the bags and the mine entryway. Leakage increased as erosion widened the gaps.

The grout bags will not conform well to an uneven surface and will not penetrate the many cracks and fissures. Efficiency may be increased by means of concrete grouting around the seal. Other mine sealing techniques such as gunite or double bulkhead seals appear to be more efficient. Costs for one experimental installation were $3,300 per seal.

FIGURE 2.8: EXPENDABLE GROUT RETAINER MINE SEAL

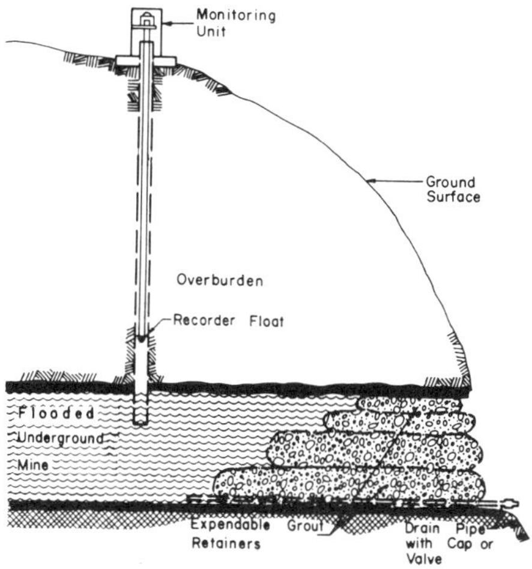

Source: EPA-430/9-73-011

Regulated Flow Seals

This technique is designed for use with treatment plants in order to maintain flow at a constant level from underground mines. The seal is used to back up excess flow within a mine during peak flow periods where complete mine inundation is impractical and treatment is required. This technique is theoretical.

Underground mine flow rates are variable and depend on the response characteristics of individual mines to seasonal rainfall variations. Near surface mines usually have sporadic (flashy) flow volumes, indicating short response times. Treatment plants are normally designed to handle the largest expected flows when complete treatment is needed. The treatment plant's capacity is then usually much larger than the average flow from the mine.

Extra costs are involved in constructing a treatment plant to handle large flows. These extra costs may be eliminated by constructing a mine seal which includes a pipe network to the plant. The mine pool will rise and fall with seasonal variations in rainfall, but the treatment plant will continue to accept average flow. This technique would also allow the treatment plant to cease operations during repairs without allowing discharges of polluted water.

Feasibility of this technique depends on the economic differential between savings in treatment plant installation and operation, and costs of mine seal installation. The mine must be capable of being sealed, and the pool capacity should be sufficient to impound water during periods of maximum rainfall. Costs must be developed on an individual application basis.

Subsidence Sealing

Mine subsidence holes can be active leak areas after a mine sealing program. Sink holes that are lower in surface elevation than the mine pool elevation are possible leak zones. These areas can be sealed using clay, concrete, or by grouting. Documented cases of subsidence sealing to prevent leakage from below are unknown. However, subsidence holes have been sealed to prevent surface waters from entering deep mines.

The subsidence holes should be cleaned of soil and surface debris with a backhoe or similar device. The underlying fractured rock can be grouted if necessary. Clay can then be compacted into the depression. Concrete would probably be more applicable than clay if high water pressures are expected.

Dry Seals

A dry seal is an underground mine seal constructed by grading earth over a mine opening, or constructing bulkheads consisting of clay or block walls. Dry seals are not meant to be used in mine openings discharging water. Their function is to prevent entrance of water and air into a mine. They have been used extensively in conjunction with air sealing programs. They should not be used where any significant amount of water pressure is likely.

These seals have only limited usefulness in water pollution control. Air sealing does not appear to be effective, therefore use of dry seals in conjunction with an air sealing program is not effective. They do have application in instances where surface water is entering an underground mine opening. A dry seal can be used to keep the surface water out of the mine.

Dry seal costs range from $100 to $500 for regrading, $1,200 to $1,500 for clay bulkheads, and $2,200 to $5,100 for masonry bulkheads.

Roof Collapse

This technique is not meant to be used as a primary mine sealing technique. It has limited use under special situations such as on a Pittsburgh coal seam underground mine that has a nonacidic clay stone (soapstone) roof containing stringers of limestone. The limestone is locked into the roof material and is not available for neutralization.

Explosive collapse of the roof would bring some of the limestone into contact with acidic underground mine waters. The physical characteristics of the clay stone are such that it would also restrict the flow of underground mine water, causing impoundment, and some acid forming strata inundation. Mine roof collapse is only a partial control technique.

Accurate mine maps, or a test boring program, are necessary to determine the best placement of explosives. Boreholes are drilled to a predetermined height above the mine roof. Explosives are set in the boreholes. The holes are then plugged to direct energy in a downward direction. This technique is only capable of a minimum amount of flooding, and should be accompanied by other water pollution control benefits such as the abovementioned neutralization. Its use is limited to areas where mine roof materials are capable of obstructing flow. Use with a blocky sandstone or shale roof would not be helpful in restricting mine flow because of the permeable nature of these materials. Roof collapse should never be used where roof materials are pollution forming, as commonly occurs in coal mines. Normal care in use of explosives is necessary to prevent damage to structures. There should not be any structures overlying or adjacent to the work areas.

Air Seals

Air seals are structures placed in discharging underground mine openings that permit water to exit from a mine without allowing air to enter (Figure 2.9). The nondischarging openings in the mine are also sealed by other conventional dry sealing techniques to prevent entry of air. Drill holes in the mined area are also plugged.

FIGURE 2.9: AIR SEAL

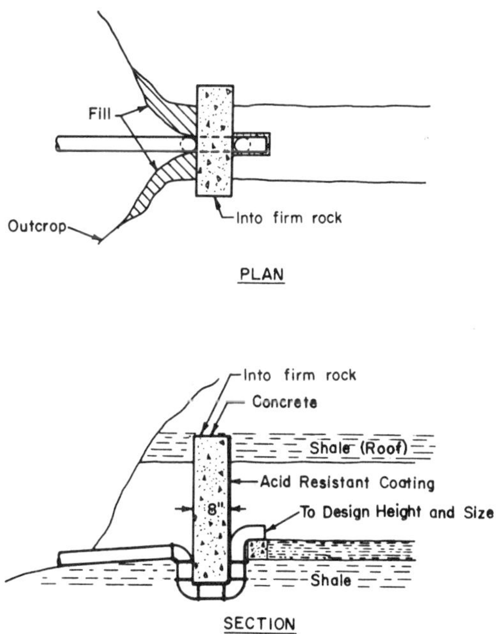

Source: EPA-430/9-73-011

An air sealing program is designed to prevent influx of free air oxygen into an underground mine. Free air oxygen is responsible for most of the pyrite oxidation, which is responsible for acid mine drainage. Elimination of free air oxygen is, therefore, the most desirable method of underground mine water pollution abatement. Many air seals were placed in eastern coal mines in the 1930s. Several air seals have been placed more recently. The seals are constructed of various materials, but their operation is based on the same principle employed by using traps in plumbing systems. There has not been much documentation of the effectiveness of older air seals. Many of these seals have been destroyed, and many of the remaining seals are discharging large quantities of pollution.

It is reasonable to conclude that air seals are not effective for two reasons. The underground mines have numerous air passages such as surface mines, boreholes, joints, fissures and mine subsidence cracks that allow passage of air into and out of underground mines. Changes in atmospheric pressure outside the mine causes a pressure gradient, resulting in air flow into and out of the underground mine. Costs generally range from $3,100 to $5,000 per seal.

Gel Material Seals

The technique involves use of commercially available grouts with a cheap filler material to remotely seal mine voids through boreholes without benefit of retaining bulkheads. One attempt to use a gel material as a mine sealing agent was made in a high flow mine entryway, and failed. This mine sealing attempt did indicate that use of gel for sealing low flow and dry mine entries may be possible.

The cost of the gel materials proved to be greater than originally estimated. The cost of this seal is not competitive with other sealing techniques. This technique has not proved feasible for use. The cost of the gel material alone is estimated at $9,000 per seal. The sealing operation would add considerably to the total cost.

Shaft Seals

A shaft is a vertical or near vertical entryway to an underground coal mine. The presence of a shaft implies that there is no coal outcrop in the vicinity. Lack of an outcrop usually increases the probability of success of a sealing program.

Discharging mine shafts are common in eastern coal fields. Coal mine shaft sealing is generally more successful than drift or slope sealing. Rock around the seal is less likely to leak because of low vertical permeability in the undisturbed coal strata.

Shaft seals are theoretically able to withstand much more pressure than outcrop seals, and large amounts of mine inundation can be accomplished. The degree of success of a shaft seal is partially dependent on depth of the mineral below the surface. Very deep underground mines can be successfully sealed. Leakage is more likely from shallow underground mines. A complete hydrogeologic evaluation is required to determine the feasibility of shaft sealing.

The shaft is first opened and cleared of debris. A suitable sealing zone within the shaft, such as a sandstone bed, is selected for sealing. Any flow from the shaft is stopped by pumping the mine pool. Miscellaneous fill is placed in the shaft up to the sealing level. The seal of clay and/or concrete is then placed. The shaft should be backfilled to the surface. A shaft seal is shown in Figure 2.10.

This technique should be highly effective in underground mine inundation when used under favorable mine conditions. The cost will be highly variable, depending on site conditions, condition of the shaft, size of the shaft and any auxiliary work required. Shaft seals will generally range in price from $7,000 to $25,000 per seal.

FIGURE 2.10: SHAFT SEAL

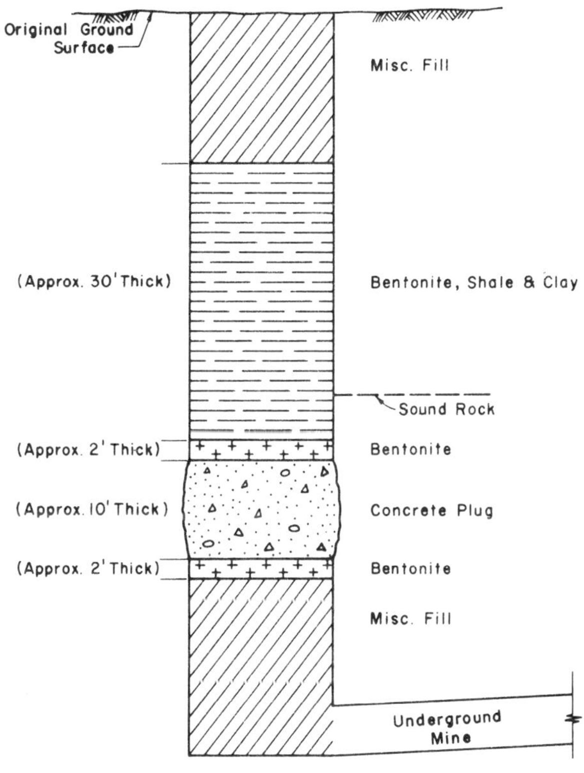

Source: EPA-430/9-73-011

LABORATORY STUDY OF SELF-SEALING LIMESTONE PLUGS FOR MINE OPENINGS

Apparatus and Procedure

This study was conducted to investigate the behaviors of various types of limestone aggregate when subjected to mine water percolation. All testing was done on a pilot scale in the laboratory. Aggregate samples were placed in square, horizontal tubes to simulate full-scale installations. The resulting model plugs were six inches square and six feet long. Synthetic mine water was supplied to one end of the test vessels and allowed to percolate through the stone. The other end was open and essentially unobstructed, allowing water to discharge freely. Water percolation was allowed to continue without interruption for up to 100 days. To evaluate the performance of these model plugs, several physical and chemical parameters were observed. Analyses were made to determine the chemical composition and particle size distribution of the stone prior to testing. During the test run, flow data and effluent water composition data were recorded. After the test run was concluded, selected plugs were subjected to chemical analysis, grain size analysis, and strength analysis.

Equipment was developed to produce simultaneously three synthetic mine waters differing only in ferric iron/ferrous iron ratios and to supply each type of water to a battery of up to 24 test vessels at a maximum rate of 1 gpm per vessel. The filled and sealed test vessels were placed on a test rack beneath the feed water supply manifolds and all piping was connected. After pressurizing the test vessels with air, the test water preparation and delivery systems were turned on to start the test run.

Flow data were recorded daily during the test run. Samples of the feed waters and vessel effluents were collected and analyzed for pH and conductivity. Weekly feed and effluent samples were collected for ferrous iron, total iron, calcium, sulfate, and hot phenolphthalein acidity determinations. After terminating the test, the vessels were opened and any volume decrease of each limestone plug was determined.

Lab Cycle I

The study was conducted in two sequential laboratory cycles, Lab Cycle I and Lab Cycle II. In Lab Cycle I, a total of 72 limestone specimens were tested. Six size ranges of each of three different limestones were tested on ferric, ferrous and ferric/ferrous synthetic mine waters. In addition, three additives were investigated. The results of these tests were used to select promising materials to be tested in Lab Cycle II.

Selection of the three limestones used in Lab Cycle I was based on a previous limestone neutralization study which showed that limestones could be classified into three groups, called Types A, B, and C limestones. Type A limestones were the most effective in neutralizing acid mine drainage, while Type C limestones were the least effective. One limestone from each group was selected for this study. Limestone No. 1809 (Type A), limestone No. 1355 (Type B) and limestone No. 1337 (Type C) were used. The following six size fractions of each stone type were tested on each of the three waters:

1" to dust (called 1 x 0)
1" to 50 mesh (called 1 x 50M)
½" to dust (called ½ x 0)
½" to 50 mesh (called ½ x 50M)
¼" to dust (called ¼ x 0)
⅛" to dust (called ⅛ x 0)

In addition to these 18 specimens (6 sizes of 3 stone types), three different additives were investigated. Limestone mixtures containing 5% bentonite, 10% flyash, and 10% air-cooled blast furnace slag were tested on all three synthetic waters. These additives were blended with both 1 x 0 and ½ x 0 size fractions of limestone No. 1809 (Type A stone). Thus a total of six specimens containing additives were tested on each water. Chemical compositions of the ¼ x 0 sizes of all three stone types were determined before testing. The results of these determinations are listed in Table 2.2. Complete particle size analyses were performed on representative samples of all 24 different stone specimens before testing. There was a considerable particle size variation in the ⅛ x 0 and ¼ x 0 size fractions of all three limestones. Limestone No. 1337 consistently contained considerably more fines than the other two stone types. Both the ½ x 50M and 1 x 50M sizes of the three stone types had similar particle size distributions.

Three test vessels were loosely filled with each of the 24 different limestone aggregate specimens for a total of 72 test vessels. The three sets of 24 specimens were tested on ferric, ferric/ferrous, and ferrous synthetic mine waters.

In well-graded materials with no particle sizes missing, the fraction of the material passing a No. 200 sieve has a great influence on the permeability. Small increases of fines greatly decrease the permeability of well-graded gravels. Granular materials with 10 to 20% passing the No. 200 sieve and placed at a medium density or greater are relatively impermeable.

Measured initial flows exhibited this effect for specimens which did not contain additives. The data show a significant decrease in permeability with a relatively small increase of fines, indicating that the specimens' gradations were responsible for initial flow behaviors.

Throughout the test run, the synthetic mine water compositions were checked and adjusted as required to maintain consistent values. The average compositions are presented in Table 2.1.

TABLE 2.1: AVERAGE COMPOSITION OF SYNTHETIC ACID MINE WATERS

	FERRIC WATER	FERRIC/FERROUS WATER	FERROUS WATER
pH	2.5	2.6	2.5
Sp. conductance	2700	2700	2850
Hot pht acidity	743	894	874
Calcium	81	78	75
Magnesium	28	22	23
Manganese	5.3	5.8	5.0
Aluminum	16	16	18
Total iron	205	209	198
Ferrous iron	10	106	197
Ferric iron	195	103	1
Sulfate	1055	1030	1122

Source: EPA-670/2-73-081

Flow and effluent composition data for all 72 test vessels were determined. This data included the following parameters: inlet head (inches), flow rate (ml/min), pH, specific conductance (mho), ferrous iron (mg/l), total iron (mg/l), calcium (mg/l), sulfate (mg/l) and hot phenolphthalein (pht) acidity (mg/l as $CaCO_3$).

When measured flow rates at measured heads were adjusted to equivalent flow rates at a six foot head it could be seen that most of the specimens reduced the equivalent flow rate of water over the test period.

Test water type had a significant effect on flow behavior. Although all flow histories showed a considerable fluctuation, this fluctuation was least severe for specimens on ferric water and most severe for specimens on ferrous water. Ferric water specimens generally had the lowest flow rates, while ferrous water specimens had the highest.

Flow behavior was also shown to be dependent on stone type. Initial flow rates were highest for stone No. 1809 (Type A) and lowest for stone No. 1337 (Type C). This was due to the initial fines content of the aggregate. For example, the ⅛ x 0 size of stone No. 1337 contained over four times as much fines as the same size of limestone No. 1809.

During the test run, however, stone No. 1809 exhibited the greatest reduction of flow, while stone No. 1337 exhibited the smallest reduction. As a result, flow rates after 50 days of testing for stone No. 1809 specimens tested on ferric or ferric/ferrous water were generally lower than for corresponding stone No. 1337 specimens. The lowest recorded flows occurred with the ⅛ x 0 and ¼ x 0 sizes, while the highest recorded flows occurred with the ½ x 50M and 1 x 50M sizes.

All the ⅛ x 0 sizes and all but one (Vessel No. 56) of the ¼ x 0 sizes maintained flows less than 300 ml/min. Flow histories for these two sizes were similar, although the ⅛ x 0 size generally exhibited lower initial flow rates. Flow histories for the ½ x 0 sizes tested on ferric or ferric/ferrous water were slightly greater than those for the ⅛ x 0 and ¼ x 0 sizes. Most of the ½ x 50M and 1 x 50M sizes were discontinued after 20 days of testing due to their high flow rates.

Specimens which had lower flow rates neutralized mine water through the stone more effectively than those with high flow rates. Also, the pH of the effluents from the specimens tended to increase slightly throughout the tests as the observed flow rates decreased. These results were expected, since the neutralization reaction is relatively slow when limestone is used, and slower flow rates provided increased detention time. The effluents typically had pH values of 6 or 7 for those specimens having flows less than 300 ml/min.

Chemical compositions of the effluents were also as expected. Ferric iron concentrations in the neutralized effluents were typically less than 20 mg/l (a 90% removal of iron) and were often less than 0.03 mg/l. Ferrous iron was also removed in many cases, but not as completely as ferric iron. Calcium concentrations were significantly higher in the neutralized effluents due to the neutralization reaction and erosion. Sulfate concentrations were essentially unchanged.

These results show that iron is precipitated and trapped within the stone, but that calcium sulfate is not. The superior flow behaviors which were observed for specimens tested on ferric water indicate that the precipitated iron had a significant effect on the permeabilities of the specimens, since ferric iron was removed more effectively than ferrous iron. Flyash and bentonite additives were shown to improve the performances of ½ x 0 and 1 x 0 stone sizes. Use of these additives, particularly flyash, provided lower flow rates and more effective mine water treatment. Performances of specimens containing these additives were comparable to performances of the ⅛ x 0 and ¼ x 0 sizes (without additives) in tests using ferric and ferric/ferrous waters. Both the ½ x 0 and 1 x 0 sizes containing flyash were more successful than smaller sizes without additives in ferrous water tests.

After the test runs were completed, the tops of the test vessels were removed and the stone specimens were visually examined. Variable amounts of stone discoloration were observed. All specimens showed some yellowish-brown or red discoloration and a thin crust on the top and sides. The discoloration was largest in the coarser stones and smallest in the stones with large percentages of fines. Although no specimens were rigidly cemented, some small blocks of lightly cemented material within 6 to 12 inches of the inlet end were observed in the more heavily discolored specimens.

A decrease in stone heights and widths was also observed. Height decreases ranged from ⅛" to 1", and width decreases ranged from 0" to ½". Since the inlet ends of the vessels were most severely affected, it is believed that stone consumption by neutralization reactions was a major cause of these volume losses. Settling of the stone upon wetting, hydraulic erosion, and compression by the pressurizing air could also be responsible. Average volume losses were determined for each specimen. The data show the following trends:

(1) Volume losses were largest for specimens tested on ferrous water and smallest for specimens tested on ferric water.

(2) Limestone NO. 1337 (Type C) showed the most volume loss, while limestone No. 1809 (Type A) showed the least.

(3) The intermediate sizes, ¼ x 0 and ½ x 0, exhibited the least severe volume losses.

(4) Bentonite and slag effectively inhibited stone volume loss.

Chemical analyses were performed on samples taken six inches from the inlet ends of all nine ¼ x 0 specimens. The results of these determinations are presented in Table 2.2. Values for the corresponding stones before testing are also included. These data show that constituent/calcium ratios generally increased as a result of testing on synthetic mine waters. Iron/calcium ratios generally exhibited the greatest increase, indicating that iron was deposited in the first foot of the plug.

In-place density, particle size distribution, compressibility, and strength parameters were evaluated for eleven of the stone plugs. Densities, compressibilities, and shear strengths were measured on 4" diameter, 6" high cylindrical specimens trimmed from the inlet ends of the stone plugs where the effects of mine water percolation were greatest.

The density of uncemented granular materials has a great influence on the compressibility, permeability, and strength of the stones. The density of granular materials is determined by the specific gravity of the particles, particle shapes, particle size distribution and the particle structure. In a loose state, particle contacts are edge to plane and edge to edge producing a structure which collapses on disturbance. In a dense state, the particle contacts are primarily plane to plane producing a strong and stable structure since the material must expand to be sheared.

TABLE 2.2: ANALYSIS OF LIMESTONES BEFORE AND AFTER TESTING (REPORTED AS WEIGHT % OF IGNITED SAMPLE)

VESSEL NO.	STONE NO.	TEST WATER	LOSS ON IGNITION	SiO_2	Al_2O_3	CaO	MgO	Fe_2O_3	S
None	1809	None	36.93	15.4	3.9	71.9	0.59	2.86	0.29
8	1809	Fe^{+3}	35.8	15.3	4.2	67.8	1.20	6.90	0.59
32	1809	Fe^{+3}/Fe^{+2}	5.7	11.1	2.0	45.1	4.03	11.52	0.27
56	1809	Fe^{+2}	29.5	25.0	20.0	49.1	1.23	9.33	0.57
None	1355	None	33.8	23.9	5.75*	62.3	1.65	2.48*	
14	1355	Fe^{+3}	32.2	23.6	6.05	59.4	1.77	4.43	0.60
38	1355	Fe^{+3}/Fe^{+2}	9.3	22.7	3.86	33.6	0.78	17.3	0.86
62	1355	Fe^{+2}	32.0	24.6	5.29	58.1	0.49	4.42	0.59
None	1337	None	46.0	1.82	0.15*	54.3	39.1	0.25*	
20	1337	Fe^{+3}	44.8	1.78	1.45	53.6	38.4	4.14	0.33
44	1337	Fe^{+3}/Fe^{+2}	44.2	5.56	1.61	52.0	40.9	5.12	0.34
68	1337	Fe^{+2}	45.1	1.42	0.73	55.7	40.6	2.06	0.18

*Data taken from previous study.

Source: EPA-670/2-73-081

These states are defined by minimum and maximum densities determined by laboratory tests, and the in-place density of a granular material is related to these limiting densities by a relative density parameter. The relative density is expressed in percent and is obtained from the following equation:

$$DR = \frac{\gamma^{d}max\ (\gamma^{d} - \gamma^{d}min)}{\gamma^{d}\ (\gamma^{d}max - \gamma^{d}min)} \times 100\%$$

where DR = relative density, %; γ^{d} = dry density; $\gamma^{d}max$ = maximum dry density; $\gamma^{d}min$ = minimum dry density. The significance of relative density values may be shown as follows:

Relative Density, %	Material Description	Type of Structure
0 to 15	Very loose	Collapsing
15 to 35	Loose	Collapsing
36 to 65	Medium dense	Intermediate
65 to 85	Dense	Dilating
85 to 100	Very dense	Dilating

The dry densities of the limestones in the test vessels were evaluated by relative density calculated from minimum and maximum densities. It should be noted that since the density affects the physical properties of the stones, the minimum and maximum densities provide a criterion for evaluating of stones with various particle size distributions. The stones with higher densities should have better properties.

The in-place densities of trimmed cylindrical specimens of the limestones subjected to mine water percolation were measured. These densities, relative density, and porosity are presented in Tables 2.3 and 2.4. The volume decrease is also shown for completeness of the density discussion. The densities were calculated from the measurements taken on trimmed cylindrical specimens from the test vessels.

TABLE 2.3: VOLUME LOSS, DRY DENSITY AND POROSITY AFTER 50 DAYS OF PERCOLATION

Stone No. & TV No.	Stone Size	Volume Loss, %	Dry Density γ_d, PCF	Dry Density Avg. γ_d, PCF	Dry Density DR %	Porosity n, %
	Ferrous Mine Water					
1809 50	1/2 x 0	38	[illegible]	[illegible]	11	47.5
	Ferric Mine Water					
			84.0		4	49.2
1809	1/2 x 0	9	94.3	91.7	30	43.0
10			96.8		37	41.4
	Ferric-Ferrous Mine Water					
			87.5		-17	47.1
1809	1/8 x 0	7	99.4	95.1	-25	40.0
31			98.5		-23	40.4
			94.5		-1	42.8
1809	1/4 x 0	7	91.6	94.2	-13	44.5
32			96.6		7	41.5
			77.2		-20	53.3
1809	1/2 x 0	11	77.8	79.8	-18	53.0
34			84.3		3	49.0
			72.2		-40	56.4
1355	1/2 x 0	15	81.8	79.5	-4	50.6
39			84.6		4	48.8
			92.0		25	44.4
1337	1/2 x 0	18	88.4	90.5	15	46.5
46			91.0		23	45.0
			90.0		20	45.4
1809	1/2 x 50	13	83.2	84.8	0	49.8
33			81.2		-6	50.9

Source: EPA-670/2-73-081

In the case of Test Vessel 58, undisturbed samples could not be obtained because of the large collapse and irregularity of the stone surface in the test vessel. The three specimens of each material are listed in order of sampling from the influent end with the center of the first specimen located about six inches from the influent, and the centers of the second and third specimens approximately 12 and 18 inches, respectively.

TABLE 2.4: VOLUME LOSS, DRY DENSITY AND POROSITY AFTER 100 DAYS OF PERCOLATION

Stone No. & TV No.	Stone Size	Volume Loss %%	Dry Density γ_d, PCF	Dry Density Avg γ_d, PCF	DR %	Porosity n, %
	Ferric-Ferrous Mine Water					
			106.8		28.7	35.5
1355	1/8 x 0	24	107.8	107.7	31.4	34.9
37			108.6		33.7	34.4
			100.0		33.1	39.7
1355	1/4 x 0	13	96.0	99.1	22.6	42.0
38			101.2		36.0	38.8
			72.7		-21.9	56.2
1355	1 x 0	17	74.8		-12.9	54.6
42			76.5		-5.9	53.8

Source: EPA-670/2-73-081

The relative densities of all trimmed specimens show the limestones at the influent end of the vessels to be loose to very loose. In six of the eleven vessels, the final densities are less than the minimum densities obtained by very loose placement of dry material resulting in negative relative densities. These negative densities indicate a large loss of material produced by the erosion of unprotected particle surfaces leaving a particle structure considerably looser and more fragile than can be obtained by physical placement. Therefore, these limestones are very loose, compressible and susceptible to structural collapse.

Finally, the settlement of the stone surfaces without corresponding increase of relative density from the loose placement density to medium density indicates stone erosion in the vessels. Particle size analyses of the stones subjected to mine water percolation all showed an increase of fines. The increase of fines is due to the dissolving of larger limestone particles and the accumulation of precipitates. This increase of fines would plug the voids in the stones and decrease correspondingly the flow of water.

The effect of density on permeability is illustrated in Figure 2.11 where permeability test results on ⅜ x 0 stone specimens prepared at different densities are presented. These data indicate a significant decrease of permeability with an increase of dry density. Thus, increasing the placement density considerably reduces the flow of water through the stone. Triaxial tests were conducted on trimmed cylindrical specimens of the limestones to determine their compressibility and strength after mine water percolation. In the triaxial test a cylindrical specimen is enclosed by a rubber membrane, confined by a lateral pressure and sheared by an axial load applied through a piston.

FIGURE 2.11: PERMEABILITY vs DRY DENSITY

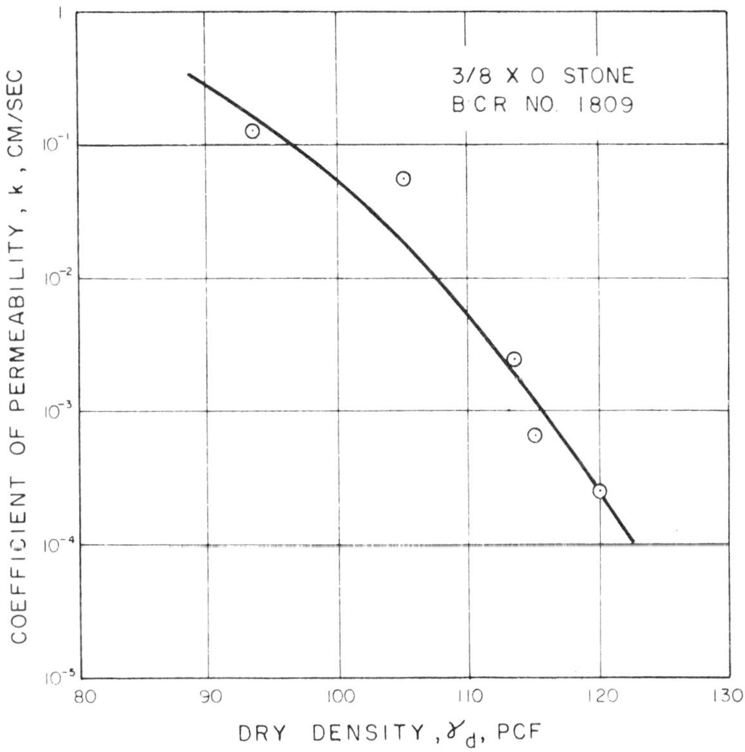

Source: EPA-670/2-73-081

Two types of tests were conducted in this apparatus. The first was a constant-diameter compression test in which the diameter was kept constant by continually increasing the lateral confining pressure during axial loading of the specimen as illustrated schematically in Figure 2.12. The need to increase the confining pressure was sensed by a lateral gauge mounted at mid-height of the cylinder. During the compression test the axial deformation, axial load and confining pressure were recorded and the axial strain, axial and confining stress and the ratio between the vertical and horizontal pressures calculated.

The second type of test was a shear strength test conducted at a constant confining pressure, illustrated in Figure 2.13. This type of test was necessary because the limestones after being subjected to the mine water percolation were found to be uncemented and, therefore, behaved as granular materials. Granular materials derive their strength from particle stresses on the failure plane. The shear strength is derived from cohesion and friction components and may be expressed as:

$$S = C + N \tan \phi$$

where S = shear strength; C = cohesion; N = normal stress on failure plane; and ϕ = angle of internal friction. The cohesion and angle of internal friction are the strength parameters and are normally evaluated in the triaxial shear test.

FIGURE 2.12: CONSTANT DIAMETER COMPRESSION TEST

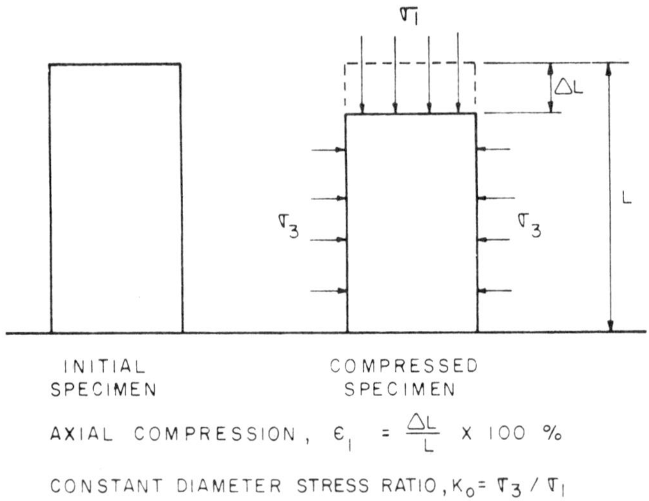

FIGURE 2.13: CONSOLIDATED DRAINED TRIAXIAL TEST

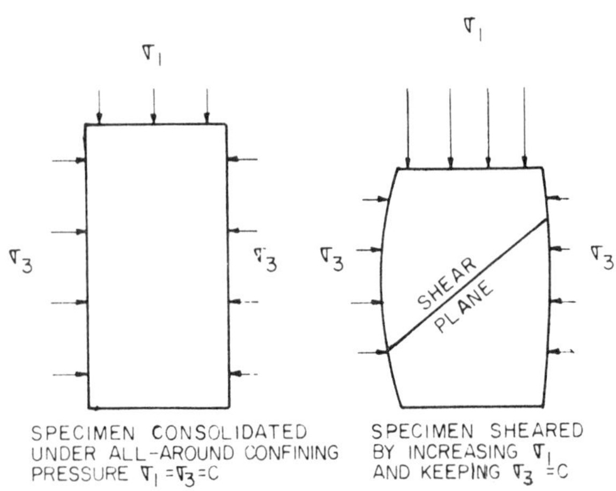

Source: EPA-670/2-73-081

A mine seal can develop a normal stress, and thus, shear strength from two mechanisms: (1) the hydrostatic pressure on the seal tending to push the seal out of the opening will tend to expand the seal and increase the confining pressure; (2) settlement of the roof will transfer part of the overburden load to the limestone seal.

Compression tests were conducted on trimmed undisturbed specimens from ten vessels and on remolded specimens from one vessel. In the latter case, a remolded specimen had to be used because it was impossible to trim a specimen from the collapsed material in Vessel 58. Since the compression data showed the limestones to be very compressible, two additional tests were conducted on remolded and compacted material prepared at a greater density than measured in the test vessels to determine the effect of density on the stiffness of the limestones. All compression test data are summarized in Table 2.5.

TABLE 2.5: COMPRESSION TEST RESULTS

Test Vessel No.	Stone Size	Type of Specimen	Dry Density, γ_d, PCF Initial	Dry Density, γ_d, PCF Final	Axial Strain at 10 TSF Load ϵ_1 %	$k_o = \sigma_3/\sigma_1$
	Ferric Water					
10	1/2 x 0	Remolded	94.3	101.7	6.4	.43
	Ferrous Water					
58	1/2 x 0	Remolded	86.8	92.0	7.5	.45
	Ferric-Ferrous Water					
31	1/8 x 0	Undisturbed	87.5	96.9	9.4	.43
31	1/8 x 0	Compacted	113.0	117.8	2.5	.34
32	1/4 x 0	Undisturbed	94.5	105.8	8.6	.44
34	1/2 x 0	Undisturbed	77.2	90.5	14.0	.46
34	1/2 x 0	Compacted	118.8	120.9	2.3	.36
39	1/2 x 0	Undisturbed	72.2	92.3	17.4	.43
46	1/2 x 0	Undisturbed	92.0	99.3	6.2	.43
33	1/2 x50	Undisturbed	90.0	102.8	9.1	.38
37	1/8 x 0	Undisturbed	106.8	117.2	7.0	.45
38	1/4 x 0	Undisturbed	100.0	111.9	8.5	.41
42	1 x 0	Undisturbed	72.7	96.9	22.5	.53

Source: EPA-670/2-73-081

The stress-strain compression data show the in-place limestones subjected to mine water percolation to be very compressible. The low stiffness of the limestones is the result of the loose placement of the stone and subsequent erosion of the limestone by mine water percolation.

Triaxial tests were conducted on two or three specimens from each vessel to determine the shear strength parameters of the limestones following the mine water percolation tests.

The results of all the tests are summarized in Table 2.6 where the average dry density, average axial strain at failure, cohesion, angle of internal friction and shear strength at a confining pressure of 2.0 tsf are given. Shear strengths developed by the stones at a given confining pressure are presented to permit a comparison of the shear strengths.

The triaxial shear test data shows the limestones subjected to mine water percolation behave as granular materials whose shear strength is a function of the confining pressure and in-place density. For a typical confining pressure of 2.0 tsf, the shear strengths were mainly a function of the density, of the stone. Some decrease of strength was observed for the materials subjected to 100 days of percolation, however, the decrease is small.

The investigation of the physical properties of limestones placed at low densities showed that they are not suitable for mine sealing. The low density produces a permeable limestone which is eroded by the mine water. Since the chemical reaction between the stone and mine water did not result in any cementation, the erosion left a very collapsible stone structure. The limestones subjected to mine water percolation were very compressible and had small shear strength.

Lab Cycle II

Lab Cycle I clearly indicated that the physical properties determining the suitability of a limestone as a mine plug, permeability, compressibility, and strength are a function of particle size distribution and density. Thus, Lab Cycle II was conducted to further investigate the effects of varying particle size distribution and placement density. In addition, additives which might aid in cementing the stone particles were investigated.

A total of twelve specimens were tested in Lab Cycle II. Ten of these were subjected to ferric/ferrous test water, and two of these were tested with South Pittsburgh city water. Commercially available ⅜ to dust (called ⅜ x 0 size) grade of limestone No. 1809 was used to prepare all specimens.

Three additives were investigated. Portland cement, calcium sulfate hemihydrate (plaster of paris), and sodium silicate were blended with ⅜ x 0 stone in 5% concentrations. These three specimens were placed in test vessels at about 30% relative density. All three were tested on ferric/ferrous water.

Four specimens containing increased quantities of limestone fines were tested on ferric/ferrous water. Minus 50 mesh fines were obtained by screening ⅜" to dust stone. These fines were added to ⅜ x 0 stone in sufficient quantity to increase the fraction of material passing a No. 200 sieve by factors of 2 and 3. Each of these two materials was placed at both 30 and 60% relative density for a total of four specimens.

A "zoned" plug was also tested on ferric/ferrous water. The first foot of stone contained 5% ferric sulfate and 15% sodium silicate. This section was intended to be a water "pretreatment" area and was not considered to be part of the actual limestone plug.

TABLE 2.6: STRENGTH PARAMETERS AND SHEAR STRENGTH FOR A 2.0 TSF OVERBURDEN PRESSURE

Test Vessel	Stone Size	Type of Specimen	Average Dry Density γ_d, PCF	Average Axial Strain at Failure ϵ_1, %	Strength Parameters Cohesion C, PCF	Strength Parameters Friction Angle, ϕ	Shear Strength Su, PSF
10	1/2 x 0	Undisturbed	9.17	15	1,600	35.5	4,300
58	1/2 x 0	Remolded	86.8	12	640	36.4	3,700
31	1/8 x 0	Undisturbed	95.1	18	950	35.3	3,900
31	1/8 x 0	Compacted	113.2	10	1,100	37.0	4,400
32	1/4 x 0	Undisturbed	94.2	19	700	35.3	3,800
34	1/2 x 0	Undisturbed	79.8	24	350	39.5	3,900
34	1/2 x 0	Compacted	118.0	6	2,700	37.0	5,400
39	1/2 x 0	Undisturbed	79.5	22	0	42.4	3,700
46	1/2 x 0	Undisturbed	90.5	23	0	41.5	3,600
33	1/2 x 50	Undisturbed	84.8	23	400	37.3	3,900
37	1/8 x 0	Undisturbed	107.7	18	0	43.8	3,800
38	1/4 x 0	Undisturbed	99.1	22	670	40.5	4,100
42*	1 x 0	Undisturbed	74.7	20	0	38.7	3,200

Source: EPA-670/2-73-081

The remaining five feet of the specimen was 3/8 x 0 stone. The entire six foot long specimen was placed at about 30% relative density. Four "blanks", 3/8 x 0 stone as received from the quarry, were tested. Two of these specimens were placed at about 30 and 60% relative density and tested on ferric/ferrous water. The other two were placed at about 0 and 30% relative density and tested on South Pittsburgh city water. As in Lab Cycle I, the test vessels were assigned vessel numbers for identification purposes.

Testing was performed in a manner similar to Lab Cycle I. Heads and flow rates were measured after 30 minutes, 3 hours, and 8 hours during the first day of testing. Effluent pH values were also recorded after 8 hours of testing. Beginning on the second day of testing (1 day after start-up) head, flow, pH, and specific conductance were recorded for each specimen on a Monday-Wednesday-Friday schedule. All specimens were tested for 50 days. Vessel effluent samples were collected after 1, 24 and 50 days of testing.

Flow and effluent composition data for the 12 specimens tested in Lab Cycle II show that all specimens tested on synthetic mine water effectively obstructed the flow of water and treated water which passed through the stone. Flow and neutralization behaviors were generally more satisfactory than those observed in Lab Cycle I.

Flow histories for specimens tested on synthetic mine water show that both increasing the placement density and increasing the fines content of the stone resulted in significantly lower flow rates. Increasing the fines content proved to be the most effective means of obstructing water flow. Flow data for the two specimens tested on tap water also showed decreasing flow rates over the test period, indicating that physical effects are at least partly responsible for observed flow decreases. The flow magnitudes, however, were much larger than for specimens tested on synthetic mine water.

These data show a discontinuity at 10 days after start-up due to an air pressure failure. Although the two vessels operated without air pressure for only a few minutes, it is believed that the stones' grain structure was permanently affected. Physical examinations, described in Lab Cycle I, were conducted on the ten test vessels from Lab Cycle II which were tested on ferric/ferrous water.

Discoloration of the limestones was observed for the entire length of the specimens containing additives, and the length of discoloration of the remaining stones was directly related to the quantity of fines and degree of compaction. The natural stone at DR = 30% (Vessel No. 81) showed the greatest discoloration while the stone with 3 x fines and placed at DR = 60% was discolored for only the first six inches from the influent end. The surface measurement of the test specimens indicated some volume decrease in all test vessels. The following trends were observed:

(1) Volume losses were largest for specimens with the highest fines content and smallest for specimens with the smallest fines content.

(2) Higher placement densities resulted in lower volume losses.

(3) Volume losses for specimens containing additives were comparable to losses for the corresponding natural stone.

These data indicate that stones must be placed at higher densities than DR = 60% to prevent excessive stone settlement. Furthermore, the degree of compaction must be increased with the percent of fines in the stone to compensate for the greater compressibility of fines. The in-place densities, relative densities and porosities of trimmed cylindrical specimens from the test vessels are listed in Table 2.7. These data were obtained after 50 days of ferric/ferrous mine water percolation.

TABLE 2.7: VOLUME LOSS, DRY DENSITY AND POROSITY OF TRIMMED SPECIMENS

Test Vessel	Material Description	Volume Loss, %	Dry Density			Porosity n, %
			γ_d, pcf	Avg γ_d, pcf	DR, %	
			95.2		6	42.4
81	Natural	7	97.7	95.2	13	41.0
	DR = 30		92.8		-2	43.9
			96.2		9	41.8
82	Natural	2	100.1	100.8	20	39.5
	DR = 60		106.1		37	35.8
			110.8		40	32.9
76	2 x Fines	16	115.0	113.9	50	30.6
	DR = 30		115.8		52	30.0
			118.0		57	28.7
77	2 x Fines	8	118.6	118.2	59	28.2
	DR = 60		117.9		57	28.7
			121.2		76	26.7
78	3 x Fines	20	116.3	116.5	67	29.7
	DR = 30		112.1		58	32.2
			117.6		69	28.9
79	3 x Fines	8	111.7	115.3	57	32.5
	DR = 60		116.5		62	29.5
			99.0		17	40.1
73	5% Cement	3	86.5	94.2	-25	47.7
	DR = 30		97.0		11	41.4
	5% Calcium		88.2		-19	46.7
74	Sulfate Hemi-hydrate	7	103.7	99.3	30	37.3
			106.0		37	35.8
	5% Sodium		77.8		-63	52.9
75	Silicate	6	86.8	85.6	-24	47.5
	DR = 30		92.3		-5	44.3
			90.4		-11	45.3
80	Zoned	8	95.7	97.7	7	42.1
	DR = 30		100.8		22	39.1

Source: EPA-670/2-73-081

Evaluating the in-place densities using relative density based on minimum and maximum densities (given in Table 2.8), the following observations can be made:

(1) Relative densities smaller than the placement densities were measured in the natural stones in Vessels No. 81 and 82. A possible explanation for the low density could be the washing out of fines in the area of sampling.

(2) Specimens from Vessels No. 77 and 78, where the stones were placed at DR = 30%, had average relative densities of 47 and 67%, respectively. This densification could have been caused by wetting of the stone combined with the confining pressure. This is supported by the large volume losses.

(3) The negative relative densities in the stone with 5% sodium silicate indicate adverse chemical reaction leading to stone erosion.

TABLE 2.8: MINIMUM AND MAXIMUM DRY DENSITIES

Stone No.	Stone Size	Description	Minimum Dry Density, pcf*	Maximum Dry Density, pcf**
1809	⅜ x 0	Natural	94.6	139.0
1809	⅜ x 0	2 x Fines	98.6	141.6
1809	⅜ x 0	3 x Fines	90.7	138.0

*Minimum by ASTM Method, D-2049

**Maximum by Modified Proctor Test, ASTM Method, D-1577

Source: EPA-670/2-73-081

Finally, the comparison of the minimum and maximum densities of the three stones indicates the stone with 2 x fines can be placed at higher densities than the other two stones, resulting in better physical properties.

The comparison of the particle size distribution of the stones before and after percolation testing indicates an increase of fines in all materials. The increase of fines is probably the result of precipitate accumulation. It can be concluded that the decrease of flow experienced in all vessels was at least partially due to precipitates plugging the stone voids.

The initial flow of the synthetic mine water through the vessels was found to be related to the percent of fines and density of the stones.

Triaxial tests were conducted on trimmed cylindrical specimens on all test vessel materials. The compression test results are shown in Table 2.9 and shear strength in Table 2.10.

These results illustrate that the compressibility and shear strength are independent of the particle size distribution in the mixes tested and are directly related to the material density. The good agreement of the behavior of the specimens from materials with additives with the natural stones indicates that the additives did not increase the stiffness of the material nor increase its shear strength.

TABLE 2.9: COMPRESSION TEST RESULTS, TRIMMED UNDISTURBED SPECIMENS

Test Vessel	Material Description	Dry Density, γ_d, pcf		Axial Strain @ 10 tsf Load, ε_1, %	$k_o = \sigma_3/\sigma_1$
		Initial	Final		
81	Natural	95.2	106.5	9.0	0.43
82	Natural	96.2	107.7	8.6	0.43
76	2 x Fines	110.8	119.8	6.2	0.43
77	2 x Fines	118.0	124.5	4.3	0.35
78	3 x Fines	121.2	128.7	4.9	0.42
79	3 x Fines	117.6	123.6	4.1	0.42
73	5% Cement	99.0	108.4	6.5	0.41
74	5% $CaSO_4$	103.7	114.6	7.2	0.37
75	5% $NaSiO_2$	86.8	100.6	11.8	0.45
80	Zoned	90.4	102.2	9.6	0.40

TABLE 2.10: SHEAR STRENGTH FOR A 2.0 TSF OVERBURDEN PRESSURE

Test Vessel	Material Description	Average Dry Density γ_d, pcf	Average Axial Strain at Failure ε_1, %	Strength Parameters		Shear[a] Strength Su, psf
				Cohesion C, pcf	Friction Angle, ϕ	
81	Natural	95.2	16	750	34.6	3,400
82	Natural	100.8	15	750	37.8	3,800
76	2 x Fines	113.9	12	1,500	39.6	4,800
77	2 x Fines	118.2	11	1,000	42.4	4,600
78	3 x Fines	116.5	16	560	43.4	4,300
79	3 x Fines	115.3	8	600	40.5	4,000
73	5% Cement	94.2		900	38.8	4,100
74	5% Ca	99.3	13	0	40.5	3,400
75	5% $NaSO_2$	85.6	18	360	33.2	3,000
80	Zoned	97.7	15	650	38.3	3,800

[a]Shear strength at a confining pressure of 2.0 tsf.

Source: EPA-670/2-73-081

Conclusions

This laboratory investigation of various limestone aggregate plugs has led to the following conclusions:

(1) The results of this study indicate that limestone aggregate plugs are a feasible means of sealing underground mines which discharge water containing ferric iron.

(2) The ⅜" to dust size of limestone No. 1809 placed at 60% relative density was the most satisfactory natural material tested.

(3) High placement densities are essential for satisfactory plug performance.

(4) Significant stone volume losses can occur when limestone plugs are exposed to acid mine water flow due to settling of the stone upon being wetted, erosion, and chemical consumption of the stone.

(5) Limestone plugs will perform best on ferric mine waters and poorest on ferrous mine waters.

(6) The Type A limestone (found in previous tests to neutralize acid mine waters better than Types B and C limestones) had the best overall performance, while the Type C limestone had the poorest performance.

(7) The ⅜" to dust grade of stone was the most satisfactory size tested.

(8) Bentonite and flyash additives improve water flow and treatment properties of permeable plugs.

(9) Bentonite and slag additives decrease stone volume losses.

(10) Increasing the fines content of commercially available stone to twice the original amount (as determined by the fraction of material which passes a No. 200 sieve) results in improved performance.

UNDERGROUND PRECIPITATION

Underground precipitation as a technique for elimination of pollution from abandoned coal mines is of interest from three standpoints: (1) filling of mine voids; (2) sealing drainage openings; and (3) continuous neutralization of effluent acid mine water. The use of underground precipitation in filling mine voids is based on forming the precipitates in the mine, thus providing an economic advantage over the cost of their return to the mine from a surface neutralizing installation.

The use of underground precipitation in sealing drainage openings is advantageous as the precipitation can be carried to the outflow points to plug the openings similar to the "blinding" of a filter cloth. Finally, underground precipitation would be a means for continuous neutralization of effluent acid mine water without the concurrent problem of sludge disposal.

Under a contract with the Pennsylvania Department of Mines and Mineral Industries, Parsons-Jurden Corporation engaged in a study of underground precipitation in abandoned mines, resulting from the reaction of mine water with hydrated lime and limestone, to prevent mine drainage pollution.

Preparations at the Mine Site

The workings of the Driscoll No. 4 mine consist of a 4½ foot coal seam that underlies an extensive hilly area. The seam dips to the northwest from the outcropping at the base of the south side of a hill that rises along the banks of Black Lick Creek at Vintondale, Pennsylvania. The mine first opened in 1906 and was considered mined out sometime in the 1950s, but it was reactivated in the early 1960s to work a new section along the western fringe of the older workings which are said to extend several miles from the portal in a northerly direction. A new portal was opened to the west of the original portal.

In compliance with newer regulations covering coal mines, three entries branched out from the portal: the No. 1 west entry, north entry, and east entry. A fourth adit, the No. 2 west entry, branched off from the north entry at a point about 25 feet from the portal and paralleled the No. 1 west entry. The north and east entries connected with the old mine workings. The No. 1 and No. 2 entries comprised the active mine.

In 1967, operation of the mine was discontinued, the pumps were removed, and the mine began to fill with water. The east entry was caved to prevent entry of personnel. The mine portal was also to be backfilled with 25 feet of noncombustible material to seal the mine against entry. However, this was not done so that the mine could be used as the field site for demonstration purposes reported here.

A field survey and a study of available mine maps led to the decision to use the No. 1 west entry for the water sealing demonstration. A rubble pile would be placed in the adit, to be sealed by the precipitate formed by pumping a slurry of hydrated lime and/or ground limestone into the mine water behind the rubble pile. Three concrete bulkheads would be required to control the water flow. It was also decided to use the No. 2 west entry and the north entry to test the possibility of continuously neutralizing acid mine water, and simultaneously eliminating the sludge disposal problem, by precipitating the mineral content from the water before it drained out and then allowing the sludge to settle into the lower portions of the mine.

A plan of the construction undertaken at Driscoll No. 4 is shown in Figure 2.14. The water level was lowered by pumping to permit repair of the portal. Steel rails and posts were installed for maximum safety. The north entry and the No. 2 west entry were propped up for safe entrance. The water level was lowered further to permit cleaning and the repair of the No. 1 west entry for a distance of 120 feet west of the portal. After the entry had been cleaned and repaired, emplacement of the rubble pile was started.

Inert material consisting of broken slate, shale, and glacial till was placed by hand, using wheelbarrows and shovels. The entry was filled and compacted to produce a pile 25 feet long (as measured at the roof of the drift). Three 2-inch steel pipes with pointed, slotted ends were driven through the pile, terminating 3, 6, and 9 feet, respectively, behind the top of the pile (Figure 2.14).

FIGURE 2.14: PLAN OF BULKHEADS, PIPING, WEIRS AND PORTAL

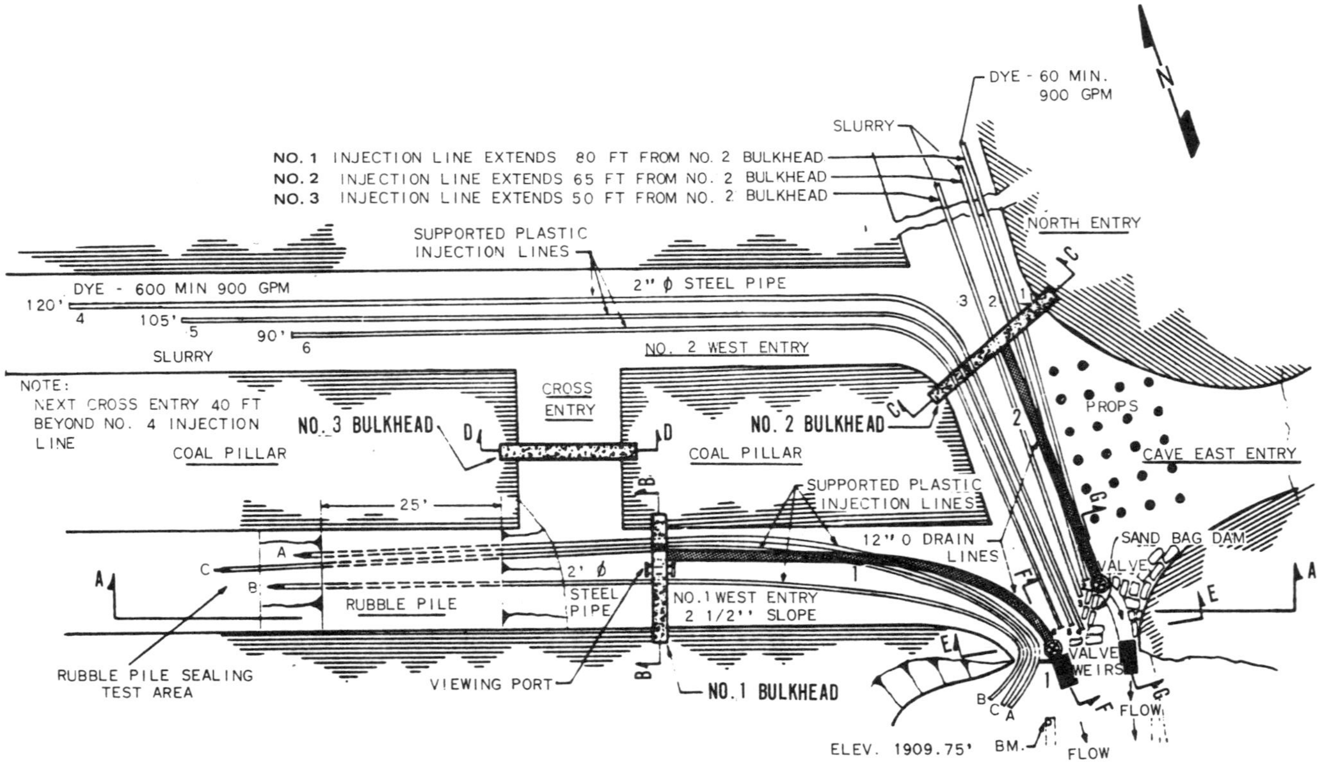

Source: EPA-670/2-73-092

Following settlement of the rubble pile, the top of the pile was refilled and the toe reinforced with sandbags. The rubble pile was porous enough to permit water to drain through.

Three 2-inch injection lines were installed in the No. 2 west entry, terminating 90, 105, and 120 feet, respectively, behind the No. 2 bulkhead. Three similar injection lines were installed in the north entry, 50, 65, and 80 feet, respectively, behind the No. 2 bulkhead. All injection lines behind the bulkheads were Schedule 40 steel pipe, supported on wooden posts near the entry roof. Prior to the injection tests, the steel injection pipes were connected to plastic pipes at a point close to the outside face of the bulkheads.

The No. 1 bulkhead was located in the No. 1 west entry and provided with a 12-inch steel drainage pipe; and No. 3 was placed in the cross entry. The three injection lines which were driven through the rubble pile extended through the No. 1 bulkhead. The No. 2 bulkhead was also provided with a 12-inch steel drainage pipe. Six injection lines were installed, three in the No. 2 west entry and three in the north entry. The No. 3 bulkhead in the cross entry was solid.

The bulkheads were located where good ground conditions existed in the entries. Eighteen inch wide slots were excavated 24 inches into the side walls, floor, and roof of each entry to lock the bulkheads in place and provide a watertight seal. Concrete reinforcing bars were placed in the excavations, and injection lines and drainage lines were installed. Wood forms were set in place and concrete pumped into the form.

Bleeder lines were placed in the back of the bulkhead excavations to bleed off air so that the entire excavated area would accept concrete. The 12-inch drain lines were extended from the No. 1 and No. 2 bulkheads out to the portal. Gate valves were installed on the ends of these lines to permit control of the flow through the two discharge points. The mine was then no longer accessible to personnel.

The 2-inch injection lines were capped temporarily to prevent water flow through them. The pumping was then discontinued, and the entry allowed to refill and drain through the two 12-inch drain lines. Some months later, when the preparations for the tests were started, it was found that a rock slide had occurred in the interim. To make the portal safe once again, it was necessary to remove several large chunks of rock from the brow of the portal and place additional timbering in the entries. The area from the portal back to the bulkheads was pumped out to permit access to the bulkhead faces. The 2-inch steel injection lines were uncapped and 2-inch plastic lines were attached and extended out to the portal.

A clay core and sandbag berm was constructed at the portal to prevent flooding this area once again and to channel the water flow from the drainage pipes. However, because of seepage from the caved east entry, it was impossible to keep this area from flooding.

One steel weir box containing a 30-degree V-notched plate was bolted to No. 1 drain pipe valve and a second was placed several feet downstream of the No. 2 drain pipe valve. The sandbag berm channeled the water from the No. 2 valve to the No. 2 weir box. Because of the large flow of water from the No. 2 drain pipe, it was necessary to cut the front plate of the No. 2 weir to produce a 12-

inch rectangular weir. A 90-degree V-notched plate was inserted during periods of low flow. Water from the No. 1 and No. 2 drainage lines discharged to an existing drainage ditch that flowed into the south branch of Black Lick Creek. The third weir, identified as the main weir, containing a 30-degree V-notched plate was placed in this ditch. A trash-catching screen was installed about 50 feet upstream of this weir to prevent interference by floating debris.

A portable gasoline-driven Shirley Mixmeter (Model 65 AP 3 x 4) was set up to prepare slurries of hydrated lime and/or pulverized limestone and inject them through the 2-inch pipes to the various flooded zones behind the bulkheads. The charging tank had a capacity of 290 gallons. This machine was used to deliver the slurries through the 2-inch lines to a space behind the bulkheads. The pH meter and the slurry metering control devices on the machine were disconnected.

The water used to prepare the slurries and carry them into the selected spaces behind the bulkheads was taken from the ditch just downstream of the No. 1 and No. 2 weirs by the suction line of the centrifugal pump. A portion of the slurry stream from the centrifugal pump was recycled back into the charging tank to maintain a constant level. The flow rate of the recycle stream was manually controlled to keep the slurry in the charging tank thin enough to flow properly to the pump suction line, approximately 5.67 weight percent of ground limestone or hydrated lime per gallon.

Tests at the Mine Site

Tests were conducted to meet two major objectives:

Test A: The demonstration of sealing the water flow through the rubble barrier by injecting lime slurries in the space just upstream of it.

Test B: The neutralization of the acid mine water by injecting lime into the space behind a bulkhead so that the resulting precipitate of ferric and ferrous hydroxide remained in the mine, and clear, neutralized water flowed out through the 12-inch drain pipe.

These tests are described below.

Test A, Part 1, Seal Outflow Through the Rubble Barrier: A series of water samples were obtained from the mine between March 25 and April 18, 1969, and analyzed for chemical content. The results of the analyses are presented in Table 2.11.

The No. 3 (main) weir, the first installed, served to establish the flow characteristics of the total water coming out of the portal. Daily readings were taken, commencing October 22. The data show that the water outflow was slowly diminishing during the period prior to Test A. A rainstorm between October 29 and 31 washed out the No. 3 weir, and readings were not resumed until November 3. Rainwater deposits did not appear at the portal until November 3 (about 5 days after the rain started), they rapidly rose to a peak of about 750 gpm, and then fell off to an average flow of about 350 gpd. For Test A, the slurry injection schedule into the No. 1 west entry was 4 hours of hydrated lime alternating with 4 hours of pulverized limestone. A high-calcium hydrated lime (with a screen analysis of 98% through 325 mesh) and a pulverized high-calcium limestone (with a screen analysis of 82% through 200 mesh; 99.6% through 60 mesh) were used. This schedule was maintained for 40 hours, during which time a total of 6,500 pounds of hydrated lime and 6,050 pounds of pulverized limestone were injected.

The following 22-hour period, injection of limestone was discontinued, and 11,000 pounds of hydrated lime was injected into the space behind the rubble pile in the No. 1 west entry to raise the pH above 8; this eventually occurred.

TABLE 2.11: MINE WATER CONTENT PRIOR TO BULKHEAD CONSTRUCTION

Date	Chemical Content (mg/1)						
	$CaCO_3$ Alkalinity*						
	P	Total	Fe	Ca	Mg	Al	Mn
March 25	2120	395	1046	220	163	163	21
March 30	2188	293	1145				
April 4	2266	312	1162				
April 9	2152	295	1101	232	119	185	21
April 14	2172	188	1126				
April 18	2228	143	1161	276	113	168	21

*P = Phenolphthalein; Total = Methyl Orange

Source: EPA-670/2-73-092

Prior to the test, water samples were taken periodically from the No. 1 bulkhead outflow; during the test, samples were taken hourly from the No. 1 weir. These were analyzed in the field laboratory for free acid, total acid, total iron, ferrous iron, and total sulfate. Results of the analyses are summarized in Table 2.12.

Readings of flow and pH at the No. 1 weir were taken hourly. The flow of water rapidly dropped to zero at about the 62nd hour of slurry injection, indicating that a plug had been formed in the rubble pile. A very small flow from the No. 1 weir started a few hours later at about 1 gpm and varied between zero and 1½ gpm thereafter. Limestone mixed with hydrated lime was injected again at about the 65th hour and continued until the 116th hour, when the injection lines plugged.

Approximately 24 hours later, one injection line was unplugged and hydrated lime was injected alone; within 9 hours, the line plugged again. When the injection lines became plugged the second time (after 149 hours), the flow from the No. 1 weir was about ½ gpm and the valve on the 12-inch pipe from the No. 1 bulkhead was closed. Because the valve was in a rather isolated area on the lower haulage track level of the portal, the closure was not noticed while efforts were underway to reopen the injection lines. The entry between the rubble pile and the No. 1 bulkhead was filled with mine water up to the invert of the 11-inch drain pipe. Thus, since the space behind the rubble pile was filled with water, the seal was under no pressure head.

TABLE 2.12: ANALYSIS OF WATER OUTFLOW

Component	Before Plugging Test 300-400 gpm (Av mg/l)	During Plugging Test 1/2-1 gpm (Av mg/l)
Free Acid	0	0
Total Acid	160	0
Total Fe	1400	24
Total Fe^{+2}	1000	5
Total SO_4	5000	4500
pH	5.1	8.8

Source: EPA-670/2-73-092

In an attempt to complete all field operations before impending bad weather forced the winter shutdown, attention was then directed to the continuous neutralization demonstration (Test B). However, while the neutralization test was being conducted, efforts to unplug the injection lines into the west entry continued and proved successful. Attempts could then be made to seal off the water from the west entry. When the No. 1 valve was reopened after being closed 168 hours, the water from the west entry No. 1 bulkhead flowed at 20 to 45 gpm, indicating that the seal had deteriorated.

Test A, Part 2, Attempt to Reestablish Seal: The equipment used for the initial plugging test was also used during the attempt to reseal the rubble pile in the No. 1 west entry. When the No. 1 valve was reopened after having been closed for 168 hours, the water issuing from the No. 1 weir box had dropped to the normal raw mine water pH level of 3.8 to 4.0.

The outflow of mine water, which was 20 to 45 gpm when the valve was reopened, increased to 90 to 200+ gpm by the time the neutralization tests were completed. This led to speculation that the great reserve of acidity in the mine water in back of the plugged rubble pile barrier had diffused into the gel and redissolved the ferrous and ferric hydrates, causing failure of the initial plug. Presumably, resumption of the injection of lime slurries would reestablish the seal. To verify this assumption, an attempt was made to reestablish the seal.

During the first test period, only hydrated lime was used; 39,300 pounds were injected continuously for 75 hours at a rate of 525 pounds per hour. The pH of the outflowing water rose to 10.2, but there was no significant drop in water flow. During the second test period, which lasted 22 hours, a mixture consisting of 4,400 pounds of hydrated lime and 5,440 pounds of pulverized limestone was injected with no noticeable effect on the water flow.

For the next 5 hours (the third test period), injection of hydrated lime was continued but pulverized limestone was omitted. At the same time, however, the production of sludge was increased by augmenting the mineral content of the water with waste pickle liquor containing 5 to 6% H_2SO_4 and 5 to 6% Fe^{++}, which was added to the hydrated lime slurry at a rate of approximately 1 gpm as the slurry was being injected into the mine workings. Addition of a total of 2,150 pounds of hydrated lime and 200 gallons of waste pickle liquor caused no noticeable effect on the water flow.

Throughout the next 9 hours (the fourth test period), a mixture of 1,550 pounds of hydrated lime and 1,280 pounds of pulverized limestone was injected without discernable effect on the outflowing water at the No. 3 weir. At this point, the Mixmeter broke down and injections could not continue until repairs were completed 29 hours later. During the repair period (Period 5), the outflowing water pH dropped to 4.7.

Injection of a mixture of hydrated lime and pulverized limestone was resumed (Period 6), but 49 hours later, after 9,800 pounds of hydrated lime and 10,560 pounds of pulverized limestone had been injected, the injection system broke down again. Again the pH rose to the 11 to 12 range.

After repairs were completed 6 hours later, injection of a mixture of hydrated lime and pulverized limestone was continued again for 17 hours, during which time 3,200 pounds of hydrated lime and 1,680 pounds of pulverized limestone were injected, again without any effect on the water outflow.

It is evident that attempts to stop the flow of water through the rubble barrier were unsuccessful even though 60,400 pounds of hydrated lime and 18,960 pounds of pulverized limestone were injected into the entry. However, the pH of the outflowing mine water rapidly increased to the 11 to 12 range when the slurry was injected and rapidly fell off to the base mine water range when the injection stopped. After the last injection period, the water flow and pH were monitored for an additional 39 hours while arrangements were being made to shut down for the winter.

Test B, Part 1, Continuous Neutralization: Test B, Part 1, the continuous neutralization of the outflowing water, was conducted behind the No. 2 bulkhead (see Figure 2.14). The water from this area of the mine flowed out of the No. 2 weir and was relatively isolated from the lime injection behind the No. 1 bulkhead. Flow from behind the No. 2 bulkhead was controlled by the No. 2 valve at the downstream end of the 12-inch pipeline just before it entered the No. 2 weir.

For the purposes of this test, the No. 2 valve was partially closed to reduce the water flow to approximately 150 gpm, and, once adjusted, was not changed for the duration of the test. However, the flow varied from 130 to 200 gpm, presumably because of changes in the head in the mine. For this test, a slurry containing 0.5 pound of hydrated lime per gallon was prepared using two Shirley Mixmeters. After 200 gallons of water were pumped into the charging tank of the first Mixmeter, the mixing arm was started and 100 pounds of hydrated lime were added. The charge was mixed until a uniform slurry was formed. A similar charge was prepared in the charging tank of the second Mixmeter. To provide a continuous supply of hydrated lime slurry during the test, as each charging tank was emptied, a new charge was prepared. The Mixmeters were used only as slurry mixers and the batches were transferred to a 55-gallon open-end drum which served as a feed tank for four small paddle pumps. Each pump was connected to one of the six 2-inch injection lines that passed through the No. 2 bulkhead into the mine.

The four injection lines used were Nos. 2 and 3 which terminated 65 and 50 feet into the north entry and Nos. 5 and 6 which terminated 105 and 90 feet into the west entry beyond the bulkheads. Each of these pumps had a capacity of 2 gpm of slurry; since they were operated together, a total of 8 gpm of slurry was injected when the tests were made.

After 39 hours, use of the paddle pumps was discontinued because the rotors were worn out, terminating the first period of the slurry injections. Operations were stopped until replacement pumps could be obtained. Both before and after the test the mine water had a pH of 3.8 to 4.5, which indicated no lasting effect of the lime injection.

Theoretically, a dosage rate of 4.0 pounds of hydrated lime per minute should raise a pH of 3.5 to 11.1. However, this did not occur; the actual mine water readings at the No. 2 weir ranged from 3.6 to 4.6 during the 39-hour period that the hydrated lime slurry was injected.

The Period 2 test was conducted with a gasoline-driven pump, which permitted the injection of the lime slurry at a higher rate than the Period 1 test; i.e., 20 gpm. The outflow of mine water was the same, and the pH, ranging from 4.4 to 4.5 during the 10-hour period between the tests, was essentially the same. Theoretically, the lime injection rate in this test was sufficient to raise a pH of 3.5 to 12. Again, however, the outflowing water failed to respond to the treatment and its pH ranged from 4.4 to 4.8 during the 26 hours of testing.

Test B, Part 2, Dye Injections: Failure of injection of the hydrated lime into the mine workings behind the No. 2 bulkhead to increase the pH of the outflow at the No. 2 weir raised the question of where the outflow was coming from. Therefore, a series of dye solution injections were made to determine the flow pattern of the water draining from the 12-inch No. 2 drain line. Both the north entry and the No. 2 west entry were tested in this manner. The dye solution was prepared by thoroughly mixing 2 ounces of methylene blue in 50 gallons of mine water. The injections were made using the 20-gpm pump, and took 2 minutes to complete.

North Entry – During the first dye test, the dye solution was injected through the No. 1 injection line into the north entry at a point 80 feet from the No. 2 bulkhead (see Figure 2.14). The control valve in the 12-inch drain line was opened completely, which gave a 900-gpm outflow of mine water. In this test, the dye appeared at the discharge of the No. 2 drain line 30 minutes after it was injected. The outflow was allowed to continue for approximately 2 hours to purge the dye from the north entry so that the dye testing could be repeated with a reduced flow. In the next part of this test, the control valve was partially closed to restrict the discharge flow to 150 gpm, and a second injection of dye solution was made under the same pumping conditions. This time the dye solution did not appear at the discharge point after 4 hours, which was long enough to drain a volume of water from a mine one-third greater than that necessary to show the dye at the higher rate of outflow.

No. 2 West Entry – For the second dye test, the dye solution was injected through injection line No. 4 into the No. 2 west entry at a point 120 feet from the No. 2 bulkhead (see Figure 2.14). As in the first dye test, the valve controlling the 12-inch drain line was left completely open and discharged mine water at a rate of 900 gpm. This time the dye appeared at the discharge 60 minutes after it was injected into the mine entries.

With the valve again partially closed to restrict the discharge flow to 150 gpm, a second injection of dye solution was made. Observation of the outflow for approximately 8 hours revealed no dye at the discharge outlet. As before, the volume of water drained from the mine at the low outflow rate exceeded the volume that showed the dye at the high outflow rate by one-third.

The dye tests indicated that the flow pattern of the water adjacent to the ends of the injection lines in the north entry and the No. 2 west entry changes when the outflow from behind the No. 2 bulkhead is changed. Obviously, when the valve is wide open, the flow at the end of the injection line is sufficient to carry the dye into the 12-inch pipe at the No. 2 bulkhead. When the valve was partly closed, the outflow was insufficient to bring the dye to the inlet of the 12-inch pipe. Presumably, this same effect was occurring when the slurry was being injected.

Thus, any effect of acid water neutralization by the hydrated lime would not be observed at the weir box for the flow rate of 150 gpm. Injection of the slurry at flow rates substantially greater than 20 gpm, to be compatible with the 900-gpm outflow rate, was not possible with the existing equipment.

Followup Inspection of Minesite, Spring 1971: A field inspection of the minesite was made on May 3, 1971, after the relatively mild winter. The weirs were still in place and no subsidence or other damage was observed at the portal. Weir and pH measurements made at the site are summarized in Table 2.13.

TABLE 2.13: FOLLOWUP WEIR DATA

Time (hrs)	No. 1 Weir		No. 2 Weir		Main Weir	
	Flow (gpm)	pH	Flow (gpm)	pH	Flow (gpm)	pH
800	15	4.2	454	3.8	560	3.9
830	16	4.3	530	3.7	545	3.9
900	15	4.1	545	3.7	560	3.9
930	15	4.3	545	3.8	560	3.8
1000	15	4.2	545	3.9	560	3.8
1030	16	4.2	545	3.9	560	3.9
1100	16	4.2	545	3.8	560	3.8
1130	16	4.2	545	3.8	560	3.8
1200	16	4.2	545	3.8	560	3.8

Source: EPA-670/2-73-092

Elevations were measured on the bottoms of the 12-inch discharge pipes, and showed that the No. 2 pipe was 0.62 foot lower than the No. 1 pipe. The values on both pipes were wide open and No. 2 pipe was blowing full. The higher pH at the No. 1 weir indicates that the lime slurry injected into the No. 1 west entry is still causing some neutralizing effect. The No. 2 and main weirs were badly corroded by the acid water.

Test Results

Sealing: After hydrated lime and ground limestone had been injected into the acid mine water behind the rubble pile for 65 hours, the flow through the emplaced rubble pile was reduced from approximately 500 gpm to zero. Thus, the initial sealing was successful.

Although the flow had been reduced to zero when the seal was first accomplished, some water began to seep at the rate of about 1½ gpm, as observed at the No. 1 weir during the test period, until the 149th hour, at which time the valve was closed. Continued injections of lime and/or limestone had no effect in reducing these seepages to zero.

An attempt was made to reestablish the seal when the No. 2 valve was reopened, but it failed. Even adding waste sulfate pickle liquor to augment the minerals in the water and produce a higher volume of precipitate failed.

As the pH of acid mine water is raised by the reaction of neutralizing agents with the acid water, dissolved ferrous and ferric ions precipitate as a sludge consisting of hydroxides or hydrated oxides. The calcium from the lime forms calcium sulfate, most probably gypsum, which is crystalline and only slightly soluble. The form and characteristics of this sludge can vary extensively, depending on such variants as water constituents, neutralizing agents, and the rate of addition of neutralizing agents.

While the sludge varies from a light, fluffy floc to a dense particulate precipitate, it is always in a nonfilterable form that would blind a filtering medium. Its nonfilterability is the basis for using the sludge to seal drainage openings, using the water flow to carry the sludge into the voids to plug them.

However, for the precipitated sludge to be effective, the size of the voids or channels that the sludge is to seal must not exceed some unknown maximum. It is certain that the precipitate, when relatively freshly formed, did indeed stop the flow of water through the channels of the rubble pile.

Flows of ½ to 1 gpm are not regarded as significant indications of seal failure; thus, it was during the time that the No. 1 valve was closed (after 149 hours) that the seal failed, and flows of 20 to 45 gpm could be observed when the valve was reopened. Two factors may have contributed to this: (1) the well-known shrinkage of gels as they age may have loosened the plugs, and (2) diffusion of acid mine water into the gelatinous plugging material may have caused resolution of the iron hydrate gel. It is entirely possible that the neutralization of the acid mine water at the face of the rubble barrier was badly impaired by the stratification of the lime slurry precipitate and its flow downslope into the mine depths. This could have occurred when the slurry fell from the ends of the injecting pipes to the floor of the entry. Since the water at the face of the rubble barriers was stagnant, once the seal was established there was no flowing force to carry or hold the injected slurry and precipitates into the channels and maintain the seals. When the seal failed later and water flows upward of 100 gpm set in, the efforts to reestablish the seal could have failed because the sandbag seal at the top of the barrier had failed and the resulting channels were too large for the gel to close. The sandbags had been in position for approximately a year and protected only the last 18 to 24 inches of the 25-foot crown of the rubble pile that had settled a few inches beneath the entry roof.

During the tests to reestablish the seal, the outflowing mine water was highly alkaline and slightly murky because of blue-green iron hydrates. This led to the hypothesis that one or more of the injection lines had been ruptured between the No. 1 bulkhead and the rubble pile, and that slurry was issuing in that zone. The rupture of the injection line could have been caused by the efforts to clear the pluggages and possibly from corrosion by the acid mine water during the 13 months or so between their placement behind the No. 1 bulkhead and the commencement of the tests.

However, it is postulated that the bulk of the gel rapidly sank to the floor of the entry and was unable to act as a sealing agent near the roof at the top of the rubble barrier, even though approximately 40 tons of material had been pumped into the mine.

Laboratory tests revealed that the sludge does, in fact, stop the flow of water through a porous sand barrier. Seals were formed within minutes, remained watertight for weeks, and withstood calculated hydrostatic heads up to 200 feet. Laboratory tests were performed using 2-inch tubes to simulate a coal mine adit that measured 4.5 by 14 feet at Driscoll No. 4. Under laboratory conditions, such problems as the settling of particles in the porous barrier and movement of surrounding rock do not affect the long-term stability of the seal.

In the field, however, these problems could become serious. The breakdown of the seal in the field was disappointing, especially since the cause of the breakdown could not be determined. While many explanations could be postulated, the sole cause cannot be determined with any degree of certainty. A combination of factors was probably to blame.

It is possible that, in trying to reopen the plugged injection lines, the compressed air that was forced into the injection lines stirred up the exposed layer of sludge that formed part of the seal. If this was the cause of the breakdown, then a fast development of the seal is undesirable because the sludge does not have the opportunity to penetrate very deeply into the rubble pile and can easily be disturbed.

It would also indicate that a lower dosage level would be desirable so as to form the sludge at a slower rate. The increased water flow for a longer period of time would then carry the sludge farther into the rubble pile. Although it would take longer to develop a seal, the seal might be more durable.

Another possibility is that the initial composition of the ferrous hydroxide sludge gradually converts to a hydrated ferric oxide, which has a higher density than the ferrous hydroxide. This would result in shrinkage cracks similar to the mud cracks that develop as a lake bottom dries up. If this was the cause of the breakdown, the need for the development of the seal at a slower rate is again indicated. The conversion of lightweight, green, fluffy floc to a reddish granular form characteristic of precipitating ferric ion, as hydrated ferric oxide, was observed in the laboratory. With the seal forming at a slower rate, more time would be available for the oxidation of the ferrous hydroxide to the hydrated ferric oxide, and additional sludge could fill the cracks as they develop.

The seal, once achieved, would not be subject to the crack formation from subsequent shrinkage. This shrinkage phenomenon was observed in the sludge settling in the bed of the drainage ditch immediately upstream of the main weir, but downstream of the screen across the ditch. Although the sludge was always submerged and never subjected to drying, cracks developed in the sludge that was deposited in the relatively quiescent pond.

It is also possible that during the injection of limestone some of the limestone does not react with the acid water, perhaps due to poor dispersion in forming a slurry, and is carried with sludge into the rubble pile to react subsequently with acid water. Carbon dioxide would be liberated and, if formed in a large pocket, could suddenly erupt, breaking the seal.

Continuous Neutralization: During the first 39-hour period of the continuous neutralization trials, the dosage level of the lime addition should have resulted in a rise in pH to 11.1. However, the pH never rose above 4.6. During the next 10-hour period, when no lime was added, the pH did not drop significantly. During the last 26-hour period, the dosage level of lime addition should have resulted in a pH rise to 12. However, during this period the pH did not rise above 4.8.

The dye tests showed that when the No. 1 valve was only partially open, and the outflow of mine water was low, the flow from the end of the injection pipe to the inlet of the 12-inch drain pipe was essentially nonexistent. Conversely, with the valve fully open, the high rate of outflow from the drain pipe created a rapid flow from the injection points to the drain pipe inlet. This shows that when the lime slurry was injected, there was essentially no flow toward the drain pipe; thus, there was no increase in pH, as anticipated.

About 3.4 times more hydrated lime was added than was necessary to neutralize the iron content of this mine water. The flow rate from the No. 2 lines was 130 to 200 gpm; thus, 585,000 to 900,000 gallons. Thus, water outflowed during the 75-hour test period (including the 10 hours between Test 1 and Test 2 when no slurry was injected). Consequently, there were between 8.2 to 12.7 theoretical water changes in the entries during the test period, which would be a change every every 5.9 to 9.1 hours. Since the dye tests did not show any flow to the drain from the injection point at this level of outflow, there must have been a significant flow away from the injection points into the depths of the mine.

During some rough bench tests in the field laboratory, it was also observed that the floc formed by the reaction of the lime slurry and the mine water rapidly settled to the bottom of the vessel in large clumps or curds. This probably took place at the end of the injection pipes located at the roof of the entries. The inlet to the 12-inch drain pipe is also at the roof where it comes through the No. 2 bulkhead. Any stratification of the lime precipitates at the bottom of the entry, back to the inlet of the drain pipe, would have approximately 2 to 3 feet before its level reaches the inlet to the 12-inch drain pipe. Thus, the flow away from the drain inlet and the rapid settling of the lime-mine water floc to the entry floors combined to migrate the lime slurry reaction products away from the drain. This would explain the negative result of the continuous neutralization tests.

The probability that this method of treatment can be successfully applied is indicated by the results obtained while attempting to reestablish the seal in the rubble pile. The pH of the discharging water, which was 3 to 4 prior to resumption of the lime injection, gradually rose to 11, then 12 during the lime injection and remained at 12 during the lime and limestone injection. When the injection was stopped, the pH dropped back into the 3 or 4 range. When the injection of a lime and limestone slurry was resumed the pH again rose to approximately 11, then gradually dropped back to approximately 4 after the injection was stopped for the last time in preparation for shutting down the test site.

Thus, while a seal was not reestablished, continuous neutralization of the acid mine water (without the iron oxidation step) was taking place and, presumably, the sludge being formed was settling back into the mine.

NEUTRALIZATION PROCESSES

The material in this chapter was excerpted from EPA 670/2-73-093 and U.S. Patents 3,516,931; 3,617,559 and 3,617,560 (see Bibliography, p 390).

Neutralization processes are designed to produce a water which will have minimum detrimental effects on the bioecological system. The treated waters must not detract from the appearance of receiving streams and be acceptable for navigational, recreational and industrial purposes. The development of potable-quality water for human consumption has not been a primary objective. These processes adjust the pH of the waters to levels above 6.0.

The treatment creates conditions which form insoluble compounds from the iron, aluminum and sometimes calcium sulfate contained in the drainage. The insolubles must be removed from the treated water. The quality criteria for such discharged waters established by the Sanitary Water Board of the Commonwealth of Pennsylvania in 1964 have served as general guides. These require that discharges to the streams of the Commonwealth have: (1) a pH between 6.0 and 9.0, (2) iron content less than 7 mg/l, and (3) no titratable acidity.

It appears these criteria should result by reacting any inorganic alkali providing it does not introduce other inimical components. The volume of such waters which may require treatment reaches millions of gallons daily, thus pragmatic considerations also necessitate the utilization of processes having minimal requirements for capital equipment, labor, process control, and operating costs. They must also offer a high degree of reliability.

These processes can be conceived as cation exchange reactions in which the hydrogen, iron, and aluminum ion species in the drainage waters are replaced by calcium, magnesium, or sodium ions. The term "cation exchange reactions" describes the replacement of the undesirable cation components of the drainage waters by more acceptable soluble ions. The detrimental components are converted to insoluble substances. In no way does the term refer to processes labelled "ion exchange" which involve a second phase, insoluble substrate. The water products resulting from neutralization treatment are not significantly improved in total dissolved solids content.

Although the number of neutralization processes that can be used to achieve these minimal quality standards is limited, the factors involved in choosing a process are more complex than commonly conceived. Once the characteristics of the product waters are defined, the process choice depends upon the loading of the drainage to be treated, its volume, logistical problems of the water sources and the treatment facility as relates to each potential reagent. The chemistry of the reactions defines the engineering criteria. The cost and availability of the various reagents establishes the pragmatic choice.

Since waters dissolve various substances as they flow through coal measures, the characteristics of the drainage differs widely with strata mineralogy. The principal mineralization results from the soluble oxidation products of the indigenous mineral pyrite. The components of concern include the following cations: hydrogen, ferrous, ferric, and aluminum. Sulfate is the predominating anion. The species in which these ions exist in coal mine drainage waters probably vary but they have not been fully defined. Furthermore, under the in situ conditions, these components are undergoing continuing internal reactions.

These reactions may be enhanced biochemically or by their contact with other strata, adding ions such as calcium, magnesium, sodium, potassium, carbonate, bicarbonate, and, to a lesser degree, manganese and chloride as well as a host of other ions at trace level concentrations. The many interactions and equilibria which exist in mineralized waters or may be developed with the introduction of inorganic alkalies create systems that may be described as "buffered." The waters considered in the study described in this chapter represent the preponderance of coal mine drainages. The generalized reactions which occur in the desired cation exchange processes are of two types:

(1) Neutralization—The available hydrogen ions in the drainage waters react with negatively-charged radicals of the reagents to form weakly-ionized species, such as water. In this chemical system, the solubility products of the iron and aluminum ionic species are exceeded, thus reducing their solution concentrations.

(2) Oxidation of Fe(II) to Fe(III)—The solubility product of ferrous hydroxide is much larger than that of ferric hydroxide. With the predominance of the ferrous species in most drainage waters, the provision of a greater hydroxyl concentration is necessary than required for ferric systems to achieve acceptable levels of iron concentration. The final pH of waters so treated may exceed the criterion for hydrogen ion concentrations. During Fe(II) oxidation, additional acid species form which represent an unacceptable pollution capability. Oxidation of the iron to the higher valence during processing obviates these problems. The Fe(II) oxidation with oxygen from air tends to proceed after neutralization. The oxidation rate increases with pH at levels above 4.0.

In these neutralization reactions, the hydroxyl ion is really the only ion species feasible to form a weakly ionized, polar compound with the hydrogen ion. The carbonate ion will react with the hydrogen ion to form carbonic acid, which will dissociate to form water and dissolved carbon dioxide. Because of the limited solubility at ambient temperatures and pressures, the carbon dioxide can be removed from the water.

The carbonates also form insoluble compounds with iron and aluminum species. Thus, the two anions, hydroxide and carbonate, are the fundamental reagent constituents in the treatment of coal mine drainage waters. They are available as compounds of calcium, magnesium, sodium, and ammonia at costs feasible for these processes. Other cations form hydroxides and carbonates, but their cost and availability eliminate them from consideration.

Ecological principles usually preclude the use of ammonium hydroxide since the soluble ammonium ion serves as a nutrient for plant and animal organisms whose growth in receiving streams is undesirable. The physiological action of magnesium ion as a cathartic, especially when associated with sulfate ion, makes its introduction into streams undesirable.

The reagents which may appropriately be considered for achieving the cation exchange reactions necessary for neutralization of mine waters are limited. These include: calcium carbonate (high calcium limestone), calcium oxide (calcined, quick, or pebble lime), calcium hydroxide (hydrated lime), calcium carbonate-magnesium carbonate (dolomite or dolomitic limestone), calcium oxide-magnesium oxide (burnt or calcined dolomite), calcium hydroxide-magnesium oxide (hydrated dolomite or partially hydrated dolomite), calcium hydroxide-magnesium hydroxide (pressure hydrated dolomite), sodium carbonate (soda ash), and sodium hydroxide (caustic soda). The oxides of calcium and magnesium meet the requirements for reagent anions (e.g., they function as hydroxides) since as anhydrides they react with water to form hydroxides.

Reagent costs, as experienced during the project studies, are cited in Table 3.1. Many other considerations become basic in choosing among these reagents. In planning a treatment system, the selection of a reagent would be the initial decision since this information forms the basis of the processes, plant design detail and economics. The selection is established by the drainage water analyses and volume, plant location and the reagent performance potential. Each of the following factors should also be optimized for the reagent: source geographical distribution, availability, transportation, handleability, cost, reactivity, chemical and physical properties and their mechanistic action within the treatment system as this affects the quality of the final water product and the characteristics of the impurity sludges.

High calcium and dolomitic limestone are naturally-occurring mineral products. The remaining reagents are all manufactured products which partially accounts for their higher cost. Quality and reactivity of high calcium and dolomitic limestone and their products vary widely with origin and manufacturing process. Calcined materials contain 5% or more inert constituents which effect costs and processes. The differences in chemistry between dolomite and high calcium materials must also be considered.

The physical form of caustic soda and soda ash (solid, flake, solution, etc.) selected must be carefully evaluated. There are major cost differences and they have an impact in determining reaction conditions, ambient storage and reaction temperatures, transportation, handling, storage and safety. Since the demand for these reagents for mine drainage treatment is in no way market controlling, consumption for this purpose must be planned around prevailing commercial products; e.g., particle sizes and qualities. The various reagents have definite limits of interchangeability within a given plant design.

TABLE 3.1: REAGENT COST DATA (1968 to 1971)

Reagent	$/ton fob source	Transportation ¢/ton mile	Delivered Cost Hollywood $/ton	Delivered Cost Hollywood ¢/lb	Reagent Cost at Hollywood[a] ¢/neutralization equivalent ¢/gal/mg/L/acidity × 10^{-6}	Delivered Cost Ratio to Limestone on Neutralization Equivalent Basis
Limestone						
Stone sizes	1.50	3.	4.50	0.23	1.92	--
Waste	0.90	3.	3.90	0.20	1.67	0.87
Pulverized		3.	8.50	0.43	3.59	1.87
Calcined lime						
Bulk	16.00-18.00	3.	21.00	1.05	4.91	2.56
Bagged	19.50	3.	23.50	1.18	5.52	2.88
Hydrated lime						
Bulk	18.25-20.75	3.	23.75	1.19	7.35	3.83
Bagged	21.25-23.75	3.	28.25	1.41	8.71	4.54
Dolomite						
Stone sizes	1.80-2.50	4.	10.00	0.50	3.84	2.00
Pulverized	7.00	4.	15.00	0.75	5.76	3.00
Calcined dolomite						
Bulk	16.00	4.	24.00	1.20	4.82	2.51
Bagged	21.00	4.	29.00	1.45	5.83	3.04
Hydrated dolomite						
Bagged	19.75	4.	29.50	1.48	7.06	3.68
Pressure hydrated dolomite						
Bagged	23.00	4.	31.00	1.55	8.56	4.46
Caustic Soda						
50% Solution						
Tanker			66.00	3.30	22.02	11.47
Drum			268.00	13.40	89.40	46.56
Flake						
Drums, Bulk			78.00	3.90	26.02	13.55
Solid, small lots			110.00	5.50	36.69	19.11
Soda Ash						
Powder						
Bulk			40.00	2.00	17.68	9.21
Bags			76.00	3.80	33.58	17.49

[a]Assumes 100% reagent purity.

Source: EPA 670/2-73-093

By-products of manufacturing processes, as "carbide lime" from acetylene manufacture, may offer cost or process advantages but their use should be carefully evaluated, including consideration of quality, uniformity, long range availability, etc.

Pulverized limestone and lime products are commonly transported in bulk utilizing pneumatic handling. Loads range from 15 to 28 tons. Air pollution by dust release during unloading can be controlled and is to be avoided. Reagent and shipment costs by bags of up to 100 pounds capacity are higher and also increases subsequent reagent handling costs. Pulverized stone products, hydrated lime and calcined lime are conveniently stored in bins.

Bin storage of hydrated lime can result in extensive water and carbon dioxide absorption within several weeks; thus long-term lime storage which can result from excessively large bins is to be avoided. In addition to bin storage, pneumatically-handled hydrated lime may be slurried (10 to 50 wt %) directly from the delivery tanker and stored as a suspension. Larger particle sizes of limestone ($>$ ⅜") have the best handleability, do not require protected storage facilities, and are not affected by the weather. Some stone products having large amounts of fines may handle differently when wet.

The total components comprising reagent costs in mine drainage treatment extend beyond the fob unit cost (as per quantity, quality and bid), to incorporate transportation, handling, storage, stoichiometry, use efficiency, etc. The reagent cost may be expressed as cents/1,000 gal water treated and compared with the theoretical value. Overall one seeks the best system efficiency which is here defined as the ratio of total treatment costs to reagent costs. For the coal operator, a more meaningful expression may be cents/ton of saleable coal produced; or for the ecologist, some broader expression as $/mi of stream controlled.

Critical factors in reagent selection and use also include the location of their origin, quality, mode and points of manufacture-distribution, mode of transportation, and availability. A proper understanding of the role of these aspects is important to the correction of the coal mine drainage problem.

DESCRIPTION OF PILOT PLANT

The concept of the Hollywood Facility was developed to evaluate and compare different chemical neutralization processes for treating coal mine drainage. Coal mine drainage waters would be pumped to the Facility, while gravity flow through the plant was sought with discharge of the effluent to natural drainage. The variability of the water sources and processes precluded optimum design for each situation. The basic unit operations would involve alkali neutralization, Fe(II) oxidation, and solid-fluid separation of the impurity sludge. Provisions included raw water storage conforming to Pennsylvania Sanitary Water Board criteria and sludge dewatering and disposal.

Capability would be included to utilize any alkaline reagent, including limestone which would require a specially-designed reactor. The thickener would involve a specially designed unit, the Infilco Densator, to permit study of sludge recycle concepts. The settling lagoon incorporated an original design for continuous sludge handling. Provision for the use of flocculant aids would be included.

Automatic plant control would be limited to pH response of alkali introduction while instrumentation would permit data collection of several process parameters, including water quality, temperature, and flow. Power consumption would be measured on the larger motors. The summarized design plan is given in Table 3.2 and Figure 3.1. The initial design permitted evaluation of only one process at a time. Data on the mine drainage sources are given in Table 3.3. In Figure 3.2 is presented the hydraulic profile for the various unit operations.

TABLE 3.2: BASIS OF DESIGN OF THE HOLLYWOOD FACILITY

Mine Drainage Water Volume		
gallons per day		500,000
gallons per hour		20,833
gallons per minute		347
Total Acidity - tons/day $CaCO_3$		2.08
Total Iron - tons/day Fe		0.21
Alkali Requirement		
tons/day, hydrated lime ($Ca(OH)_2$)[c]		1.62
Oxygen Requirement		
for iron oxidation pounds/hour[d]		2.52
Estimated Air Requirement		
ft^3/min, 45 minute retention		22
Thickener or Settling Pond Sludge Resulting from Process		
Solids content, percent by weight		1.0
gallons/1000 gallons water treated	a.	50
	b.	10
gallons/day	a.	25,000
	b.	5,000
gallons/hour	a.	1,040
	b.	208
Dry Solids in Sludge		
pounds/1000 gallons water treated		5
tons/day		1.25
pounds/hour		104
Densator-Thickener Dimensions (Diam. X SWD) feet		25 x 20
Drum Filter		
Solids loading pounds/square foot/hour	a.	5.0
	b.	15.0
1-6 foot diameter by 7 foot long rotary drum-belt vacuum filter Filter area, square feet		125

a. Based on sludge from normal settling.
b. Based on dense sludge.
c. Assumed to have 95% available CaO.
d. Assuming all iron in ferrous state.

Source: EPA 670/2-73-093

FIGURE 3.1: SCHEMATIC DRAWING OF TREATMENT PLANT

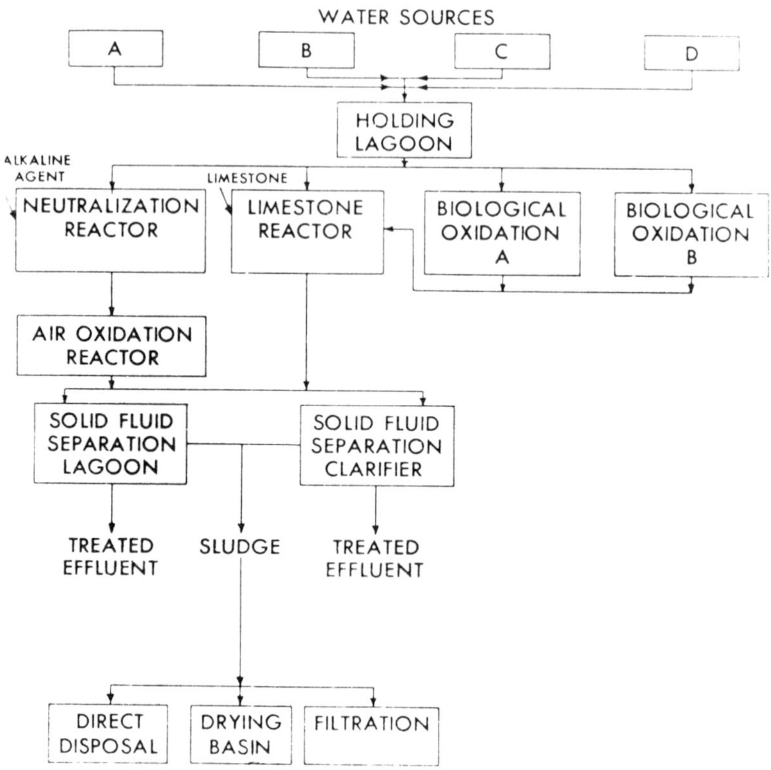

Source: EPA 670/2-73-093

TABLE 3.3: SOURCES OF MINE DRAINAGE ESTIMATED VOLUMES, CONSTITUENTS AND CHARACTERISTICS

Source	Proctor No. 2	Tyler Run	Proctor No. 1	Bennett Branch
Volume - mgd				
Maximum	0.063	0.600	0.288	not available
Average	0.025	0.119	0.199	30
Minimum	0.009	0.003	0.127	3
pH				
Maximum	3.81	4.30	3.49	4.31

(continued)

TABLE 3.3: (continued)

Source	Proctor No. 2	Tyler Run	Proctor No. 1	Bennett Branch
Average	3.05	3.35	3.05	3.43
Minimum	2.70	2.91	2.60	2.80
Acidity (as $CaCO_3$) - mg/l				
Maximum	8,112	420	1,271	523
Average	6,700	330	780	278
Minimum	5,343	150	316	39
Total Iron - mg/l				
Maximum	3,604	90	199	105
Average	2,090	27	134	53
Minimum	1,516	2	99	11
Ferrous Iron - mg/l				
Maximum	2,300	90	199	105
Average	1,780	26	133	53
Minimum	1,360	2	99	10
Aluminum - mg/l				
Maximum	770	131	196	45
Average	725	100	141	30
Minimum	680	51	93	22
Calcium - mg/l				
Maximum	252	108	144	56
Average	150	77	102	40
Minimum	60	40	40	20
Magnesium - mg/l				
Maximum	129	114	132	46
Average	80	83	80	42
Minimum	3	58	10	39
Sulfate - mg/l				
Maximum	6,174	712	1,127	514
Average	5,150	640	1,080	385
Minimum	4,062	536	990	248
Temperature - °C				
Maximum	26	25	21	28
Average	19	16	15	18
Minimum	8	3	4	3
Specific Gravity	1.11	1.09	1.10	1.10

Source: EPA 670/2-73-093

FIGURE 3.2: HYDRAULIC PROFILE OF THE TREATMENT FACILITY

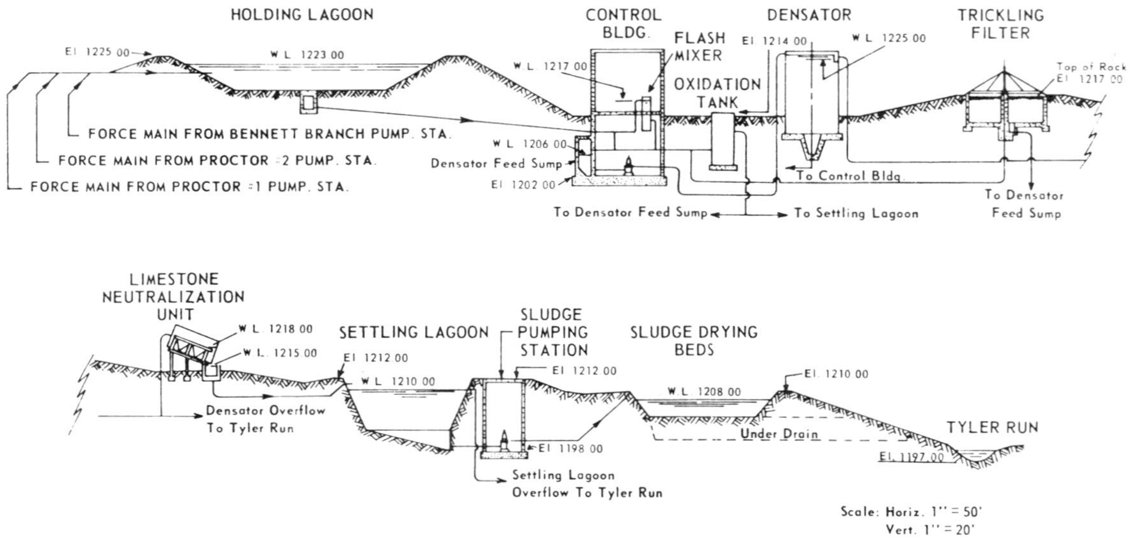

Source: EPA 670/2-73-093

NEUTRALIZATION TREATMENTS

With Hydrated Lime

The process most commonly utilized to treat coal mine drainage mixes the water with calcium hydroxide, raises the pH and forms insoluble iron and aluminum compounds. The reaction mixture is subsequently oxygenated converting Fe(II) to Fe(III). The suspended solids are then separated, discharging the treated water.

The studies, considering each of the four waters available, utilized this procedure with two modifications of solids separation: settling lagoon (Process 10) and thickener (Process 11). The process conditions and flowsheet for Process 10 are detailed in Table 3.4. In these tests the controlling pH probe was located in oxidation tank No. 2. The process conditions and flowsheet for Process 11, where the pH probe was located in the flash mixer, are given in Table 3.5. The reagent was introduced as an aqueous slurry containing between one and five percent calcium hydroxide. The system permitted the lime slurry to be controlled and introduced either by a manual valve or by a proportioning valve which responded to a setting of the pH probe.

TABLE 3.4: PROCESS 10–HYDRATED LIME

Water source	Average flow, gpm	Run duration hrs	Water treated gal	Test No.
Tyler Run	182	47.5	520,912	10-70
	247	67.5	1,001,060	9-70
Bennett's Branch	153	48.0	441,400	3-70
	240	48.0	693,600	2-70
	317	90.0	1,716,650	1-70
Proctor No. 1	112	18.0	120,636	8-70
	182	24.0	263,200	4-70
	251	60.0	904,400	5-70
	298	48.0	859,570	6-70
Proctor No. 1 and Bennett's Branch	342 (290+52)	48.0	984,960	7-70

Process Flow:

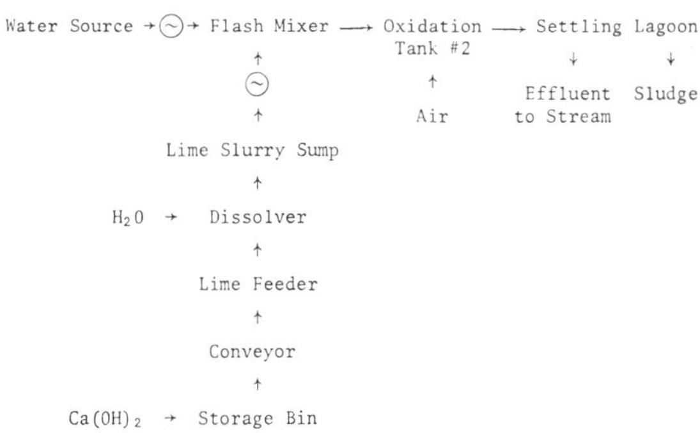

Source: EPA 670/2-73-093

TABLE 3.5: PROCESS 11–HYDRATED LIME

Water Source	Average Flow, gpm	Run Duration hrs	Water Treated gal	Test No.
Bennett's Branch	103	45.5	281,440	71-71
Proctor No. 1	96	26.8	154,700	70-71
	100	35.0	210,000	38-71
Proctor No. 2	87	22.8	119,470	66-71
	89	19.5	104,980	56-71
	97	5.3	30,860	68-71
	100	19.8	119,140	67-71
	106	24.0	153,310	69-71
	134	7.0	56,370	57-71
	178	15.5	165,830	65-71

Process Flow:

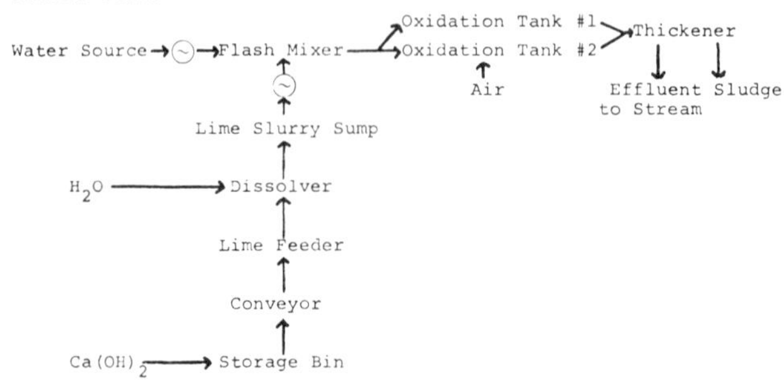

Source: EPA 670/2-73-093

A constant manually-controlled slurry addition was utilized when treating large volumes of the more highly mineralized water with the concurrent addition of lime slurry through the proportioning automatic valve. This procedure yielded minimal pH variation within the system. The manual valve alone did not provide reliable, constant slurry flow. The lime slurry feed system was effective in providing a continuous, controlled flow completely automatic from lime storage bin. No difficulties were experienced with lime scaling in the lines or hang-up in the bins. Due to intermittent operation and prevailing high humidity, the electronic controls of the feeder required frequent servicing.

Some carbonation of the lime slurry occurred in the lime slurry sump, especially in those operations where the lime consumption was low and the slurry was prepared at infrequent intervals. This introduced an uncertainty in reagent measurements, especially with short runs. The test program necessitated that some batches of hydrated lime remain in the storage bin for weeks. Although no handling or moisture absorption difficulties developed, there were major levels of reagent carbonation which changed the character of the alkali reagent, and its reaction mechanism. In operating treatment facilities, the lime storage capacity should not exceed one month's requirements.

The addition of lime (or any of the reagents) as an aqueous slurry avoids delays in wetting the solid. The slurry mixes with the water more rapidly. The solubilized lime reacts on contact while those solid particles remaining are more rapidly dispersed for solution and reaction. It appears desirable to add the reagent at the most dilute level feasible. The total reagent requirements establish the limits. This rapid mixing and reaction permits a more responsive action of the controlling pH probe. Localized excesses of reagent are less likely to occur thus avoiding unwanted side reactions. Reagent addition control is most versatile when a slurry is employed. Large clumps of unwet lime are more likely to exist with dry reagent feeding and to remain together leading to possible coating and to loss of unreacted reagent.

The lime slurry was prepared with either house water or treated effluent. The preparation of a concentrated (10 to 20%) reagent slurry with good quality water (as in the mixer-dissolver), to be diluted with the water being treated, is indicated. The dilution would become the first of several partial neutralization stages. In the Hollywood design this could be accomplished in the lime slurry sump. The level of dilution would vary with the acidity of the feed water but the final slurry probably should never exceed 0.1% by weight. This procedure would offer the best control to maintain the desired uniform pH and avoid excess alkali consumption by unwanted reactions.

In the Process 10 tests, the pH control was maintained at the lowest pH possible, consistent with meeting final iron specifications. In some cases the acidity requirements were not maintained unless significant reagent excesses were utilized. In those cases (as Test No. 8-70), the alkalinity developed in the effluent was less than that indicated by the excess reagent. This suggests that other reactions, probably involving aluminum and magnesium, were consuming reagent. This observation is important in considering reagent consumption in respect to theoretical requirements and treatment objectives. The goal to reduce iron concentration and develop some alkalinity involves aluminum and magnesium concentration secondarily.

The reaction of further lime with aluminum hydroxide beyond its minimum solubility level (above pH 6.2) appears to lead to its resolubilization as the aluminate. This means less removal of aluminum from the waters and higher lime consumption. Despite theoretical considerations, magnesium does appear to react at unexpectedly low pH levels, possibly precipitating and reducing magnesium concentration in the effluent. This consumes lime beyond the level related to established goals. The desirability of these effects requires a management decision: use of minimum reagent (reduced costs) versus less aluminum and more magnesium removal.

This situation was more evident in Process 11 tests, where the pH probe was located in the flash mixer. Two probe locations in this reactor were utilized. The pH variation was found to be greater (more excess reagent utilized) when the probe was located near the flash mixer overflow (Tests 66-71, 67-71, 68-71, 69-71 and 71-71), as opposed to a location in the same compartment where the alkali was added (Tests 38-71, 56-71, 57-71 and 65-71). The excess of reagent consumption tends to increase with the acidity of the raw water and with smaller flows where reaction time was longer.

The Process 10 tests generally produced satisfactory effluent meeting the specifications: pH >6.0, slight excess alkalinity (several tests resulted in low acidity values), iron <4 mg/l, and a clear effluent with suspended solids <12 mg/l. The reagent consumption decreased in efficiency with increasing acidity, ranging from 54 to 192%. The adjustment to levels near theoretical was feasible but was difficult during these short duration runs. The sludge production also varied with the total acidity of the feed water. The settled sludge (two hour level) ranged from 0.14 to 0.52 w/v %, the dry solids from 0.23 to 2.84 lb/1,000 gal and the sludge volume from 1.2 to 14.5 vol %. The sludges displayed Type 1 settling rate behavior (see "Sludge Control") which tended to gel upon aging. Settling rate data are given in Figures 3.8 and 3.9, while summary sludge settling characteristics are given in Table 3.15. The Fe(II) oxidation was satisfactory at the prevailing retention times, ranging from 0.56 to 30 min/ppm Fe(II).

The Process 11 tests were also satisfactory, with more emphasis on higher acidity waters, but at lower flow rates due to the thickener size. The effluent pH was maintained at higher levels than the Process 10 (settling lagoon) tests, always exceeding 7.0. The alkalinity values ranged to 345 but were generally near 14 mg/l. The soluble iron was consistently less than 1 mg/l. The effluent was sparkling clear without the use of flocculants. The reagent consumption ranged to 283% of theoretical with control becoming more difficult with the higher acidity waters. Adequate adjustment was obviously feasible.

In Proctor No. 2 waters the sulfate content was reduced by 22%, as was magnesium by nearly 50%. The two hour sludge content was similar to the previous tests (0.42 to 0.69 w/v %) but increased to 2.6% after 10 hours. The dry solids were higher with Proctor No. 2 waters ranging to 14.2 lb/1,000 gal treated water as was sludge volume extending to 26 vol %. The oxidation retention time was increased but proved to be adequate at levels as low as 0.37 min/ppm Fe(II). The oxidation-reduction potential ranged between 162 and 350 mv. The minimum oxidation rate extended to 338 mg/l/hr in these runs. In contrasting the two unit operations for solid-fluid separation (settling lagoon versus thickener), as applied with this reagent, there are no obvious differences which affect the chemical treatment.

With Calcined Lime

Tests with calcined lime followed the flowsheet used with hydrated lime with the replacement of a slaker for the dissolver-mixer. (see Table 3.6). Each of the four waters was treated in the flow range from 101 to 325 gpm. The slaked reagent was diluted to the one to five percent level for pumping to the flash mixer. The average percent concentration of $Ca(OH)_2$ for runs 6-71 through 11-71 were, respectively, 2.8, 0.93, 1.70, 2.30, 1.31 and 1.18. The slurry addition was controlled as described with hydrated lime.

TABLE 3.6: PROCESS 10–CALCINED LIME

Water Source	Average Flow gpm	Run Duration hrs	Water Treated gal	Test No.
Tyler Run	101	23.5	143,220	16-71
	199	37.0	443,770	14-71
Bennett's Branch	109	45.3	297,500	8-71
	197	62.5	741,490	7-71
	289	60.0	1,040,560	6-71
Proctor No. 1	107	21.5	138,030	9-71
	196	45.3	534,960	10-71
Proctor No. 2	103	40.3	250,590	11-71
	195	29.2	342,900	13-71
	199	41.0	490,210	12-71
	200	50.0	601,330	15-71
Proctor No. 1 and Proctor No. 2	325 (260+65)	29.3	571,560	17-71

(continued)

TABLE 3.6: (continued)

Process Flow:

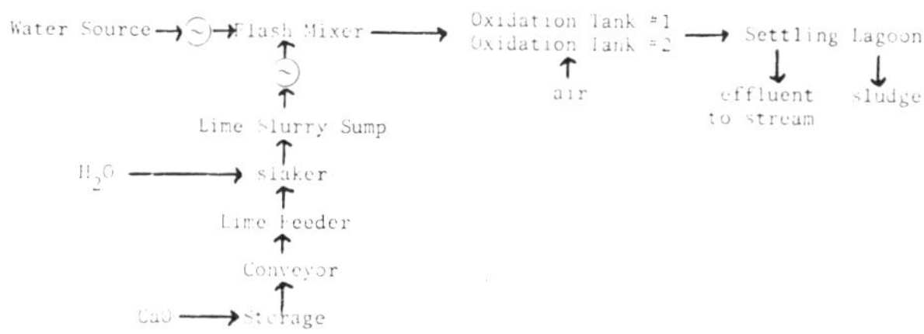

Source: EPA 670/2-73-093

The pH probe controlling slurry addition was moved from the oxidation tank to the first compartment of the flash mixer after Test 7-71. For Test 15-71 it was moved to compartment No. 3 near the overflow pipe. As with the hydrated lime tests, proximity of the control probe to the slurry feed point tended to result in greater alkali consumption. The difficulty of pH control (as evidenced by range of pH observed) tended to increase as retention time decreased, as shown in Table 3.7.

These values are extremes and do not reflect the stable control attained which was normally less than ±0.5 pH units. These variations with calcined lime were greater than hydrated lime, reflecting a slower level of reactivity. The feed water was 8° to 17°F colder during these studies than the hydrated lime tests which would also decrease reactivity. Otherwise there were not any distinct operational differences observed. Calcined lime is the most commonly used reagent other than hydrated lime.

The storage, conveying and slaking of calcined lime gave a satisfactory response without mechanical difficulties. However, it required additional attention to remove the unreactable grit and to maintain slaker operation. Consideration of known variations in slaking characteristics with different calcined limes and attainment of optimum slaking conditions was beyond the scope of this study. The nonslaking grit represents unburned limestone. The 4 to 8% loss of grit, its handling, equipment maintenance as well as its capital costs minimize the cost advantage of this reagent. The grit products not only vary with the calcined lime product but with the operation of the slaker. In Test 12-71 the grit production was 9.2% while in Test 13-71 it was 5.2%.

The slaker must be kept at hydration reaction temperatures (>130°F); thus it lacks the versatility of intermittent operation feasible with the hydrated lime dissolver. The slaking temperature ranged between 102° and 210°F with an average of 143°F. Failure to maintain a proper slaking temperature tends to result in loss of alkalinity with the grit and a less reactive reagent. The lime consumption must be at least adequate to maintain continuous operation of the smallest slaker commercially available (about 300 lb/hr and a production of 12 to 25 lb/hr of grit). Only in a few tests was this rate approached. Calcined lime is favored cost-wise as the reagent requirements of the system increase.

TABLE 3.7: SUMMARY OF pH CONTROL VARIATION CALCINED LIME TREATMENT—PROCESS 10

Test no.	Oxidation tank response time minutes	Flow rate GPM	pH Range	
			Flash mixer	Oxidation tank
17-71	<5	325	8.0	7.7
7-71	10	197	6.8	1.9
10-71	minimal	196	1.6	2.5
12-71	10-15	199	1.4	2.3
15-71	15	200	1.9	2.3
14-71	15	199	4.0	4.1
13-71	15-20	195	1.0	1.2
9-71	20	107	3.2	2.8
8-71	<30	109	7.2	6.1
11-71	60	103	1.9	0.9
16-71	60	101	4.0	2.5

Source: EPA 670/2-73-093

The observations during the hydrated lime tests were confirmed regarding advantages of stepwise addition of the alkali to prevent localized overtreatment and subsequent reactions with aluminum and magnesium. These responses tend to consume excess reagent. There was no indication of moisture or carbon dioxide absorption by the calcined lime during storage. Calcined lime must be utilized as a slurry to achieve an adequate utilization and reaction rate.

During these runs the pH control was adjusted to maintain a level in excess of 7.0 to avoid any residual acidity. This control, believed to be typical of industrial practice, leads to considerable excess reagent consumption. The utilized excess increases with raw water acidity and tends to increase at lower flow rates with longer reaction times, ranging from 77 to 272%. These excesses were not detectable as alkalinity in the effluents. Those tests showing very high reagent consumption used a raw water having relatively large concentrations of aluminum and magnesium.

Despite the relatively high pH levels maintained and the excess reagent utilized, the settling lagoon effluents tended to develop small levels of acidity. The iron content was characteristically 1 mg/l or less. The effluents were sparkling clear with suspended solids seldom exceeding 10 mg/l.

As with the hydrated lime tests, the sludge production increased with total acidity of the feed water. The solids level of two hour settled sludge was somewhat higher (0.20 to 0.95 w/v %) than with hydrated lime, as was dried solids production (0.55 to 19.1 lb/1,000 gal). The settled sludge volume was consistent with the hydrated lime tests, varying between 1.3 and 29.6 vol %. The settled behavior of the sludges was typical Type 1 excepting those developed from Tyler Run and Bennett's Branch waters which responded as Type 2. These sludges tend to gel upon aging. See Figures 3.8 and 3.9 for typical settling curves and Table 3.15 for sludge settling data on page

The Fe(II) oxidation was adequate at the retention times employed [0.19 to 196 min/ppm Fe(II)]. The largest minimum oxidation rate observed was 416 mg/l/hr. The measured oxidation-reduction potentials were somewhat higher (increased pH levels) than observed with hydrated lime extending from 198 to 414 mv.

To summarize the responses shown by hydrated lime and calcined lime: there are no obvious deviations. There is the very critical cost difference between them but this appears advantageous only when the reagent consumption is large enough to accomodate the additional equipment and maintenance necessary to slake the calcined lime.

With Caustic Soda

The tests using sodium hydroxide to treat drainage waters were carried out with the conventional "yellowboy" procedure, Process 10, (see Table 3.8). Each of the four waters was treated at flow rates ranging from 87 to 292 gpm. As with the calcined lime studies, the water temperatures were low, frequently near freezing. The reagent was supplied as a 50% solution in 55 gallon drums. The undiluted reagent was added to the flash mixer and controlled manually through a variable-speed, positive displacement pump to a head tank with return overflow. The head tank provided improved flow control at very low flow volumes.

TABLE 3.8: PROCESS 10—SODIUM HYDROXIDE

Water source	Average flow, gpm	Run duration hrs	Water treated gal	Test no.
Tyler Run	91	48.0	263,230	1-71
Bennett's Branch	87	30.0	157,280	2-71
	199	48.0	573,900	3-71
	292	24.0	421,380	4-71
Proctor No. 1	88	36.0	190,570	5-71
	204	21.5	263,160	18-71
	255	24.0	368,250	19-71
Proctor No. 2	87	13.5	70,880	20-71
	195	25.3	296,110	21-71
	227	27.0	367,760	22-71

Water Flow:

Source → (~) → Flash Mixer ⟶ Oxidation Tanks 1 & 2 ⟶ Settling Lagoon → Effluent to Stream; → Sludge

Flash Mixer ← Reservoir ← (~) ← NaOH (drums) 50% solution

Oxidation Tanks 1 & 2 ← Air

Source: EPA 670/2-73-093

This reagent is dangerous to handle, requiring gloves and eye protection. The drums utilized to ship sodium hydroxide are not pressure tested. Safety considerations prohibit the use of air pressure to discharge the solution from the drums. This reagent freezes at relatively high temperatures (54°F), a factor which frequently required the drums to be heated to allow transfer of the reagent. Commonly, shipments were received in a solid condition. The reagent drums tend to develop a concentrated slurry in the bottom which can block the delivery system, create difficulties in total recovery of the reagent from the drum and lead to some deviation in concentration which complicates treatment control.

The reagent requirements were estimated from acidity and flow of the raw water with operating control adjustments made manually from pH measurements in the flash mixer and oxidation tanks. Automatic addition of caustic soda solution is quite feasible. During tests at large flow volumes (200 gpm) of highly mineralized waters, the consumption of reagent was so great that a full time helper was necessary to change drums.

In operations utilizing more than one drum per shift, the labor costs to handle the reagent supply become excessive. In such instances large, expensive, storage facilities would be required. These requirements (capital costs and labor) for delivery, storage and handling of this reagent further add to its high utilization costs. However, the use of a purchased, aqueous reagent does minimize equipment and labor requirements for reagent preparation.

The reagent effectiveness and control capability was satisfactory while treating the lower mineralized waters (Tyler Run and Bennett's Branch). However, in the tests with the more severely polluted waters the tendencies observed with hydrated and calcined lime toward excess alkali consumption resulting from aluminum and magnesium reactions become pronounced. Utilizing manual control, the pH variation in the oxidation tanks with flow and acidity did not show the trend as did lime products under automatic control. The pH deviations were smaller (0.77 to 4.27).

Controlled, stepwise addition of caustic soda as described with lime product reagents is indicated. The extreme difficulty of maintaining a pH below 7.0, with acceptable iron and acidity levels in the effluent, was evidenced by these tests. Although the ease of control with sodium hydroxide is excellent, there is a danger of overtreatment with subsequent damage to receiving streams. This potential hazard requires constant personal monitoring or elaborate automatic monitoring and control equipment.

The reagent consumption tended to be less than theoretical (66 to 125%) with the low acidity waters while maintaining a pH in the low six levels. Under these conditions a small acidity value remained although final iron concentrations were satisfactory. The consumption excesses tended to increase with flow rates, contrary to the experience with lime products. With the higher acidity waters, the reagent consumption increased drastically (82 to 175%) in seeking to maintain an alkalinity value in the effluents.

These large excesses beyond the measured alkalinity levels must represent side reactions involving aluminum and magnesium. In Test 19-71, the final effluent persisted with an acidity level (7 to 22 mg/l) despite a reagent consumption of

131%. The iron content of the effluent was consistently satisfactory. The effluents were sparkling clear, ranging between 4 to 22 mg/l suspended solids with turbidity levels between 2 and 10 jtu.

As with lime products, the sludge production increased with acidity of the feed waters. The density (v/w %) of settled sludge (2 hours) was somewhat lower than with lime products: Tyler Run, 0.15; Bennett's Branch, 0.25 to 0.30; Proctor No. 1, 0.19 to 0.28; and Proctor No. 2, 0.61 to 0.76. The dry solids production (lb/1,000 gal) was definitely lower than with lime products: Tyler Run, 0.37 compared with 0.55; Bennett's Branch, 0.72 to 0.88 compared with 0.23 to 1.67; Proctor No. 1, 2.16 to 3.28 compared with 2.7 to 4.9; and Proctor No. 2, 10.1 to 12.8 compared with 11.7 to 19.1. The settled sludge volume (%) was greater than with lime products for the low acidity waters (3.0 to 3.5) but less voluminous with the high acid waters (14.0 to 20.0), compared with a maximum of 29.6 with calcined lime.

The sludges displayed settling behavior described as Type 1 or 2 depending on the raw water acidity. These sludges also tended to gel on standing. Table 3.15 provides sludge settling data. The shortest retention time utilized in the oxidation tanks, providing 0.28 min/ppm Fe(II), was adequate. The oxidation rates observed were similar to those with lime products. The oxidation-reduction potentials measured in the oxidation tank were somewhat higher than with lime reagents, especially in the lower acidity waters (128 to 540 mv).

The sulfate content of Tyler Run and Bennett's Branch effluents was reduced to less than 25% of the feed water. The conductivity of the effluents was doubled (171 to 258%) in the low acidity waters, with a smaller increase of Proctor No. 1 effluent (113%).

The treatment of coal mine drainage with caustic soda is characterized by its rapid reaction rate and quick responses, necessitating careful control. It is an expensive reagent, dangerous to handle, which may require special storage facilities. Its use eliminates reagent preparation equipment. The weight of sludge produced is somewhat less than formed with lime. This is attributed to the lack of precipitation of calcium sulfate. The sodium hydroxide added to the several waters as a neutralization reagent would result in a sodium concentration between 45 and 742 mg/l.

Tests 1-71 and 2-72 gave treated water effluent levels of 59 and 49 mg/l respectively in contrast to sodium additions of 57 and 63 mg/l. Sodium levels found in the raw waters ranged between 4 and 49 mg/l. Otherwise the overall response is similar to that observed with lime product reagents. Its greatest advantage lies with emergency use to prevent discharge of unacceptable quality waters.

With Soda Ash

The reactions with sodium carbonate were carried out with its addition to the flash mixer as a dilute solution prepared in the dissolver. The high solubility of soda ash ($7^{0°}$ to $45^{100°}$ g/100 ml) permits concentration control. Although caution must be exerted, the handling danger is less than experienced with caustic soda. Unless fed as a solid, dissolution equipment is necessary. Each of the four waters was treated in flows ranging from 105 to 288 gpm. The water temperatures ranged from 38° to 45°F, excepting Proctor No. 2 (50°F), whose waters are collected in the mine. The system was controlled by a pH probe located in the first compartment of the flash mixer. Following iron oxidation, the solid-fluid separation was made in either the settling lagoon (Process 10) or in the thickener (Process 11). Process conditions and flowsheet for Process 10 are shown in Table 3.9 and for Process 11 in Table 3.10.

TABLE 3.9: PROCESS 10–SODIUM CARBONATE

Water Source	Average Flow, gpm	Run Duration hrs	Water Treated gal	Test No.
Tyler Run	276	24.0	398,150	25-71
Bennett's Branch	100	22.8	137,869	30-71
	144	14.3	123,800	27-71
	196	25.0	294,880	29-71
	288	24.0	415,000	26-71
Proctor No. 1	106	23.8	152,370	32-71
	169	24.3	246,731	31-71
	222	45.5	607,300	28-71
Proctor No. 2	105	21.8	137,860	33-71
	187	12.8	144,200	34-71

Water Flow:

Source → (~) → Flash Mixer → Oxidation Tanks 1 & 2 → Settling Lagoon → Effluent to Stream; Settling Lagoon → Sludge

Flash Mixer ← (~) ← Reagent Sump ← Dissolver ← Feeder ← Sodium Carbonate

H_2O → Dissolver

TABLE 3.10: PROCESS 11–SODIUM CARBONATE

Water source	Average flow, gpm	Run duration hrs	Water treated gal	Test No.
Proctor No. 2	202	7.0	84,880	35-71
	253	22.3	338,610	37-71
	256	14.8	227,445	36-71

Water Flow:

Source → (~) → Flash Mixer → Oxidation Tanks 1 & 2 → Thickener → Settling Lagoon → Effluent to Stream; Settling Lagoon → Sludge

Flash Mixer ← (~) ← Reagent Sump ← Dissolver ← Feeder ← Sodium Carbonate

H_2O → Dissolver

Source: EPA 670/2-73-093

The reaction control, as evidenced by pH measured in the oxidation tanks, was excellent with the extreme values occurring in less than 20 minute intervals. The improved uniformity results from the highly buffered system. In maintaining these pH levels, the response of the electrodes in the flash mixer led to the addition of excess reagent, developing conditions favoring unwanted reactions. The excess reagent did not remain available as free alkalinity. Even with the very large reagent excesses employed, the buffering effect did not allow the pH to increase to unacceptable levels. Improper dosages of soda ash can develop high pH levels, but these are less likely to occur than with caustic soda. The pH ranges in the flash mixer and oxidation tank for these tests were:

	pH Range	
Test No.	Flash Mixer	Oxidation Tank
25-71	6.0 - 8.5	7.0 - 8.2
26-71	7.5 - 9.6	7.3 - 8.1
29-71	5.0 - 10.0	6.8 - 7.0
27-71	7.8 - 10.1	7.3 - 7.8
30-71	6.4 - 10.6	7.0 - 9.0
28-71	8.2 - 9.6	7.8 - 8.3
31-71	5.8 - 9.8	6.6 - 8.2
32-71	6.3 - 9.8	7.6 - 8.4
34-71	7.6 - 9.2	7.2 - 7.6
33-71	6.6 - 9.0	7.3 - 8.3
36-71	6.9 - 8.6	7.5 - 7.7
37-71	3.6 - 9.6	6.9 - 8.0
35-71	6.0 - 9.5	6.3 - 8.0

The reagent consumption experienced during the plant tests ranged from 146 to 313%. The excess increased with flow rates but showed no particular trend with acidity levels, which probably reflects the delayed response of the buffered system to pH changes. The effluents did maintain an alkalinity value but the percentage reagent utilized which was related to this contribution was small and large excesses of reagent went into the treatment per se, as shown below. The reagent representing effluent alkalinity decreases with increasing acidity of the feed water.

Test No.	% Reagent Feed Representing Effluent Alkalinity	% Reagent Consumed for Treatment Based on Theoretical
25-71	63.2	82.9
26-71	17.1	259.1
29-71	28.2	149.8
27-71	23.0	164.6
30-71	14.5	141.6
28-71	12.2	235.4
31-71	11.4	246.2
32-71	14.9	124.4
34-71	7.2	272.3
33-71	4.0	256.1

The iron removal response was excellent with the effluents having consistently less than 2 mg/l soluble iron. The effluents were clear with suspended solids ranging between 3 and 26 mg/l and turbidity levels between 2 and 20 jtu.

The sludge formed with soda ash settles to greater densities than sludges resulting from lime products but does not approach densities achieved with limestone. The two hour settled densities ranged from 0.37 to 2.1 w/v %. The level of sludge production is equivalent to that found with caustic soda (0.25 to 19.4 pounds dry solids per 1,000 gallons water treated) and less than the lime product sludges. The corresponding sludge volume was decidedly less than caustic soda and lime product sludges (0.3 to 11 vol %) but was not as favorable as the response from limestone. The sludges may be described as exhibiting Type 1 behavior but have better mobility and are less inclined to gel. Typical settling rate curves are shown in Figures 3.8 and 3.9 with other sludge data shown in Table 3.15.

The retention time provided for iron oxidation was excessive, ranging from 0.78 to 72 min/mg/l Fe(II). The fastest oxidation rate was at least 230 mg Fe(II)/l/hr. The oxidation-reduction potentials were favorable within the limits of 84 to 383 mv. The low iron levels in the effluents also reflect proper oxidation. The use of sodium reagent in excess resulted in substantial increases in dissolved solids as reflected by conductivity. The average increase of about 450 micromhos was substantial (from 42 to 254%). The sodium concentrations which would be developed in the water with the reagent excesses employed ranged from 55 to 1,713 mg/l. This response was similar to that experienced with caustic soda treatment.

The treatment response of Proctor No. 2 waters when using the thickener (Process 11) for solid-fluid separation confirmed the other studies using the settling lagoon. The thickener settling operation was equivalent to similar tests with lime product treatment. The reagent consumption was excessive (223 to 292%). The effluents were alkaline with 1.4, 28.0, and 13.1% (Tests 36-71, 37-71, and 35-71, respectively) of reagent added occurring as alkalinity in the effluents.

The actual excess reagent consumed in the reaction was uniform for the three tests: 219.8, 210.0, and 194.8%. The consistent effluent pH levels were between 7.56 and 7.71 reflecting good reagent control response. The pH variation during the three runs was 2.50, 2.86, and 0.90. The iron levels were satisfactory with an unexplained exception in Test 36-71.

The two hour sludge densities were less than the Process 10 tests (0.56 to 0.75 weight per volume percent) as was the sludge production (5.1 to 6.9 lb dry sludge per 1,000 gallons) and sludge volume (1.1%). This response is inconsistent since the raw water loadings were similar.

The oxidation retention times were less than the Process 10 studies [0.52, 0.49, and 0.39 min/mg/l Fe(II) respectively for the three tests] but were adequate. In all of the tests utilizing soda ash with the settling lagoon, the effluent specifications were achieved; however, the reagent consumption was excessive. Of the excess alkali, only a portion occurred as effluent alkalinity. The remainder apparently being consumed in other reactions, favored by the buffered conditions. The sludge characteristics are somewhat more favorable than those resulting from lime products or caustic soda, especially in resulting sludge volumes. The process drastically increases dissolved solids in the effluents.

With Dolomitic Reagents

There were three groups of studies made with dolomitic reagents considering the hydroxide, the oxide anhydride and the carbonate. Some process variations were also introduced. All four waters were treated at flow rates from 31 to 139 gpm. Process conditions and flowsheet for the hydrated dolomite studies are given in Table 3.11. The prepared hydroxide slurry was reacted in the flash mixer followed by air oxidation and solid-fluid separation in the thickener. The effluent was discharged through the settling lagoon. The pH control probe was located in the flash mixer with a set point of 7.0. The slurry preparation did not differ from the use of hydrated lime. Calcined dolomite was utilized in the same flowsheet as the hydroxide, replacing the dissolver-mixer with a slaker.

TABLE 3.11: HYDRATED DOLOMITE PROCESS CONDITIONS

Water Source	Average Flow, gpm	Run Duration hrs	Water Treated gal	Test No.
	Process No. 11			
Proctor No. 2	83	13.5	67,900	59-71
	105	11.3	71,700	58-71
	106	21.3	135,740	60-71

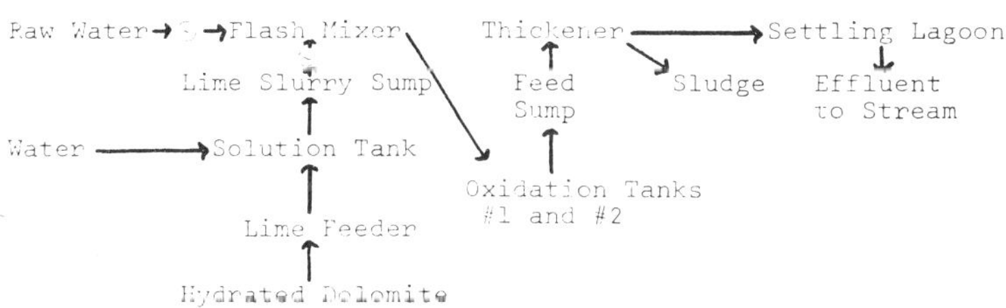

Source: EPA 670/2-73-093

The calcined dolomite responded differently than the anhydride lime during slaking. The slaking temperature was difficult to maintain; the solid feed resisted wetting. A solid crust tended to develop in the slaker reaction chamber and formed a thick mix which moved with difficulty to the grit separator. Much dilution water was required to separate the grit. The hydration may have been inadequate, producing a less reactive reagent. This behavior was related to the inert, insoluble MgO component, since only the calcium portion of the compound slakes under atmospheric pressure. The dolomite was fed as larger-sized particles (Grade 4) to the rotary reactor. In some tests, it was supplemented with pulverized limestone.

The dolomitic hydrate yielded a satisfactory effluent with the highly mineralized Proctor No. 2 waters. The reagent consumption was reasonable although in one test a 146% consumption occurred. Even this level was less excessive than experienced with high calcium reagents.

The pH control of the reaction was good, maintaining a level above 7.5. The effluents had an excess alkalinity, which accounted for only 1.3 and 1.6% of the reagent fed in Tests 58-71 and 60-71, respectively. When the reagent was introduced in large excesses, it did not appear as water alkalinity. The iron removal and effluent clarity were excellent. The settling lagoon effluents showed a reduced magnesium content, but the calcium levels rose nearly four-fold. It appears that magnesium was being removed from solution.

The settled sludge densities (0.58 to 1.3 w/v %) were consistent with those resulting from hydrated lime treatment, as were the sludge production levels (11.4 to 16.2 lb dry solids/1,000 gal). The sludge volumes were somewhat greater (to 33 vol %) but were subject to reduction under compression. Settling rate curves are shown in Figure 3.9, while other sludge data are given in Table 3.15. The settling behavior of these sludges respond as Type 1, tending to gel. The sludge composition had unusually high calcium but low magnesium and sulfate content. Sludge drainage rates were also consistent with other lime product reagents. There is strong evidence for magnesium-sulfate-complex association, indicating possible advantages for magnesium reagent in treating high sulfate-containing waters.

The retention time utilized during iron oxidation [0.61 to 0.90 min/mg Fe(II)/l] was probably excessive. The oxidation rates exceeded 263 to 297 mg/l/hr, while the oxidation-reduction potential ranged between 202 and 229 mv. The settling lagoon maintained a pH consistent with the oxidation tank effluent.

The calcined dolomite also gave satisfactory responses with all four waters using low flow rates. It appears that stoichiometric dosages of calcined dolomite are feasible, assuming reaction with only the calcium portion of the compound. The reagent consumption levels were consistent, with near theoretical values being attained for tests on low acidity water and high acidity water at a low flow rate. However, one test utilized 186% theoretical reagent while another was atypical (605% excess), being unexplainably out-of-control. These last two responses must relate to reaction time in the flash mixer, pH probe location or possibly to unobserved mechanical problems.

In one test, utilizing the control set point of 7.0, reaction control difficulty was experienced throughout the test with the pH in the oxidation tanks rising to 9.5 for over three hours. The system was mechanically and chemically responsive. The control point was lowered in increments to 5.3 with the oxidation tank effluent pH slowly declining to 8.0 and finally to 7.5 near completion of the test. The reagent feed rate also approached theoretical with the control change. The test illustrated control problems which could be costly in unattended plants and lead to dangerous overtreatment with reactive reagents.

These tests showed greater variation in the alkalinity of the oxidation tank effluent than with the hydrate reagent. Each test having reagent excesses did not develop an alkalinity. The two attaining an effluent alkalinity represented 15.6 and 4.2% of the reagent fed. The iron reduction was satisfactory. Reactions with these reagents continued throughout the process as there was a rise of pH levels in the settling lagoon effluents.

The settled sludge density confirmed the values from the hydrated dolomite tests, as did the sludge production level. The settled sludge volumes were less than resulted with hydrated dolomite and more consistent with lime product reagents. The sludges showed Type 1 settling behavior and would gel. They are high in calcium and magnesium but generally lower in sulfate content. The filtration data generally agree with that from lime product sludge. Sludge drainage rates indicate their bulk densities to be slightly lower than hydrated lime sludges.

The iron oxidation levels were satisfactory at the retention times utilized [0.76 to 82 min/mg Fe(II)/l], covering the range 3 to 303 mg Fe/l/hr. The oxidation-reduction potential ranged between 197 and 293 mv.

The dolomite tests indicated this reagent to be less responsive than high calcium limestone, but capable of developing higher pH levels in the effluent. With the time allotted to study this reagent, it was not possible to explore all the parameters of the rotary reactor and stone properties (only one particle size was used) which control the reaction. The reagent provided adequate treatment for Tyler Run waters in the rotary reactor at a flow of 89 gpm. The feed water had little iron ($<$2 mg/l) which made iron oxidation unnecessary.

The reactor effluent maintained a pH above 7.0, a reduced iron content, and an excess alkalinity. The alkalinity added to the effluent represented 58% of the neutralizing requirement. The estimated reagent consumption was excessive since equilibrium conditions were not established. This unreacted reagent was not solubilized but reported to the sludge. With appropriate control, reasonable consumption could be expected. Under these conditions reagent cost would not exceed 2¢/1,000 gal.

The sludge behavior responded as those produced in limestone reactions (Type 3) maintaining a slight turbidity. The sludge settled rapidly to a high density (9.1 w/v %) and small volume (0.4%). The sludge production was 3.0 lb dry solids per 1,000 gallons of water. This sludge contained excess dolomite fines. This sludge formed from a reactor effluent containing 600 mg/l suspended solids. The settling lagoon effluent did not change in quality from the reactor effluent, indicating little continuing reaction. This response was different than observed with high calcium limestone.

Dolomite shows a reduced neutralization reactivity level in respect to high calcium limestone. Waters with acidity levels above 300 mg/l may be treated satisfactorily but only with great increases in reaction time. The reaction times necessary with relatively insoluble carbonate reagents is extensively controlled by those parameters which establish contact between available reactive solid surface and diffusion levels control carbon dioxide concentrations. The conditions maintained for these tests utilizing the rotary reactor were inadequate when utilizing dolomite but were satisfactory when applied with high calcium limestone. The use of supplemental reagents, as pulverized limestone, can be helpful in improving these reaction times.

The evaluation of dolomite-based reagents indicate their potential use in treating coal mine drainage but point to major limitations. In Pennsylvania, the higher cost of these reagents (mainly transportation costs) would discourage their use; however, in locations where the reagent supply sources are closer to the treatment plant, their use may be indicated. Under good control, a near stoichiometric consumption of the hydrated and calcined products can be expected based upon reaction of the calcium component.

The treatment effluent will meet requirements but overall reaction control appeared more difficult to attain. The effluents tended to maintain a slight turbidity which can be controlled with flocculants. The natural dolomitic carbonate was less reactive than high calcium limestone but did permit attainment of higher pH levels. The use of mixed reagents offers advantages.

With Limestone

The tendency to oversimplify processes to treat coal mine drainage (CMD) has led to overemphasis of the ratio of reagent cost to overall treatment economics. Yet the selection of the reagent establishes the system and to a significant extent controls total treatment costs. The theoretical 1.92 cents per neutralization equivalent for limestone, with its 2.56-fold advantage over quicklime and 3.83-fold advantage over hydrated lime has continued to be viewed advantageously. Despite this and other advantages, pragmatic process considerations have prevented the more extensive use of limestone. These studies sought to develop a perspective of this situation.

Chemical Aspects of Limestone Treatment of CMD – In the general equations written for limestone neutralization of CMD, the water solubility relationships existing for the reactant and products are too often ignored:

$$CaCO_3 + H_2SO_4 \longrightarrow CaSO_4 + H_2CO_3$$

$$3CaCO_3 + Fe_2(SO_4)_3 + 6H_2O \longrightarrow 3CaSO_4 + 2Fe(OH)_3 + 3H_2CO_3$$

$$3CaCO_3 + Al_2(SO_4)_3 + 6H_2O \longrightarrow 3CaSO_4 + 2Al(OH)_3 + 3H_2CO_3$$

The small amount of $CaCO_3$ (0.014 g/l), which dissolves in water, reacts rapidly with mine water. The continued neutralization reactions are governed by the limestone solubility rate which varies with the physical properties of the stone, the presence of dissolved salts, and the carbon dioxide partial pressure. There is also some reaction at the solid limestone surface which is governed by surface area, surface reactivity, diffusion rates of reactants and products, and reactant concentrations. As the pH rises above 4.5, the equilibrium which yields an excess of hydroxyl ions becomes increasingly important.

In this same pH region, the solubility products of the hydrated ferric and aluminum compounds are exceeded and their precipitation begins. Ferrous compounds are not precipitated until higher pH levels develop. Calcium sulfate, another reaction product, also has limiting solubility and plays a unique role in this process. Gypsum tends to deposit and grow, though its solubility may not be exceeded. Localized concentrations of gypsum, especially on the limestone surface, are undoubtedly high. Carbonic acid, subject to dissociation, acts as a buffer to maintain a weakly acid environment. Carbon dioxide can be stripped from such waters allowing the pH to rise.

These equilibria suggest overall slow, continuing responses, especially when fine (near-colloidal) calcium carbonate particles are present. They also indicate why mechanical effects tend to become controlling.

The overall chemistry of this system is complex, especially the solution chemistry of iron between pH 2 and 10. Of concern are: the solubility products of iron, calcium and magnesium salts (sulfates, carbonates, bicarbonates and hydroxides) as well as carbon dioxide and oxygen; the interactions of three physical states (and the response of partial pressures, complexes, ionic strength, particle size, nucleation); and reaction conditions (temperature, contact times, concentrations, oxidation state, equipment configurations).

The sludge characteristics are determined by the reagent and processes used in the treatment. Limestone treatment appears to result in an optimum situation forming a sludge which is rapidly settled, dense, and readily dewatered. The treated water tends to maintain a persistent turbidity after solid-fluid separation which clears very slowly. This haze is attributed to the presence of near-colloidal limestone particles, with the particles eventually settling or solubilizing.

It is also feasible that part of the turbidity originates from colloidal silica in the limestone. The mass of sludge produced during treatment relates to the pollution loading of the raw waters, primarily the iron and aluminum content. In some high sulfate ($>$1,400 ppm) waters, gypsum may also form. Other sludge contents result from insoluble impurities in the reagent and from excess reagent. The inefficient use of limestone increases the mass of sludge to be handled and total process costs.

There is no reason to anticipate limestone requirements in excess of the stoichiometric value: 8.3453×10^{-6} lb of calcium carbonate/mg/l acidity/gal of water treated. This value must be adjusted to consider the stone purity and the efficiency with which it is utilized. The latter is dependent upon a myriad of parameters to be indicated from these studies.

Considerations in the Use of Limestone — The chemical factors of concern include: reactant solubility rates, precipitation and neutralization rates, iron oxidation rates (either pre- or postneutralization), and composition of the sludge, which determines suspended solids settling rates, and sludge dewatering rates. These factors determine equipment selection and sizing, hence capital costs. The equipment basically establishes reagent efficiency and operating costs. Among the advantages in the use of limestone as a reagent are:

(1) Lowest cost neutralization reagent—about 0.002 cents per part acidity per thousand gallons.

(2) Minimal sludge volume is developed—usually less than one-third volume from more soluble reagents.

(3) The limestone-produced sludges have high solids content.

(4) The sludge tends to be granular and crystalline, making dewatering operations less costly and more efficient.

(5) It is impossible to overtreat the water, leading to minimum control problems.

(6) Reagent availability is extensive over a wide geographical distribution.

(7) Limestone storage is favorable. The larger sizes do not require protective enclosures. Pulverized stone, which requires bin storage, handles well and is nonhygroscopic.

(8) Simplicity of plant design.

(9) Reduced plant maintenance and operation requirements.

(10) Reduced safety hazards relative to reactive reagents.

(11) Reaction capability indicate stoichiometric reagent requirements

need not exceed 20% theoretical.

There are limitations and disadvantages in the use of limestone. Among them are:

(1) Use is limited by the ferrous iron concentration in the feed water. The limiting pH attainable, near 8.3, establishes maximum iron oxidation rate.

(2) Reactivity is lower than more soluble reagents. Surface coating of the reagent by-products reduces reactivity.

(3) Settled sludge has clay-like character which may create handling difficulties.

(4) The actual sludge weight produced (on a dry basis) may be greater than with lime products since stone tends to be more impure. Slower reactivity and lower use efficiency may introduce unreacted reagent to the sludge.

(5) Near-colloidal-sized particles of stone and silica create turbidity in plant effluents.

(6) The limestone-produced sludge decreases in volume very little under compression.

(7) The carbon dioxide formed during reaction must be removed to provide a stable pH.

(8) Rotating reactors are noisy. They require power to turn the drum and are inefficient in comminution.

The response of mine drainage waters to limestone treatment suggests three types of feed waters, grouped by ferrous iron content. They are listed in Table 3.12. The ferric iron acidity concentrations of CMD do not appear to be limiting criteria. In other tests, not detailed here, it was established that mine drainage containing up to about 100 mg/l of Fe(II) could be satisfactorily treated in the rotary limestone reactor at maximum flow without a separate iron oxidation stage. The stoichiometric oxygen requirements for the oxidation of this level of iron concentrations are closely met by the air saturation level in water, assuming no replacement by additional oxygen to the solution. However, at higher Fe(II) levels, acceptable iron concentrations in the settling lagoon are not attained.

Various limestone reactor modes have been considered. Two modes of reacting limestone with CMD have been utilized which differ primarily in feed stone size—pulverized or larger sizes (+¼ inch). The pulverized reagent is mixed with the CMD in a flash mixer in near stoichiometric quantities. Long reaction times are required with vigorous agitation to maintain particle suspension. The equipment should be housed. This mode tends to result in poor reagent utilization.

Larger particle-sized reagents are introduced, in excess, to autogenous, attrition reactors. Only rotary reactor designs seem feasible. The reaction time is short. The reactors provide vigorous movement of stone and water but do not require weather protection or housing. Much of the water movement results from gravity flow. Reactor capabilities must permit desired reaction levels but avoid excessive loss of unreacted stone.

Reagent cost favors the use of larger-sized stone. Although the reactivity increases with decreasing particle size, the extent each particle responds lessens. This phenomenon is associated with particle surface coatings by reaction products and, the ease in maintaining the surface active. With the commercially-available pulverized stone sizes, reaction times in excess of twenty minutes become essential to minimize loss of unreacted stone.

TABLE 3.12: POTENTIALITIES FOR LIMESTONE TREATMENT OF MINE DRAINAGE WATERS

Ferrous concentration mg/l	Response to limestone treatment
0-100	Effective treatment may be achieved without pre- or post-neutralization iron oxidation.
100-500	May be effectively treated but requires post-neutralization aeration and significant reaction-retention time. Pre-neutralization oxidation can reduce iron II concentrations.
>500	Potential treatment is uncertain with experience to date, unless combined with pre-neutralization ferrous iron oxidation to achieve the above ferrous levels.

Source: EPA 670/2-73-093

The power requirements between the two modes are utilized differently. With the pulverized stone, the power required to maintain particle suspension must also move large volumes of water which serves minimal useful functions. With the larger stone reactor systems, the power is more effectively utilized to create and maintain reactive surfaces.

The rotary limestone reactors perform two principal functions, the relative extent of each has not been defined: (a) maintain freshly exposed, reactive surfaces by attrition to remove insoluble reaction products; and (b) achieve an autogenous comminution of the stone to produce fines of great reactivity. Additional functions also occur: (c) contacting the neutralization reactants to enhance reaction rates at the solid-liquid interfaces which includes wetting of solid surfaces; (d) contacting the oxidation reactants [dissolved Fe(II) and oxygen]; (e) contacting air (oxygen) and water to maintain an oxidizing environment; (f) removing carbon dioxide from the water; (g) controlling the reaction-retention time; (h) transporting the water through the reactor; (i) providing a suitable transport and distribution of the stone in the reactor to optimize the other functions.

The capability with limestone treatment to achieve an effluent with an alkalinity value is the main control requirement. The occurrences which do not achieve an alkaline effluent or tend toward inadequate reagent utilization may be modulated, but tend to have counteracting responses. They may be enumerated as follows. In pulverized stone systems: increase reagent addition; decrease stone particle size; use stone of greater purity and reactivity; and increase reaction time. In larger-sized stone systems: increase rotation speed; increase stone loading of reactor; add attrition devices to reactor; use smaller top size stone; use softer stone; and use stone with higher alkali equivalent.

Should these steps fail to be adequate, supplemental additions of a more reactive reagent may be employed after the limestone reaction. A greater reagent efficiency may result by achieving only partial neutralization with limestone and completing the treatment with hydrated lime slurry.

Even with an adequate effluent alkalinity, a depressed pH may result from carbon dioxide solution. This gas should be stripped from the effluent by subsequent mixing or aeration. During this gas stripping, the alkalinity and pH of the waters may decrease but an excess alkalinity level must be maintained. Should an acidity develop, inadequate available alkalinity has been introduced to the water. Should an excess ferrous iron concentration persist, the following actions may be considered:

Increase oxygen transfer rate. This dissolved oxygen level in the effluent should be maintained at least 5 ppm.

Increase reaction-retention time. Fe(II) oxidation rates are relatively slow at the pH levels prevailing under these conditions.

Increase pH of the system by the addition of further limestone or other alkali supplement.

Add a chemical oxidant such as chlorine gas or calcium hypochlorite.

Decrease ferrous iron concentration of feed water by dilution or preneutralization oxidation.

With the attainment of suitable water quality after separation of the sludge, effluent turbidity may persist. This can be controlled with polymeric flocculants if necessary, but it is usually nondeleterious.

Plant Operating Observations – Limestone neutralization studies were made with the rotary reactor employing feed stone with varied top sizes from ¼" to 3" (see Figure 3.3). In some tests pulverized limestone was also added to the reactor as a slurry. In other studies pulverized stone slurry was reacted in the flash mixer and the Densator (see Figure 3.10). Tests were made with high calcium as well as dolomitic stone from several geographical origins and were applied to the several waters available.

The rotary reactor was designed to meet the multiple functions cited. Vigorous movement of water and stone was achieved within the reactor by horizontal lifters which enhance autogenous comminution, mixing, and gas exchange. The rotational movement of the drum carried the stone and water load to about "ten o'clock" before discharge.

The 2 mm wedge wire screen, covering the discharge aperture, was effective in preventing loss of all but the finest size stone particles. The most excessive loss of unreacted stone, as well as potential sanding problems in pipes, was prevented by this screen. Some of the fine particles which do escape were trapped in the effluent sump while others are carried with the effluent flow, providing additional reaction time while reaching the settling lagoon. Physical characteristics of lime-

FIGURE 3.3: SCHEMATIC OF LIMESTONE REACTOR

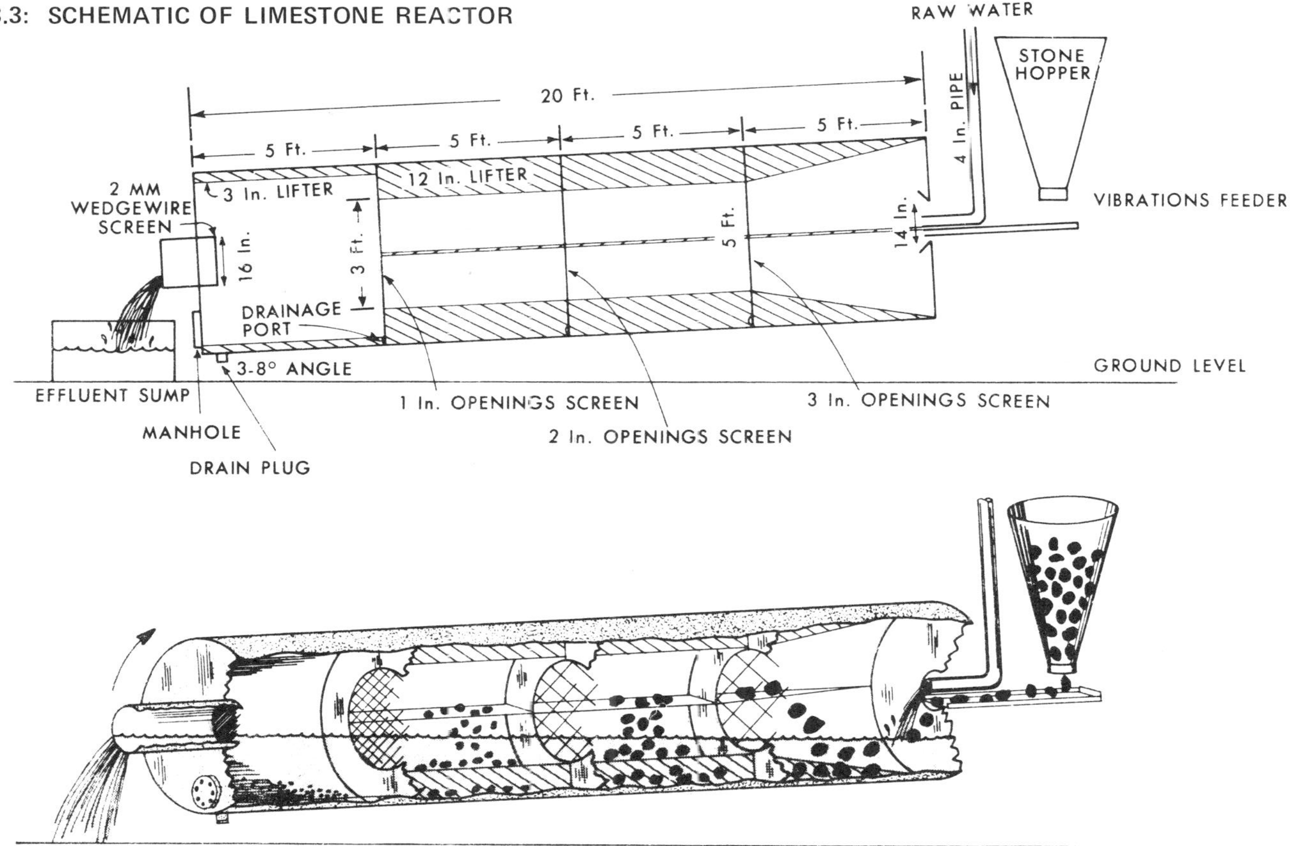

Source: EPA-670/2-73-093

stone and dolomite sludges found in the bottom of the settling lagoon, as the water level was dropped, are given in Table 3.13.

TABLE 3.13: CHARACTERISTICS OF LIMESTONE SLUDGE FROM SETTLING LAGOON

		Sample 1	Sample 2	Sample 3
Sample location Inches from weir		96	104	124
Bulk density - gm/ml		1.167	1.139	1.171
Solids Content Wt. % @ 105°C		35.4	30.2	37.9
		Size Content Direct Weight Percent		
Size (mesh)				
passed	retained	Sample 1	Sample 2	Sample 3
	100	2.2	1.4	4.7
100	200	23.2	39.1	26.3
200	325	24.5	47.9	25.2
325	400	19.2	8.4	27.4
400		30.9	3.2	16.4

Source: EPA 670/2-73-093

The larger particles would be expected to settle in the first region of the lagoon, with decreasing diameters as distance from the weir increased. The particle sizes were small and their distribution relatively uniform. The bulk density and solids content were surprisingly high—showing a significant density increase under compression in comparison with 24 hour settled sludge (Table 3.15). X-ray diffraction patterns of these sludges showed calcite, dolomite, quartz, and opal, in addition to ferric oxyhydroxides. No difficulty was experienced from the buildup of nonreactive inerts in the reactor.

The systems used with pulverized stone permit no control over loss of unreacted reagent; the stone reacts, settles in the reactor or is carried with the water to the next unit operation. In two years of intermittent operation of the rotary reactor, there was no evidence of internal wear and corrosion. Rubber, ceramic, or other abrasion-resistant liner material would further extend reactor life. The control anticipated by varying the rotary reactor slope was ineffective due to inadequate design of the feed end of the reactor. The "open" design, which permits the extension of a vibrating feeder and the influent pipe into the reactor, also permitted excessive splashing from the reactor at low slope angles and high flows.

The result was potential corrosion, and limitations in the useful limit of fluid flow and horizontal reactor angle. The horizontal reactor angle effects stone movement within the reactor, but has negligible control over water retention time. The angles available favored excessive movement of the stone to the exit end of the reactor. The reactive capacity of the device is optimized only if the stone charge maintains a uniform distribution throughout the length of the unit. The experience attained with this unit suggests a horizontal, or even a shallow reverse angle to be preferable. The slope variation capability provided in this research reactor is not needed in an industrial unit. A minor design change to a "closed" feed system would circumvent these difficulties and enhance reactor performance.

The median retention time in the rotary reactor appears at about 150 seconds regardless of rotation speed, flow rate (up to 250 gpm), or reactor angle (3° to 4°). The retention time in the pulverized limestone reactors is a function of the reactor size and flow rate.

The belt drive of the reactor created difficulties in "training" the belt. Changes in temperature, humidity, precipitation and load caused the belt to creep, resulting in unreliability and undue operator attention, thus limiting the testing program. A chain drive is preferable.

The power requirements for the rotary reactor are distributed between those required to turn the base reactor, to move the water, and to abrade the stone. The power requirements associated with the water movement do not appear to vary appreciably with flow. They increase with stone loading and change rapidly with reactor rotation speed between 9.6 and 16 rpm.

With this unit, the power costs range between 6 and 14¢/hr or between 0.4 and 2.3¢/1,000 gal of water for the conditions cited. Power consumption by the reactor reflects current (amp) levels drawn. These levels were monitored to protect the motor from overload and also served as a rough indicator of reactor stone loading. The range observed was from 12 to 24 amp. More effective power transmissions would be expected with a chain drive.

The effective capacity of the rotary reactor is related to achievement of each of the necessary functions in response to the water and stone characteristics. Most critical is the reactor capability to provide adequate reactive stone surface which is a function of reactor stone load, of stone size distribution, of stone physical attrition, of stone distribution within the reactor, and of the stone consumption.

The reactor must operate with excess stone present, but at an optimum and constant load. The power capability of this reactor will carry a load in excess of 1.5 tons of stone. Since power consumption varies with stone load, the optimum load must be limited to process requirements. The stone feed rate must balance the required stone consumption and must relate to the stone feeding system. Stone consumption results from reaction and the discharge of unreacted fines, with the latter determining reagent utilization efficiency.

With the introduction of a maximum load of stone to the reactor, the stone will distribute itself rapidly (within minutes) throughout the reactor length, governed by movement of different size particles through the three screens. The stone tends to move with the water flow toward the effluent end. Reactor slope control assists in establishing this rate; but appears to be effective only at angles less than 3° from the horizontal.

The potential stone distribution within the reactor has extremes ranging from complete loading in section 1 (with +3" stone) to complete loading in section 4 (with 1" stone). The reactor design did not provide for exclusive use of pulverized stone due to the very low retention time (less than 3 minutes) which would result in excessive losses of reagent. In the first condition, attrition and comminution will be maximal in the first compartment, but stone surface area will be minimal, with three-fourths of the reactor being ineffective.

With rotation time, the rapid attrition of the sharp edges of the large stone permit particle transfer through the screens to subsequent reactor compartments with the particles assuming smooth surfaces and round edges. A gradual, even distribution of stone throughout the reactor is approached. Larger total surface area of the stone is then achieved. However, the attrition-comminution rate decreases as the initial stones abrade to smooth surfaces and redistribute themselves to a reduced volume level throughout the reactor. With batch loading, it is during this initial period that conditions permit the greatest loss of unreacted stone. The attrition rate is also a function of particle size since the mass of larger particles falling will offer the greater impact for breakage.

Thus, with increasing distribution and decreasing compartment loading in the reactor, the loss of unused stone decreases. Subject to the water flow and acidity loading, treatment will remain satisfactory with increasing stone surface area until the steadily decreasing attrition-comminution rate reaches a level that the total available reagent surface becomes inadequate. The production of a reactor-stone consumption level in excess of 3 lb/min was attainable. This would meet the requirements of CMD water having 2,000 ppm acidity at a flow rate of 300 gpm. This rate is beyond the design and hydraulic capacity of the reactor with the existing end-feed design.

The other loading extreme (-1" stone which moves rapidly to reactor section 4) would also limit reactor volume use and, due to greatly reduced attrition, would limit production of reactive stone surface despite large surface area. The proximity of this section to the effluent port would provide functional retention time of seconds, leading to inadequate reaction and excessive stone loss.

The optimum stone loading operation would require a uniform stone distribution throughout the reactor, controlled size distribution in each reactor section, and a continuous, uniform stone feed rate of specified size distribution. Ideally, the feed rate would precisely conform to the theoretical reagent requirements of the feed water. These conditions will vary with stone properties, especially hardness. Design of a reliable reactor feed system is feasible and would limit labor requirements to filling the stone feed hopper and maintenance.

Stone was introduced in the rotary reactor via a vibrating feeder from an 1,800 pound hopper. The hopper was loaded by a mobile conveyor belt. The stone feed rate to the reactor was adjustable by varying the amount of vibration and the size of the hopper diaphragm. The reliability of feed control varied extensively with stone size distribution and moisture content. Mixed sizes (-3") were least reliable, while narrow size ranges (especially ¼" to 1") were most satisfactory.

Limited moisture content of the stone tended to resist flow while operations during rain gave enhanced flow. Some difficulties were experienced from freezing during intermittent operation in cold weather. The stone feed equipment design is optimum for a given product with a limited range of feed rates, and will have decreased reliability with deviation from these specifications. The wide range of stone feeds and rates considered increased the control difficulties.

The Hollywood unit has been applied satisfactorily to CMD volumes in excess of 200 gpm, with higher levels limited by the reactor feed design. Evidence suggests a capacity of at least 400 gpm is entirely feasible for this size of vessel. Treatment capacity could be increased by enlarging the drum diameter and reaction response by increasing drum length.

The economic, operating, and design criteria basic to the selection of the limestone treatment in contrast to other reagents for a given water is most complex. Figure 3.4 illustrates the alternate routes for the limestone-water reaction.

FIGURE 3.4: ALTERNATE UNIT OPERATIONS FOR LIMESTONE NEUTRALIZATION

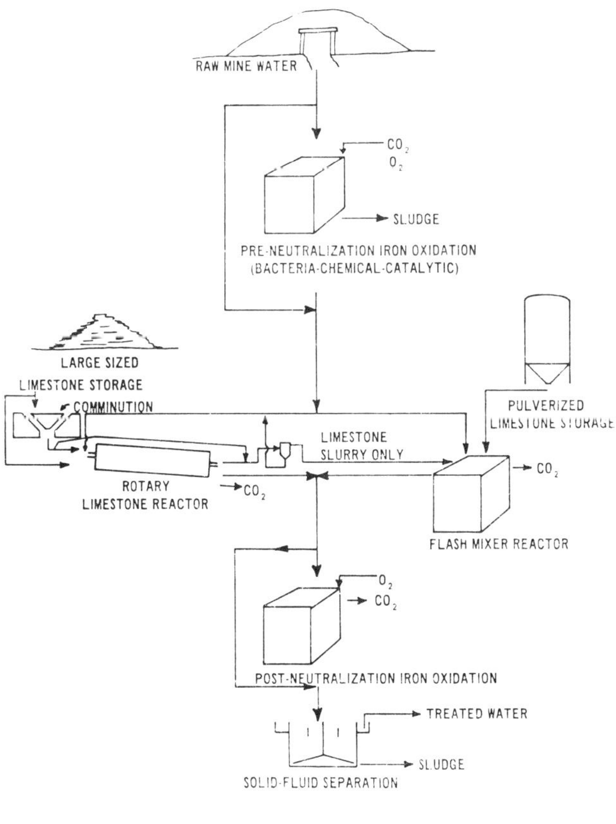

Source: EPA 670/2-73-093

The more significant design parameters for limestone treatment given by unit operation follow.

Reagent control: Consumption rate, lb/hr; specifications for limestone grade, size, and hardness; type of storage (bins or piles); feeders and hopper design; all weather provisions.

Limestone neutralization unit: Mode of operation (mixer or rotary mill); working volume and excess capacity allowance; weight and volume of limestone charge; pulp retention time vs ferrous iron content; rotation or mixing requirements; gas exchange capability (oxidation and CO_2 removal); supplementary reagents required, e.g., $Ca(OH)_2$; materials of construction; cold weather provisions.

Iron oxidation unit (CO_2 elimination): Type (tank or pond); function, pre- or postneutralization oxidation, aeration or bacterial; aeration capacity required by Fe(II) concentration; retention time required.

Settling lagoon(s): Settling rate of solids; loading rate (feed flow rate and % solids); range of rising velocities; sludge bulk density; sludge storage capacity; provision for sludge removal; baffles; multipond systems; cold-weather provisions.

Thickener: Settling rate of feed slurry (including recycle); optimum feed solids content; maximum underflow solids content; allowable rising velocity range, gpm/ft^2; ratio of recycle volume; operating characteristics, w/wo flocculant; feed flow rate; temperature effects.

Filter: Solids filtration rate, lb/24 hr/ft^2 filter area; liquid filtration rate; gpm/ft^2 filter area; type filter, filter area, filter medium; feed load, lb solids/24 hr/1,000 gal CMD/ppm acidity; filter aid (precoat) requirement; flocculant requirement, lb/1,000 gal feed slurry; percent solids in filter feed; percent solids in filter cake; materials of construction; vacuum requirements.

Sludge drying basin: Bed design including porosity and drainage rates, area and depth; percent solids in feed slurry and "dried" solids; settling rate of feed slurry; mass of "dried" solids/1,000 gal feed water; precipitation and evaporation rate of the immediate area.

Rotary Limestone Reactor Data – The capability of the limestone neutralization process was demonstrated under severe conditions utilizing highly ferrous Proctor No. 2 waters. Water quality specifications were attained. Reasonable reagent use efficiency was experienced. Any coal mine drainage of lesser pollution loading will also respond satisfactorily. The Fe(II) content was biochemically oxidized prior to neutralization and the waters were neutralized in the limestone reactor. The overall flowsheet is detailed in Figure 3.5 and the supporting conditions and analyses given in Table 3.14. The bacteria inoculation is used only to initiate the biochemical oxidation system and need not be maintained.

The weight of limestone was evaluated volumetrically. The weight of given bulk volume of reagent was established. Stone was added through a vibrating feeder with nearly constant manual attention. The suspended solids in the effluent of the limestone reactor for this test were higher than normal, the value usually being closer 2,000/mg/l.

The limestone reactor sludge solids settled rapidly, with one-third final sludge volume being achieved within 3 minutes and nearly all of it within 10 minutes, without the use of flocculants.

FIGURE 3.5: BIOCHEMICAL IRON OXIDATION—LIMESTONE NEUTRALIZATION SYSTEM

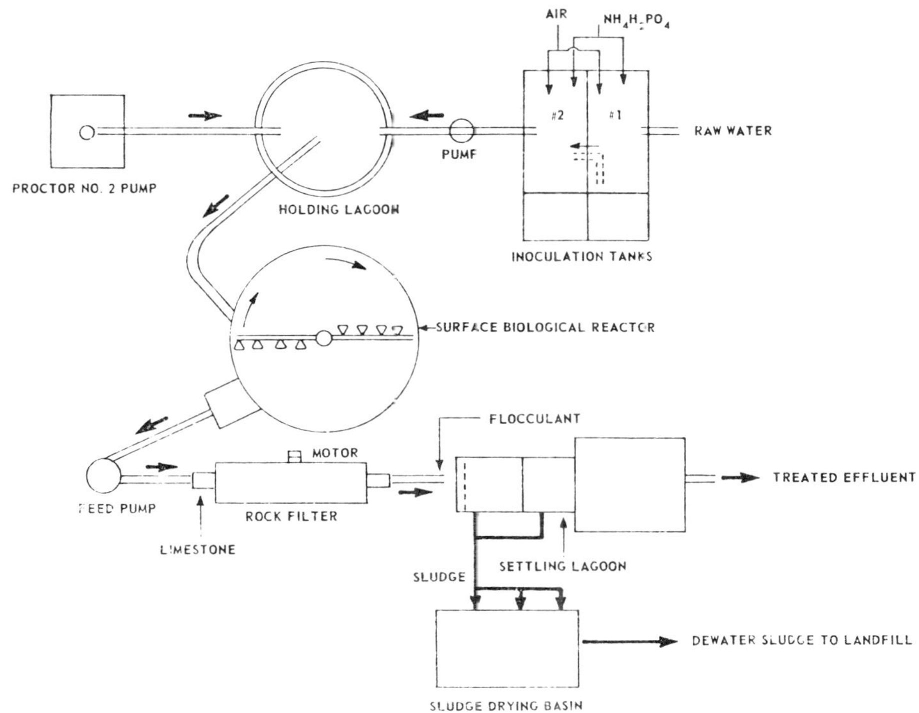

Source: EPA 670/2-73-093

TABLE 3.14: NEUTRALIZATION OF CMD BY LIMESTONE IN THE ROTARY REACTOR

Test Period 9/2-10/71

Reactor Conditions

Speed - 13 rpm; Power consumed - 7.88 KWH/Hr. @ 19-23 amperes; Slope - 4°

Limestone Reagent

Seam - Valentine (south western end)

Location - Oak Hall, Centre County, Pa.
Neidigh Bros. Limestone Co.

Chemical Analysis

Moisture = 2.0%

Dry Basis:
CaO = 50.0%
$CaCO_3$ = 85.0%
$MgCO_3$ = 1.5%
R_2O_3 = 0.5%

Feed Rate - 60 lbs./hr
continuous feed

10% excess theoretical

Cost
Stone $1.33/ton
Transportation 1.95/ton (65 mi.@ 3¢/ton/mi.)

Hardness

Los Angeles Rattler Test - B Grading
100 revolutions - 5.6% passing
500 revolutions - 25.7% passing

Size Analysis:

Screen Size Passed	Screen Size Retained	Direct Wt.%	Cumulative Wt.% Retained	Cumulative Wt.% Passing
	1/2 inch	39.0	39.0	100.0
1/2 inch	3/8 inch	10.0	49.0	61.0
3/8 inch	8 mesh	29.5	78.5	51.0
8 mesh	60 mesh	16.3	94.8	21.5
60 mesh	200 mesh	2.7	97.5	5.2
200 mesh		2.5	100.0	2.5

Process Flow Analyses

Source - Proctor No. 2
Flow - 60 gallons/minute
Process Temperature - 18°C
Theoretical limestone requirements - 0.0127 lb./gal.
Theoretical requirement of stone reagent utilized - 0.0152 lb./gal. or 0.90 lb./minute.
Flocculant addition - 5 ppm Atlasept 1A1 as 0.2% wt. solution.

	pH	Acidity mg/l $CaCO_3$	Alkalinity mg/l $CaCO_3$	Fe II mg/l	Fe III mg/l	Al mg/l	SO_4 mg/l	Cond. μ mhos.	Sus. Solids mg/l
Feed Water	2.8-3.2	1,520[c]		374	56	230	2346	2490	
Limestone Reactor									
Influent	2.9-3.2	1.500		26	183				
Effluent	6.2-6.6		128	20	35				4,232[a]
Settled Solids									19.6% wt./vol.
Settling Lagoon									
Influent	6.40			6	1				
Effluent	6.48		78	<1	<4				
Settled Solids									9.5% wt./vol.[b]

Sludge Control:

Assuming transfer of about 2,600 gallons of sludge at 9.5% wt./vol. solids daily from settling lagoon to sludge drying basin over period of one hour, would add about 3 inches of sludge depth (assuming 1400 ft^2 area in a compartment). With the three compartment sludge drying basin, the transfer cycle would be: daily transfer of three inches of sludge for two consecutive days to each compartment.

Day No.:	1	2	3	4	5	6	7	8	9	10	11	12
Compartment No.:	1	1	2	2	3	3	1	1	2	2	3	3

Experience shows this would result in 1.50 inches/6 days of sludge level at solids content between 15 and 45 wt. % solids depending on drainage rate and weather: i.e., upon dewatering the settled solids reduced to about 1/4 their original volume. This sludge accumulation rate would require removal of sludge from the drying basin about every 144 days representing about 600 $yds.^3$ or 450 ton (about 110 $lb./ft^3$). The sludge drying basin was emptied with a 2 $yd.^3$ bucket on a model 922 caterpiller unit on cats. Two trucks with a total capacity of 15 ton carried the dewatered sludge about 1/2 mile for landfill. The sludge drying and volume reduction continues, reaching a final volume between 25-33 percent of the dewatered sludge and a water content related to the ambient conditions (1-10%).

[a]0.035 lb. dry sludge/gallon or 2.12 lb./min.
Settled sludge volume - 3.0 vol. % in 10 minutes.
[b]0.03 gal. sludge/gal. process influent or 1.8 gal./minute.

Composite sample during sludge transfer, value represents dilution with transfer water. Samples taken from lagoon bottom have exceeded 30 wt. %.

[c]Modified procedure - (pH 7.3) - 1,433 mg./l.

Source: EPA 670/2-73-093

These limestone-produced sludges (Type 3 settling behavior) do not exhibit a clear settling interface and their settling rates can best be observed by measuring sludge buildup. The supernatant water tends to be cloudy, but its clarification rate and level can be enhanced by the use of high molecular weight flocculants. A very high flocculant dosage was employed during this test, seeking indications of sludge balling with carbonate sludges, as was attained with hydroxide sludge. Positive results were attained which suggests that a bottom feed-type of solid-fluid separation operation would be especially effective for this type of sludge.

In such an approach, the settling sludge blanket serves to collect and filter the flocculating colloidal particles resulting in an extremely clear effluent. The bottom fed weir to the settling lagoon provided sufficient retention time to settle the flocculated sludge and allowed a very clear effluent to enter the settling lagoon. This operation was effective to the degree that, had a proper pump been available to remove the settled sludge from the bottom of the weir, the settling lagoon would have been completely superfluous. Such a procedure would have major advantages in both capital and operating costs.

The response to other limestones was satisfactory but it varied according to particle size, purity and hardness. Dolomite also responded satisfactorily but was more difficult to control having reduced reactivity and greater hardness thus limiting maintenance of adequate available reactive surface. The use of pulverized limestone (-200 mesh rock dust) was helpful as a supplemental feed to the rotary limestone reactor to complete the reactions and to add excess alkalinity to the effluent. However, beyond requiring a separate feeding system, control of its addition must be exceptionally tight to prevent use of excess reagent.

Proprietary Processes

Gravity Flow Lime-Limestone Process — The process of *J.J. Birch; U.S. Patent 3,516,931; June 23, 1970; assigned to Barnes & Tucker Co.* offers the advantage of being very inexpensive in capital equipment, requiring no clarifiers or the like. Aside from a very minimum of treating equipment, the operation is carried on in a concrete tank and earthen ponds. Furthermore, the process permits very large through-puts, the plant being described serving to treat two million gallons of acid mine water per day.

One of the advantages of the process is that it can be carried out under gravity operation. Another advantage is that the labor cost is a minimum. The plant will operate without attendants for considerable periods and when attendants are needed they are only concerned with such features as supplying chemicals, servicing equipment, and removing the floc formed.

Referring to Figure 3.6, acid mine water enters from an intake **20** and is withdrawn through a pipe or channel **21** into a mixing tank **22**, suitably a concrete septic tank. Quick lime or hydrated lime from a storage hopper **24** is fed by a suitable power-operated feeder to introduce it at a controlled rate into the tank. In the tank the lime forms a slurry mixed by a propeller mixer driven by a motor. The lime slurry passes off at the top of the mixing tank through a pipe or conduit **30**.

FIGURE 3.6: GRAVITY FLOW PROCESS

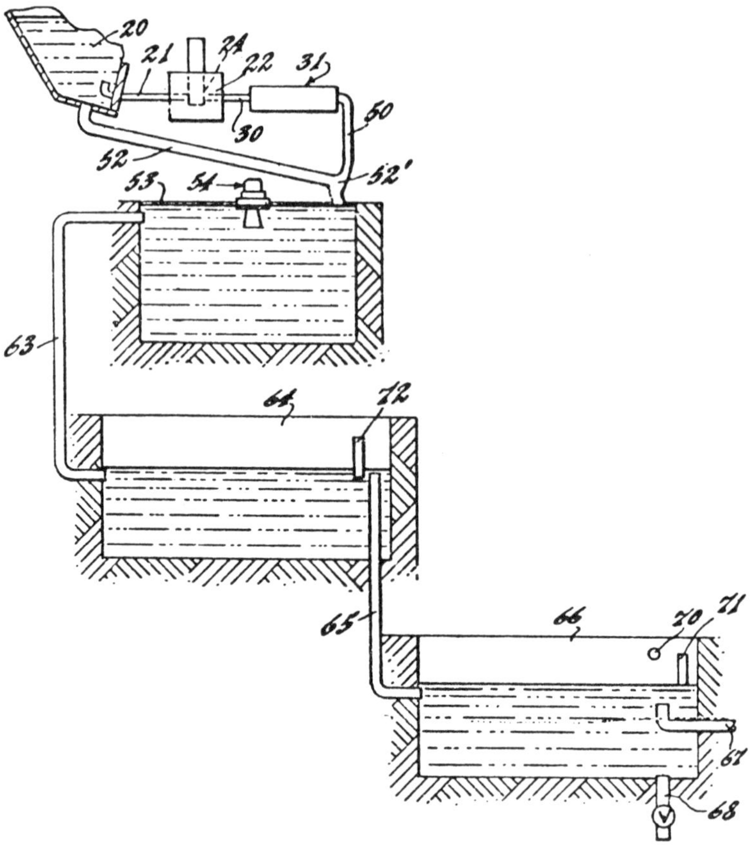

Source: U.S. Patent 3,516,931

At a suitable point below the mixing tank there is provided an attrition mill **31** having a cylindrical housing, an inlet end and an outlet end. The attrition mill is conveniently made from a coal drying kiln or cement kiln removing the burners as no heat is applied in it. The attrition mill is arranged horizontally without any slope. The cylindrical housing has journal bands near the opposite ends which support it rotatably on pivotable rollers. At a convenient point near the center, a gear is affixed around the outside of the housing and the mill is rotated by a pinion driven by a drive.

The inlet end has at the center an opening into which enters an extension of pipe **30** bringing slurry from the mixing tank **22**. This pipe does not fit closely into the end of the attrition mill but has a clearance between the end of the mill and the outside of the pipe through which air enters. In operation the clearance may be of the order of ½" to 1" at each side, on a mill of 4' diameter.

To permit charging additional limestone when required, a limestone hopper feeds by a screw conveyor into the center of pipe **30**, leaving ample room around the limestone feed for entrance of slurry. At the outlet end the attrition mill has a plate or ring extending to a position near the center where an opening is provided covered with a fine screen which prevents pieces of limestone from leaving the mill. A discharge pipe discharges slurry that passes through a screen into pipe **50**. Limestone lumps in the interior of the attrition mill are desirably of a coarse size, and too large to leave through the screen of 20 Tyler standard mesh per linear inch. From the mill water leaving by pipe **50** joins a main stream **52** of raw water from intake **20** at point **52'** and is discharged into aeration pond **53** provided with aerator **54**. The aerator may be of the float type.

The aeration pond in a plant handling two million gallons of acid mine water per day should be of about 400,000 gallons capacity. From the aeration pond, the water containing floc ready to precipitate is conducted through pipes **63** from a point near the top of the aeration pond to a settling pond **64** where the water is allowed to lie quiescent in order to settle out the floc. In a plant of the size of two million gallons treating capacity per day, the settling pond should be of about 4,000,000 gallon capacity.

From the settling pond, water from a point near the top is carried over by pipes **65** to an equalization pond **66**. In the case of a plant of about two million gallons per day capacity, the equalization pond should be about 4,000,000 gallons. The final treated effluent from the equalization pond passes by pipes **67**, conveniently about half way up the height of the pond to a suitable stream. There are also provided a bottom pipe **68** provided with a normally closed valve for draining the equalization pond and a top overflow pipe **70**.

The settling pond has near the outlet an automatic pH probe **72** and the equalization pond has at a point near the outlet an automatic pH probe **71**. These probes and their automatic controllers can, if desired, be set to maintain a particular pH level and add to or reduce the feed of lime into the mixing tank as required to maintain such a level.

Typical analysis of the final treated water as it leaves the equalization pond and enters a stream is as follows: pH, 7.2; total acidity, 0; total alkalinity 40 ppm; and total iron, 1.5 ppm.

Split-Stream Limestone Process – M. Deul and E.A. Mihok; U.S. Patent 3,617,560; November 2, 1971; assigned to The U.S. Secretary of the Interior found that the limitations inherent in previous processes for the limestone neutralization of acid wastewaters may be overcome in a relatively simple manner. Instead of using a limestone abrasion or grinding means as a reactor through which the total of wastewater flow is passed, it has been found that splitting the wastewater flow and passing only a minor portion of that flow through the reactor results in significant process advantages. This process modification can increase the neutralization capacity of a plant by as much as 300 to 600%.

The flow of acid wastewater through the grinding means is limited to a low-level which maximizes the production of extremely fine limestone particles. The slurry of particles thus produced is then mixed and reacted with the major portion of the wastewater stream. Extremely rapid (on the order of 5 seconds or less) reaction then occurs.

Referring to Figure 3.7, a stream of acid water waste **1** flows into surge storage means **2**. Provision of surge storage capacity, which is not always necessary, is highly desirable in those cases where the wastewater flow rate is fluctuating or intermittent. In the case of acid mine waters containing ferrous iron some oxidation of ferrous to ferric iron will also occur during storage.

A controlled flow of wastewater is drawn from the storage means and is split into two streams, **3** and **4**. Stream **3**, comprising only a small fraction of the total flow, is passed into limestone grinding means or mill **5**. This mill may comprise any conventional type of grinding mill provided that it is capable of producing a very finely divided product having a median particle size of less than about 10 microns. Make-up limestone feed is added to the mill by means of hopper **6**.

FIGURE 3.7: SPLIT-STREAM LIMESTONE PROCESS

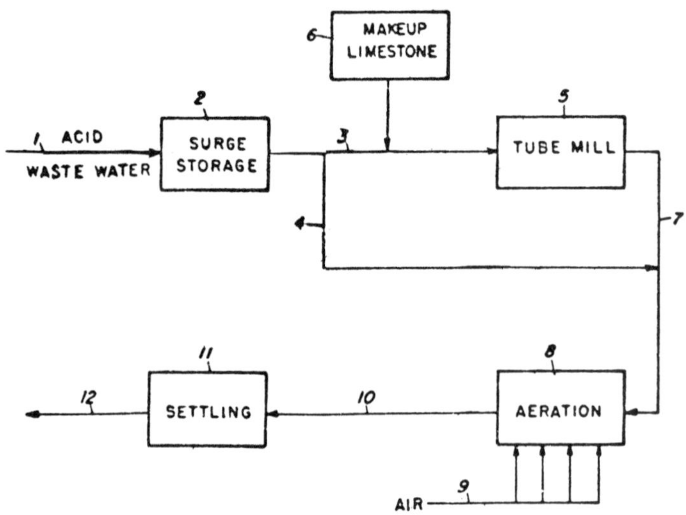

Source: U.S. Patent 3,617,560

A particulate limestone slurry stream **7** is removed from the mill and is mixed with the major portion of the wastewater stream by-passing the mill via conduit **4**. It is preferred that mixing means such as riffles or baffles be provided in conduit **7** after its juncture with stream **4** so that the neutralizing limestone slurry is thoroughly mixed with the acid water waste stream. A mixing time as short as 5 seconds is sufficient for substantially complete reaction to occur.

The combined limestone slurry and acid wastewater streams are then passed to aeration means **8** having air supply means **9**. The aeration means may comprise any conventional devices capable of intimately contacting water with an oxygen-containing gas. A combination of surface aeration and air sparging was found to be adequate. The treated effluent stream entering aeration means typically will have a pH in the range of 5 to 6. This low pH is due almost entirely to dissolved carbon dioxide. Sparging with as little as one-third cubic feet of air per gallon of effluent is sufficient to strip much of the dissolved carbon dioxide, thus raising the pH to the range of 6.5 to 7. Aeration also contributes to efficiency of iron removal in those wastewater streams containing ferrous iron by oxidizing the iron from the ferrous to the ferric stage. The aerated effluent stream is then passed via conduit means **10** to settling means **11**. A neutralized and clarified effluent is then discharged from the process by means of conduit **12**.

SLUDGE CONTROL

Any treatment removing the undesirable ions from mineralized waters creates a by-product which is troublesome and costly to control. In neutralization processes applied to coal mine drainage, a sludge is formed from the separated water-insoluble compounds.

Sludge control may incorporate three stages: solid-fluid separation of the sludge, sludge dewatering, and sludge disposal. Two techniques may be satisfactorily employed in the solid-fluid separation: settling lagoons and mechanical clarifier-thickeners. Lagoons are most commonly employed. Although the choice is usually determined by land availability and process economics, there are other more subtle differences which should not be ignored. The process economics do not consistently favor either practice.

Solid-fluid separation in a settling lagoon may provide limited sludge storage or permit permanent disposal. Limited storage requires sludge transfer but uses less land. Settling and disposal in the same lagoon necessitates large land areas, but involves minimal design and construction. The method can be quite satisfactory. This approach is the least acceptable aesthetically. Other detailed design, construction, and operation considerations are also necessary.

In contrast, mechanical clarifier-thickener systems require the least land area and provide the least sludge storage. They are more versatile in process control by permitting continuous, automatic sludge removal. Their mechanical nature does not necessarily imply higher costs.

Further dewatering of the settled sludge prior to disposal may be indicated by its voluminous nature, low percentage of solids, and tendency to gel. It may involve some form of in situ evaporative drying or direct water removal by a drying basin, by a filter, or by a centrifuge. The settled sludge usually may be pumped through pipes to the disposal area or to a tanker truck for haulage to the disposal site requiring multiple handling. In either case, handling large volumes of water increases costs. However, after dewatering the solids may be moved by dump truck. Ultimate disposal is usually made into abandoned mines (deep and surface) or to landfill. No broad, feasible use for the impurity solids has been developed.

The settled sludge is difficult to handle and dewater. When collected by pump suction, the gelled or flocculated solid particles do not readily disperse. They move at slower rates than the supernatant water which leads to "rat holing". This phenomenon makes removal of settled sludge from a lagoon bottom difficult and costly. Some sludges tend to gel, while others achieve a plastic clay-like character which resists any flow.

The important physical properties of the sludge are: settling rate, settleability (an arbitrary concept combining settling rate and final settled volume which also incorporates compression phenomena), bulk density, and flowability. All relate to the hydrated character of the sludge. The sludges settle to a solids content between 0.5 and 14% by weight. Most are in the lower range. When dried or frozen, the highly hydrated character of the sludge is changed.

These hydration properties are not restored upon rewetting. The physical properties of sludges are governed by chemical composition and precipitation processes which tend to be reasonably uniform for a given drainage water and treatment process. However, the properties vary widely with the neutralization reagent employed. Limestone shows the greatest deviation from the other reagents, producing a rapidly settling sludge.

The volume of the settled sludge increases with iron-aluminum concentration of the raw water extending to more than 30 vol %. The sludge volume level obviously relates to disposal problems and costs; there are advantages in achieving higher density sludges. In addition to control by the initial precipitation reactions, a greater sludge density may be achieved by recycling a portion of the settled sludge to the initial reaction chamber. A very dense, rapid settling magnetic sludge has been described (see Catalytic Oxidation Process, p 192) when the iron is converted to a magnetite structure.

The factors which relate density, flowability, and surface character of sludges to their composition and formation, have not been adequately elucidated. Obviously the crystal composition, size, surface area, and the nature of its surface charges all affect the sludge behavior. Sludges formed in the pH range of 6.4 to 7.2 appear to settle most rapidly. This pH range also permits use of minimum reagent levels.

Sludge Settling Rates

Three types of settling rate phenomena have been observed for the CMD slurries developed at Hollywood.

Type 1: The sludge settles rapidly and uniformly with a distinct interface and a clear supernatant liquor. This response results from reaction with strong alkalies (hydrated and pebble lime, sodium hydroxide, and sodium carbonate) when the mine water has a typical or high iron and aluminum content. The settling rate curve has a classical shape, i.e., the sludge volume decreases sharply during the first 10 to 30 minutes, asymptotically approaching a relatively constant terminal volume within 24 hours. See Figures 3.8 and 3.9.

Type 2: This behavior occurs when very low iron- and aluminum-containing waters are treated with strong alkalies. Tyler Run waters containing 1 ppm iron and 30 ppm aluminum are illustrative. The amount of precipitate formed is quite small and resists flocculation. The flocs tend to remain diffused and form a nearly stable suspension. Complete settling may not occur for several days. Even then the settled sludge might not exceed 0.1 wt % solids. The water may retain a slight turbidity. Rates are difficult to evaluate.

Type 3: A two-phase solids system is developed. The major portion (to 90%) of the solids settle very rapidly and can best be measured as they collect on the bottom of the test device.

FIGURE 3.8: SLUDGE SETTLING RATE—PROCTOR NO. 1 WATER

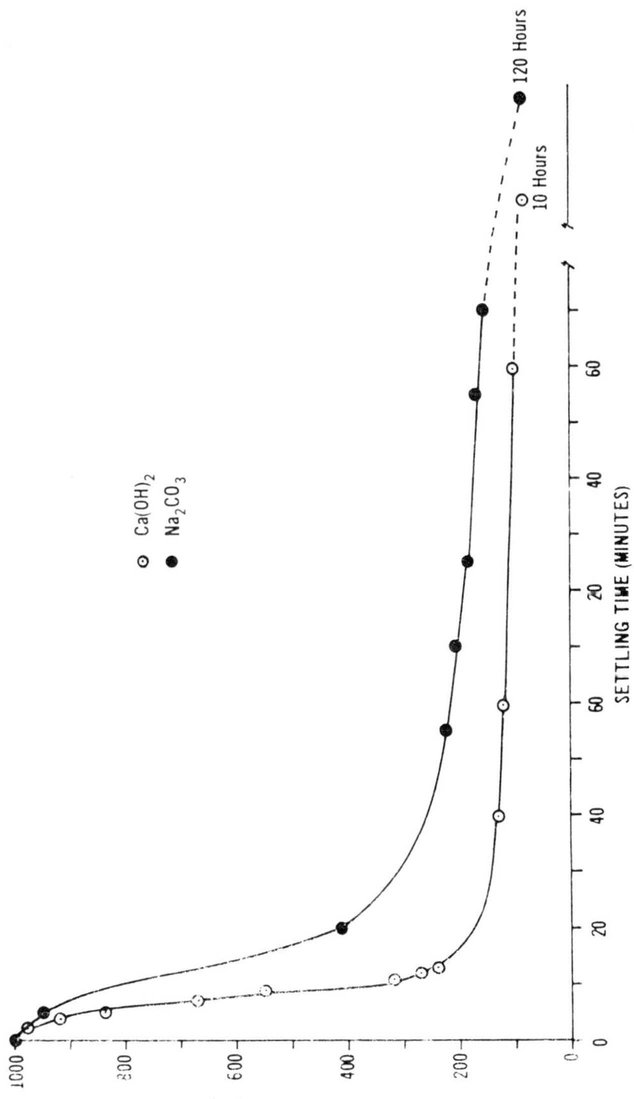

Source: EPA 670/2-73-093

FIGURE 3.9: SLUDGE SETTLING RATE—PROCTOR NO. 2 WATER

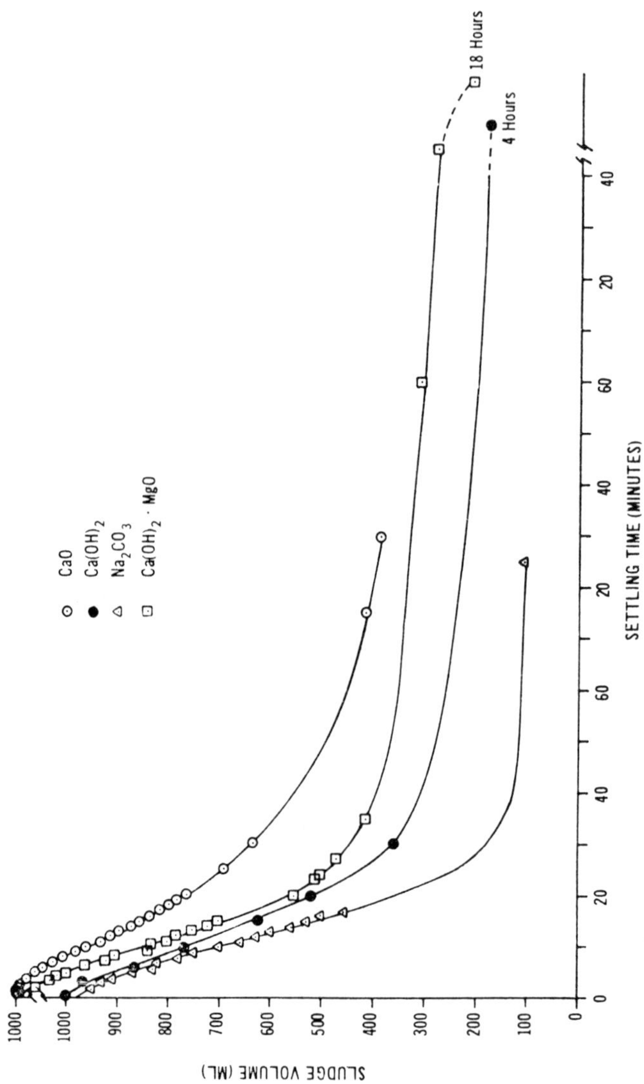

Source: EPA 670/2-73-093

The particles which are collected are not gelatinous flocs and do not appear to "grow" significantly in size. This buildup is an asymptotic reciprocal of Type 1 behavior. The bottom collection usually approaches its maximum level within 10 minutes. This volume is used as the sludge volume. This buildup rate can be measured but with less certainty. The values reported are conservative. The supernatant liquid is turbid. This settling behavior appears to be independent of pollution loading of the feed drainage.

Sometimes an interface will form slowly to yield a clear supernatant. This behavior is similar to Type 2 slurries, requiring about 24 hours to settle. In other cases the solution will decrease in turbidity but remain cloudy. This behavior results with limestone neutralization. The residual turbidity is attributed to the presence of near-colloidal solids. The behavior has also been observed with sodium carbonate and sodium hydroxide when continuing reactions (both neutralization and iron oxidation) are experienced. When observed with these reagents, the behavior is a symptom of improper reaction control.

Solids separation from all three types of settling behavior can be improved by the use of flocculants, by sludge recycling to increase the total solids content of the suspension, or by "filtering" the floc through a deep sludge blanket such as can be maintained in a thickener. The sludge blanket in a thickener can be improved by weighting it with loading agents such as pulverized limestone, fine coal and silica.

A summary of sludge settling data from various plant tests are presented in Table 3.15. No data resulted from the sludges exhibiting Type 2 behavior. The observed settling rates ranged from 1.2 to 5.4 ft/hr. The higher values resulted with limestone treatment but yielded a turbid supernatant liquor. Plant data did not indicate rate trends among the other reagents. The sludge settling rate seemed to approach a maximum with increasing iron-aluminum levels in the 100 to 200 ppm range then decreased.

At low metal levels floc growth was slow while at levels about 200 ppm the growing flocs approached compression and settled more slowly. There was no distinct trend measured during plant tests in settling rates with pH (6.3 to 8.3) variation, if the neutralization-oxidation reactions were complete. However, laboratory tests indicate maximum rates near pH 6.8. Continuing reactions literally "muddied" the water. The density of the settled sludge does not directly reflect either metals content of the raw water or settling rates of the sludge.

Use of Polyelectrolytes as Settling Aids

Polyelectrolytes (usually polyacrylamide polymers of various molecular weights and structures) have found universal application in enhancing the settling rates of all types of slurries, including most coal mine water sludges. They enhance floc growth and decrease turbidity. They are available in a range of ionic charges, molecular weights, and other characteristics which govern their performance. There is no evidence that these polymers increase final sludge densities either in bench or plant tests. This was evident in sludge from Tyler Run waters which have low suspended solids containing predominantly aluminum hydroxide. At very high polymer dosage, a "balling" phenomenon occurs with these sludges which may have the practical response of increasing density. The polymers do not decrease residual soluble iron levels in treated waters as has been suggested.

TABLE 3.15: SUMMARY OF SLUDGE SETTLING DATA FROM HOLLYWOOD PLANT TESTS

Water Source	Total Acidity mg/l	Fe^{++} mg/l	Al mg/l	Final pH	Alkali Used	Sludge Settling Rate ft/hr	Suspended Solids in Neutralized Water mg/l (aver.)	Sludge Settled Volume % of original	Solids in Settled Sludge wt. %	Calculated Solids in Settled Sludge (T.A. + $2Fe^{++}$)
Tyler Run	114	4	15	6.26	NaOH	--	50	2.2	0.23	--
Bennett's Branch	120	29	15	6.51	NaOH	--	91	--	--	--
Bennett's Branch	206	35	17	6.95	CaO	--	96	4.5 + T	0.21	0.61
Bennett's Branch	206	35	17	6.95	CaO	--	96	2.5 + T	0.38	1.10
Proctor No. 1	502	80	80	8.30	CaO	3.49	350	10.0	0.35	0.66
Proctor No. 2	1,707	420	210	7.46	CaO	1.02	1,365	29.6	0.46	--
Tyler Run	78	1	15	7.00	Soda Ash	--	--	--	--	--
Bennett's Branch	86	10	5	8.03	Soda Ash	--	37	1.0	0.37	1.06
Proctor No. 1	397	50	50	7.38	Soda Ash	2.96	290	5.0	0.58	1.00
Bennett's Branch	79	18	10	8.00	Soda Ash	--	55	0 + T	--	--
Proctor No. 1	328	57	57	7.60	Soda Ash	3.95	252	3.0	0.84	1.47
Proctor No. 1	302	52	75	7.78	Soda Ash	1.87	178	3.0	0.59	1.35
Proctor No. 2	1,398	241	205	7.31	Soda Ash	1.78	2,318	11.0	2.11	1.71
Proctor No. 2	1,174	118	59	7.38	Soda Ash	3.00	--	8.0	--	1.76
Proctor No. 1	405	69	62	7.25	$Ca(OH)_2$	4.72	848	8.0	0.06	0.68
Proctor No. 1	1,380	317	191	6.75	Limestone	5.40+T	1,692	3.0	5.63	6.69
Tyler Run	104	<1	12	7.75	Pul.Limestone	4.68+T	104	0.2	5.2	5.2
Proctor No. 2	1,401	322	191	7.75	Pul.Limestone	4.10+T	1,356	2.0	6.78	10.3
Proctor No. 2	1,305	250	191	7.1	$Ca(OH)_2$	1.47	1,370	18.0	0.76	1.0
Proctor No. 2	1,360	266	191	8.2	Dol.Hydrate	1.95	1,990	18.0	1.11	1.05

NOTE:

"+T" - Indicates supernatant liquor was turbid.

Source: EPA 670/2-73-093

Since their usage does aid uniform plant operation and reduce the size of equipment, an evaluation of several types of polyelectrolytes was made. Previous pilot studies involving several preparations found nonionic Superfloc 84 and Nalcolyte 67 effective at dosages of about 10 ppm.

Representatives of several polyelectrolyte producers carried out evaluation of their flocculants on fresh sludge at the pilot plant. Although the use of these formulations is simple, the techniques must be carefully established. Considered were: selection of the most effective reagent, preparation of the reagent solution, and reaction parameters as dosage, dilution and mixing procedures.

These polymers consist of long length molecules. Their response is governed by the reaction of the polymer and the particle surface, thus contacting methods between the polymer and the suspended particles are critical. Excessive agitation (as passing a solution through a high speed mixer or pump impeller) can shear polymer molecules and destroy their effectiveness.

The data presented are not intended to compare products nor imply that other products may be unsatisfactory. High molecular weight polymers can be effective flocculants at dosage levels of 0.5 to 2.0 ppm of neutralized mine water. Those polymers designated nonionic or weakly anionic were most responsive. The settling rates with the polymers ranged from 4 to 16 ft/hr. The optimum dosage level increased with the suspended solids level.

In recycle sludge systems, flocculant dosage levels should be based on the ppm polymer/lb of dry solids. It cannot be assumed that polymers associated with settled recycle particles remain effective. Excessive agitation can destroy polymer-formed flocs. The best flocculant and dosage for any particular sludge must be established experimentally, although the data presented here should limit the evaluation. No obvious sensitivity of the polymers to pH was indicated over the range 6.0 to 9.0 normally experienced. Flocculants recommended by their respective companies were: Superfloc 127, Decolyte 930 and Atlasep 1A1.

Primary Solid-Fluid Separation

Thickeners: Gravity settling of suspended solid particles is divided into clarification and thickening. The former term is applied to dilute pulps as illustrated by Type 2 mine drainage sludges. The devices utilized are characterized as to means for removal of the settled solids. Unit sizing is the primary consideration. The size is established empirically, based on influent volume and particle levels and characteristics. Particle settling rates establish the surface area required to provide the desired water rise velocity (ft/hr or gal/min/ft^2). The clarified effluent rise velocity must be less than the settling rate of the slowest moving suspended particle. A rise velocity, which is much lower than the particle settling rate, may be required with some slurries to provide adequate retention time for floc growth.

The depth of the thickener must comprise four zones: the clear supernatant, the feed region, and the transition and the compression zones. The clarified zone contributes a safety factor against discharge of turbid waters (usually at least two feet) while the compression zone aids in producing a more dense sludge. Observations indicate the advisability of introducing the feed slurry from some mine waters into the transition zone to gain a filtration effect of the settling sludge blanket. At the bottom of the compression zone, a slow moving rake moves the sludge to a removal port and prevents nonuniform, excessively dense sludge buildup.

Hollywood Thickener – The device utilized is atypical in its design as a thickener since in addition to this application, studies of the sludge recycle concept were to be made with this unit. The unit was ruggedly constructed and was not prone to mechanical problems. It performed both types of functions satisfactorily; provided means for good control, sampling and observation capability. The calculated flow rate parameters are given in Table 3.16, while related sludge parameters are given in Table 3.17.

TABLE 3.16: THICKENER FLOW RATE PARAMETERS

Zone	Zone capacity gallons	Retention time minutes 100 gpm	Retention time minutes 200 gpm	Retention time minutes 350 gpm	Flow velocity feet per hour 100 gpm	Flow velocity feet per hour 200 gpm	Flow velocity feet per hour 350 gpm	Gallons/minute/ft² 100 gpm	Gallons/minute/ft² 200 gpm	Gallons/minute/ft² 350 gpm
Primary (Downflow)	3,730	37.3	18.6	10.7	8.46	16.9	29.4	1.05	2.10	3.68
Secondary (Downflow)	3,730	37.3	18.6	10.7	8.46	16.9	29.4	1.05	2.10	3.68
Compression Zone	13,550 1,130[a]	See Table 3.17 for Compression Zone Data								
Solids Separation (Upflow)	29,200	292.0	146.0	83.2	2.06	4.12	7.2	0.253	0.506	0.885
Total	51,340	513.0	257.0	147.0						

[a]Sludge Collection Zone

Source: EPA 670/2-73-093

TABLE 3.17: SLUDGE RATE PARAMETERS FOR 25 FOOT THICKENER

	Influent Solids Level mg/l								
	500			1000			2000		
	Flow Rate - gpm								
Solids in Underflow %	100	200	350	100	200	350	100	200	350
	Retention Time-Compression Zone-hours[a]								
2	98	49	28	49	25	14	25	12	7
4	196	98	56	98	49	28	49	25	14
6	295	147	84	147	73	42	73	37	21
8	392	196	112	196	98	56	98	49	28
	Sludge Generation (Equilibrium Blow Down) Rate-gpm								
2	2.5	5.0	8.75	5.0	10.0	17.5	10.0	20.0	35.0
4	1.25	2.5	4.38	2.5	5.0	8.7	5.0	10.0	17.5
6	0.83	1.66	2.91	1.67	3.33	5.8	3.3	6.7	11.7
8	0.63	1.25	2.19	1.25	2.50	4.4	2.5	5.0	8.8

(continued)

TABLE 3.17: (continued)

	Influent Solids Level mg/l								
	500			1000			2000		
	Flow Rate - gpm								
	100	200	350	100	200	350	100	200	350
	Dry Solids Production Rate								
lbs/min.	0.42	0.84	1.46	0.84	1.68	2.92	1.67	3.34	5.85
lbs/day	605	1210	2102	1210	2420	4204	2420	4840	8408
$lbs/ft^2/$ 24 hrs	1.25	2.47	4.29	2.47	4.94	8.58	4.94	9.88	17.16

[a] Calculated from compression zone volume of 14,700 gallons and bottom area of 490 ft^2 Sludge depth from bottom of secondary zone - 5 feet.

Source: EPA 670/2-73-093

The physical unit was a modified Infilco "Densator," High Density Solids Contact Treating Plant. Bottom rakes were added to permit the operation of the unit as a conventional clarifier-thickener. A detailed drawing of the unit is shown in Figure 3.10. The system provided for controlled pumping of sludge from the bottom well to the Densator feed weir, to the sludge storage sump for subsequent dewatering, or to the sludge drying basin.

The recycle sludge rate could be monitored by a magnetic flow meter. Alkali slurry could be pumped to the secondary reaction zone of the Densator in response to a pH sensor. This probe sampled water from the secondary reaction zone. The system permitted premixing either recycle sludge and raw water or recycle sludge and alkali before further reaction. The above-ground installation was convenient for regional sampling of the Densator.

The single unit was designed to provide neutralization, sludge densification and clarification for a complete treatment facility. The sludge densification capability was most important. The test sought to demonstrate the potential of the Densator as a multifunction unit producing a densified sludge with good dewatering characteristics.

In this thickener, the primary influent zone serves as a feed well, regulating flow surges and providing additional reaction time. The secondary zone permits flocculant addition and provides slow agitation for floc formation. This design permits the introduction of the feed slurry below the sludge blanket into the transition zone, if desired.

These atypical aspects of this thickener's design reduces the overall capacity. In the solids separation zone (feed solids and transition regions) the downflow settling of the floc proceeds along with the upflow rise velocity of the clarified water. The feed well extends ten and one-half feet below the overflow level. This provision permits an upflow sludge filtration region of eight feet when the sludge blanket is maintained two feet below the overflow.

FIGURE 3.10: INFILCO DENSATOR DESIGN

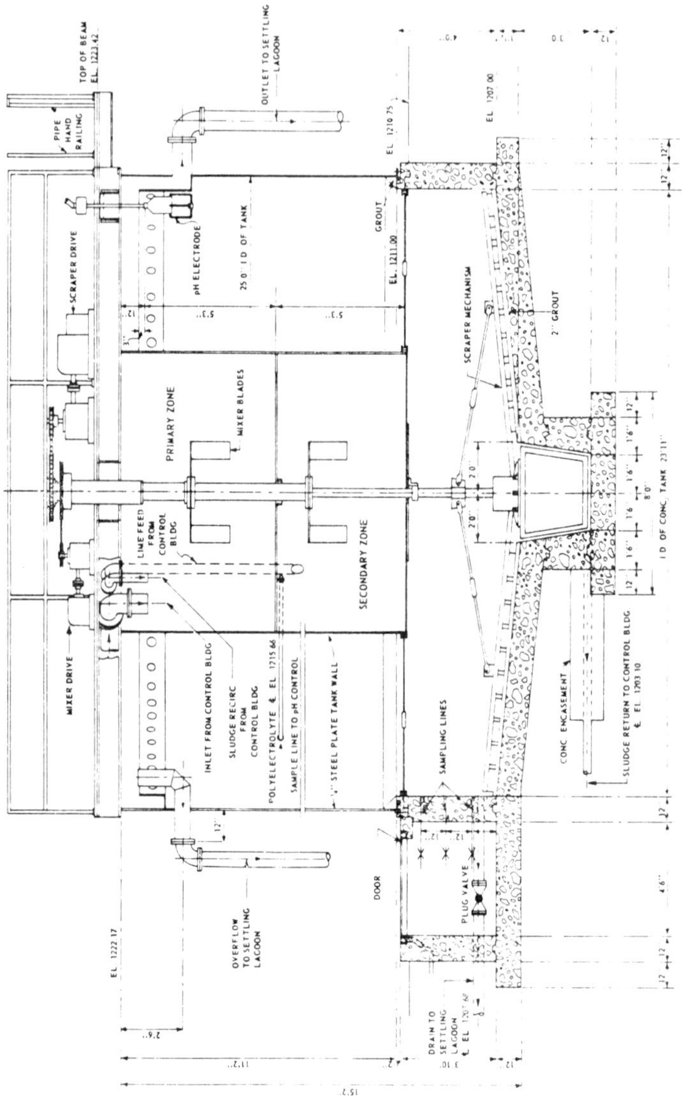

Source: EPA 670/2-73-093

When the sludge level is carried above the bottom of the secondary zone wall, a desirable upflow filtration of fines through the blanket occurs and can function to remove near-colloidal material from the rising water column. The thickener was satisfactorily operated under these conditions during several limestone neutralization runs resulting in good effluent clarity. This actual performance deviated significantly from bench scale settling tests with the same feed which failed to yield clear effluents, even with some flocculant additions. The use of thickener designs which include deep feed wells thus permitting upflow sludge blanket clarification is deemed to be an important discovery.

The upflow velocities of this thickener ranged from 2.06 ft/hr at 100 gpm to 7.2 ft/hr at 350 gpm. The measured settling rates of freshly precipitated sludges ranged between 1 and 6 ft/hr with a maximum of 16 ft/hr with flocculants. More typical sludge settling rates were nearer 2 ft/hr. This size thickener would function satisfactorily with flows near 100 gpm; however, when flocculants were used the rates could be expanded about three times and still maintain a clear effluent. The functioning of the primary and secondary zones were quite adequate for the maximum flow (350 gpm) to which the unit was subjected. In some tests, with Type 2, slow settling sludges, and with the use of pulverized limestone as a reagent, excessive turbidity in the effluent did occur.

Densator Performance as a Thickener – The Densator was operated as a thickener. All data utilized the following flowsheet. Limestone reactions introduced a pulverized slurry into the flash mixer since the rotary limestone reactor effluent could not be introduced to the thickener.

```
Raw Water  → Flash Mixer  → Oxidation Tanks  → Thickener
                                              ↙          ↘
Alkali Slurry ——↑      Dewater ←  Underflow          Effluent
                                                        ↓
                                                 Settling Lagoon
                                                   of Stream
```

Good clarification was maintained at flow rates less than 125 gpm when 1 to 2 ppm of flocculant were used. Flocculant additions were essential. Higher flow rates presented difficulty in keeping the sludge blanket level below a 2 foot depth from the overflow. This suggests rising velocities greater than about 3 ft/hr should not be exceeded.

Sludge thickening was erratic with underflow slurries ranging from 0.7 to 7% solids. The lower levels resulted from low acidity waters at high flow rates. A typical underflow solids content was about 2%. The more dense sludges developed when pulverized limestone was used as the neutralizing agent. This experience was consistent with the results of bench settling tests. Other observations from the operation of the Densator as a thickener are indicated below.

(1) Pulverized limestone, as a neutralizing agent, often resulted in turbid overflows which were difficult to control. Pulverized limestone produced a greater quantity of neutralized solids in the influent than with a similar alkaline equivalent of hydrated lime.

(2) Reagents with slower neutralizing reactions rates, as dolomitic hydrate and limestone, continued both neutralizing and oxidizing reactions in the thickener.

These continuing reactions were apparent from decreases in ferrous iron concentration and increases in pH in the thickener effluent. A thickener can be expected to provide important secondary reaction time to insure specification effluents, but with possible turbidity problems.

(3) A sludge level maintained above the bottom of the feed well provides for particle entrapment during upflow through the sludge blanket. This entrapment improves effluent clarity when difficult dispersions occur, as in limestone neutralization.

(4) The thickener bottom rake action produces a dewatering action on the settled sludge. Some underflows attained a solid content of 5 to 7% after 24 hours of settling time, whereas the same slurry, settled under quiescent conditions, never exceeded 2% solids after 48 hours settling time.

(5) Waters having low levels of pollutant loading, produce low settled solids density and require large unit areas. In plant design, serious consideration should be given to the use of a second stage thickener possibly with flocculant addition, to attain a more dense sludge product.

(6) Sludge removal control followed either slow continuous pumping or timed-interval blow-down, utilizing automatic pneumatic valves in the system. The test schedule was too short to establish optimum sludge storage and blow-down considerations.

Design Parameters for Thickeners – The thickener studies provided information for establishing the design criteria for effective clarifying and thickening of mine slurries. An optimum design would provide the following features:

(1) Clarity of overflow with all types of slurries.

(2) An underflow of high density solids suitable for secondary dewatering.

(3) Provision for sludge recycling to achieve underflow solids content of at least 5%.

(4) Provision for second stage thickening, if necessary.

(5) Provision for upflow of suspensions through sludge blanket if colloidal solids are present.

(6) Provision for settled sludge surge capacity to permit economical sizing and operation of secondary dewatering devices or disposal.

(7) Provision for flocculant addition to improve and control sludge settling rates.

Design parameters required to attain the above conditions were established and are listed in Table 3.18.

Summary and Conclusions for Thickener Operation – Pilot plant experience confirmed engineering predictions that a thickener is the key primary dewatering process for any plant involving the treatment of flows greater than 750,000 gpd with a neutralized suspended solids content of 1,000 mg/l or greater. The precipitated solids handling and disposal problems arising from such waters require the superior process control that only a thickener can provide. This is an essential requirement for an effective, low-cost treatment process and is the logical alternative for settling lagoons. The great advantage of a thickener over the settling lagoon is the greater efficiency and the less costly sludge handling by making secondary dewatering feasible.

TABLE 3.18: OPTIMUM DESIGN PARAMETERS FOR THICKENER OPERATION

1.	Rising Water Velocity - gpm/ft^2	0.2 maximum
2.	Solids Settling Rate - ft/hr	2.0 minimum
3.	Unit area ft^2/T/day	300 minimum (may exceed 10,000)
4.	% Solids in Underflow	5.0 minimum
5.	Sludge Recycle Provision - % Circulating Load	5 to 30
6.	% Solids to Feed Well (Combined raw feed plus recycle sludge)	1 minimum
7.	Flocculant Dosage - ppm influent	1 - 2
8.	Depth of Feed Well Below Sludge Level - ft	2 minimum
9.	Sludge Depth Below Feed Well - ft	6 minimum
10.	Settled Sludge Retention Time - hrs	72 minimum

Source: EPA 670/2-73-093

Settling Lagoons: The objectives of the settling lagoon studies were to observe: lagoon performance during year-round operation as a water clarifying process; flow and turbulence within the lagoon; secondary water quality changes within the lagoon; and the response to atmospheric factors as temperature, precipitation, and wind. Further, information was desired on the nature of sludge accumulation and changes in sludge properties which control its handling by gravity flow and by pumping.

Simplicity, convention and apparent low cost have encouraged the use of this unit operation. However, maintained and reliable performance of lagoons is found to involve many complex functions, all of which are seldom optimized. Among the functions frequently sought, in addition to solid-fluid separation, are: uniform effluent quality based on buffering response of large volumes, providing treated water for other uses, effluent quality monitoring, retention time to complete reactions, and the control to densify, dewater, store, and even dry the sludge. The particular functions required will determine the design.

The lagoon system types may be grouped: clarification of low level solids requiring minimum sludge storage, clarifier-thickener providing for temporary sludge storage and intermittent removal of sludge (most common approach), total sludge control without sludge removal (requiring very large multiple lagoons, and secondary sludge storage including permanent disposal.

Providing capability for intermittent sludge removal usually limits the lagoon size. The process factors which determine lagoon design are: flow rate, unit sludge volume, slurry settling rate, sludge characteristics, and requirements for secondary dewatering of the sludge. Weather conditions, including heavy rainfall, wind, temperature layering and freezing, can drastically affect operation of a settling lagoon but these effects may be partially counteracted by design.

Lagoon operation based on continual sludge deposition is not only subject to decreasing retention time but currents and particle suspension which lead to particle discharge in the effluent. However, such trends in lagoon design modification increase costs and lead to mechanical sludge removal and peripheral effluent discharge; just a step removed from a thickener.

Basis for the Hollywood Settling Lagoon Design – The lagoon design incorporated a bottom, gravity flow sludge removal capability. Some of its major design features include:

(1) Three-compartment, sectionalized construction permitting operation of the influent section as a thickener, with the following sections acting as clarifying and polishing operations.

(2) Full lagoon-width influent weir minimizing velocity and providing a uniform feed stream at the surface leading to effective utilization of the entire lagoon. This greatly reduced turbulence and bottom agitation at the feed end.

(3) Transverse baffles separating compartments but also minimizing currents, surface windage effects, and short circuiting. The bottom elevation of the baffles defined the design sludge storage volume.

(4) Tapered sides and bottom with transverse sludge channels (in first two compartments) allowing bottom sludge removal with minimal disruption of the settling functions or development of turbidity. The slopes assisted the sludge flow to the outlet.

(5) Smooth, impervious, bituminous-covered lagoon bottom assisting movement of the settled sludge along slopes.

(6) Sludge pump station transferring the settled sludge to secondary dewatering operations.

(7) Lagoon effluent structure releasing clear water from surface of lagoon, providing for monitoring flow and quality, permitting transfer of treated water for plant use and allowing variation in water depth.

The lagoon retention times at different flow rates are plotted in Figure 3.11. The changes in retention time with filled sludge storage areas are also shown. Major changes in retention time with flow rate occur at low flows while small retention differences result at high flow rates.

Settling Lagoon Operating Observations – The settling lagoon was essentially in continuous use during the study as the final settling and water polishing unit operation. This provided observation of long-term operating performance and operating information, especially during cold weather.

Water Handling: Very few water handling problems arose with the lagoon except during severe winter weather. Under normal operating conditions, a minimum clear water depth of 5 feet was maintained in the feed compartment while hand-

FIGURE 3.11: SETTLING LAGOON CALCULATED RETENTION TIME

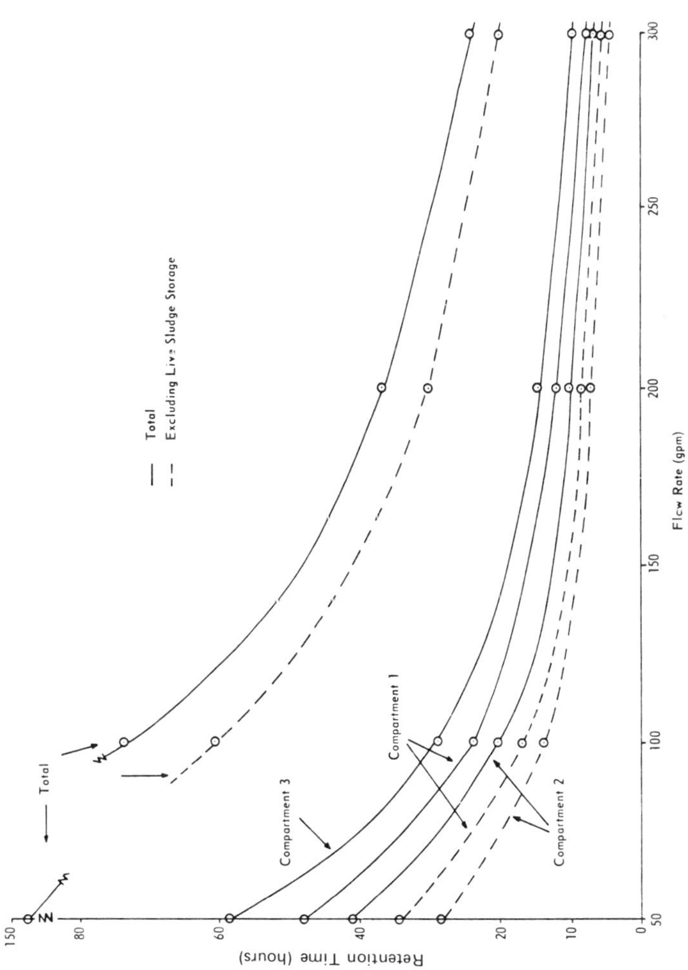

Source: EPA 670/2-73-093

ling maximum plant flow of 350 gpm containing up to 2,000 ppm of neutralized solids. Retention time was more than adequate even with sludge storage exceeding design capacity. Turbid effluents occurred only when situations unrelated to settling phenomena existed. The baffle and lagoon bottom design controlled short circuiting difficulties. No sludge inversions or floating islands of sludge were experienced. The clarity of lagoon effluents under these conditions was satisfactory for discharge.

A major increase in turbidity of the lagoon effluent was associated with heavy rains. Lesser increases in turbidity occurred during rapid sludge removal periods. The heat loss from the quiescent water during severe winter weather permitted ice cake formation on the lagoon surface. Under these conditions, a limited water flow was formed as a narrow channel between the ice, growing down, and the sludge, building up.

This situation impeded clarification. Under these conditions, the water velocity increased, while the retention time and sludge settling rates decreased. Sludge accumulation was gradually extended into the final compartment and the effluent turbidity increased. Continuous or frequent sludge removal from the bottom minimized this problem, although adequate sludge storage should be provided. If ice prevents surface removal of the sludge, the increased compression, aging, and gelling of the sludge further degrades its flow characteristics.

The adjustable height overflow pipes discharged only the clearest effluent in the lagoon. They were maintained nearly vertical to maximize retention time. There was no apparent advantage from changing their position from center of the effluent structure. The pipes were moved only to lower the water level prior to sludge cleanout operations.

The full-width influent weir compartment was found to be excessively deep which permitted settling of larger or rapid settling particles in the weir. However, the design objective to achieve a low influent velocity and initiate rapid sludge settling functioned well, even during freezing temperatures. The only effluent clarity difficulties occurred during the use of carbonate reagents or when the preceding neutralization or oxidation reactions were incomplete.

Sludge Handling: Bottom sludge removal from the lagoon was at times troublesome. The observations regarding sludge handling characteristics included:

(1) The sludge moved satisfactorily from the bottom channels of the settling lagoon to the sludge drying basin by gravity, prior to its gellation, although it tended to resist flow from the sloped bottoms to the channel. The positive displacement pump in the sludge pumping stations was used only during major lagoon cleaning and in transferring sludge to the Control Building for filtration. The limestone-produced sludge is most difficult to move due to its density. The plant operators were reluctant to transfer sludge from the lagoon to the drying basins at controlled slow rates, seeming to prefer more rapid flows. This psychological reluctance, unexplained, extended beyond the press of other duties. Low velocity flows are preferable since the tendency toward "rat-holing" increases with flow rate.

(2) The mobility of an "aged," settled sludge, when covered by several feet of water was poor. It would not flow freely from the sloped bottom into the

sludge channels and hence through the effluent pipes by gravity flow or from the pump. This indicates that the angle of repose of the submerged sludge is greater than 30°, and slopes approaching 45° are preferable. A slow moving screw with deep blades within the bottom channel would be helpful in directing the sludge toward the exit ports to maintain maximum sludge density.

(3) Mobility of aged sludge on the sloped surface of a drained lagoon was satisfactory when movement was initiated by a squeegee or hose water pressure. It tends to move as a large block. On quiescent compression and aging, the sludge forms a gel which has thixotropic tendencies. The gellation behavior develops within 48 hours under sludge and water pressure compression. There was no evidence of this gellation phenomena in the settled sludge in the thickener. Apparently the slow movement of the sludge by the thickener rake prevented this gel formation. A detailed understanding of this aging and gel formation is needed to assist lagoon design and operation.

A two- and three-foot thick layer of gelled sludge in one section of the lagoon could be moved into the sludge channel by two men in 4 to 5 hours using squeegees and low pressure hoses. Blocks of sludge were cut with shovels and would slide down the sloping bottom to the sludge channels, provided the bottom was kept thoroughly wetted by a small flow of plant water from a hose. Such settled sludge would normally range between 5 and 15% solids. This procedure was rapid and minimized sludge dilution. There was no evidence that the bituminous bottoms significantly enhanced sludge movement, but they did maintain stable dimensions and resist erosion or mud formation during sludge removal.

The side walls of the lagoon were earthen. Although no significant erosion problems developed with the compacted clay lagoon walls, sludge removal from these areas was more difficult. An occasional small stone from the lagoon walls would block the check valve of the sludge pump. Total bituminous side and bottom surfaces are definitely advantageous and recommended. Under no conditions should a stone rip-rap be employed below water levels in a lagoon.

(4) Transfer of gelled sludge from the settling lagoon to the drying beds by gravity flow was unsatisfactory. This sludge form tends to rat-hole quickly, allowing excessive amounts of clarified supernatant to move with the sludge and complicate subsequent operations.

(5) The sludge had adequate flow characteristics in lines and through pumps when movement began. The sludge transfer lines were flushed with clear water after ceasing sludge transfer to prevent scaling or further gellation. No line blockage was experienced.

(6) Sludges produced with hydroxide-type alkalies were the least dense and did not adhere very tightly to the lagoon bottom. In contrast, settled limestone slurries were dense, sticky, and clay-like in character and adhered tenaciously to the bottom surface.

(7) The distribution of sludge among the three compartments of the settling lagoon could be related to the sludge settling rates. Sludges settling at rates in excess of 2 ft/hr seldom passed into the second compartment. When the sludge was not removed and the first compartment approached its sludge capacity level (bottom of baffles), the settled sludge was carried into the second compartment.

During one three-month period without sludge removal, some sludge collected in the final polishing compartment.

Slower settling, Type 2 sludges, tended to drop more uniformly throughout compartments one and two with little deposition in the final compartment. The very rapidly settling sludges, such as those from limestone treatment, accumulated on the sloped bottom of the first compartment, a few feet from the weir. Only as the accumulation in this region increased, did the sludge storage extend beyond the sludge removal channel.

The sludge removal practice sought to maintain low sludge storage levels in the first compartment. There were distinct differences in the sludges settled in the three compartments. The settling in the second compartment was less dense and tacky with smaller, lighter floc. These sludges showed lesser tendencies toward gelling, rat-holing and were easier to move. They maintained a smaller angle of repose.

Settling Lagoon Performance Data — The settling lagoon functioned well as a clarifier in dewatering. The turbidity values of the effluent ranged between 1 and 5 jtu with suspended solids between 2 and 13 ppm. The ditch returning the plant effluent to Tyler Run showed no staining after the year's program. Satisfactory water quality was indicated by several generations of animal and plant growth in the third lagoon compartment. These included toads, frogs, salamanders, and insects. Trout and bass transferred to this section of the lagoon found ways of accumulating excessive-levels of sludge in their gills, although they survived well in the treated waters transferred from the lagoon. Deer and bear frequently drank from the lagoon.

The slow settling floc of Type 2 sludges gave the highest effluent solids level, although not necessarily the highest turbidity. The suspended solids and iron levels of these effluent waters met specifications at the highest flows hydraulically possible through the plant. The lagoon was operated once continuously without sludge removal for a sufficient period of time to consider its "live" sludge storage capacity. The longest service (over a period of three months) was equivalent to 42 days of continuous operating time at an average influent flow rate of 175 gpm. Assuming the formation of 12,500 gpd of settled sludge (5% of original volume) and a live sludge storage volume of 77,800 gal, the capacity of the lagoon would be only six days of settled sludge production.

At the conclusion of the long operating period, sludge had accumulated nearly to the influent weir with about 3 feet of clear water in the first two compartments and about 2 feet of settled sludge in the third or polishing compartment. The clarity remained satisfactory. This observation suggests that sludge compression had occurred and a live sludge load capacity, based upon 48 hour settling volume, is a conservative but realistic estimate for design purposes. A considerable safety factor beyond the one month sludge storage recommended by Pennsylvania guidelines is achieved. Due to difficulties which can develop during subfreezing weather, a two month sludge storage volume seems preferable.

The continuous or frequent interval removal of sludge from a settling lagoon is definitely advantageous. When sludge was permitted to accumulate in the lagoon and its removal was carried out, many difficulties were experienced. Removal would have been most difficult during freezing winter conditions. Labor requirements to manually assist sludge removal, even with described bottom design, was high. It appears that a settling lagoon is not an optimum unit operation in a mine water treatment flowsheet except for plants with small capacity or when treating waters with low solids loading.

Secondary Dewatering of Sludges

The solids content of any feed slurry to a mechanical dewatering operation is important. In general, feasibility increases and costs decrease with increased feed density. The physical and chemical nature of the sludge and its solids content are determined by the reagents and processes which precede the secondary dewatering. Thus any dewatering process evaluation must involve a coordinated process flowsheet.

Vacuum Filtration: Continuous vacuum filtration is a principal industrial dewatering unit operation. The filtration tests, all made on thickener underflow products, sought to establish technical and economic feasibilities for settled mine water sludge.

The technical feasibility of filtration with hydrous iron oxide sludges has been demonstrated in the processing of waste pickle liquors in the steel industry and with sewage sludges. Important properties of mine water sludges affecting filtration are: The particle size and the zeta potential of the solids; the pH and temperature of the slurries; the floc size, as enhanced by the use of flocculants or degraded by agitation of high-shear pumps; the solids content of the feed slurry; and the chemical composition of the solids, including the presence of impurities as excess neutralizing agents.

Filtration operational techniques and economics must be related to subsequent handling and disposal of the sludge filter cake and the filtrate. Among these factors are: clarity of the filtrate, and moisture content of the filter cake, which will determine its bulk handling characteristics, the equipment required for transferring, costs of transportation and ultimate disposal.

Plant Operation—Equipment Used — The filtration tests used a precoat rotary drum filter with diatomaceous earth (Celite 545) as a filter aid. The Eimco filter was 6' in diameter by 6' in length, having 113 ft^2 of filter area. The filter is of acid-proof construction and has a variable speed drum drive, a variable speed precoat knife-cutting mechanism, a variable rate slurry feed pump, and a slurry flocculant conditioning system.

Bench tests utilized a 0.1 ft^2 test leaf. Some leaf tests were made with strongly flocculated slurries but no precoat with quite promising results. These tests could not be confirmed with the continuous filter because internal filter modifications would have been necessary for its conversion to a nonprecoat configuration permitting a blowback cake discharge.

Operating Parameter Limits — The parameter operational ranges of the filtration unit are listed in Table 3.19. The values shown under "Maximum Operation Range Experienced" indicate the parameter spread observed while the "Typical Range" values could be maintained under steady-state condition without mechanical operating difficulties.

TABLE 3.19: OPERATIONAL PARAMETERS FOR FILTRATION UNIT OPERATION HOLLYWOOD PILOT PLANT

Parameter	Maximum operating range experienced	Typical range
1. Filter Slurry Feed Rate - gpm	5-40	5-20
2. Percent Solids in Feed	0.8-15	1-3
3. Dry Solids Feed Rate - lbs/min	0.35-50	1.26-4.7
4. Filter Drum Speed - rpm	0.22-0.85	0.29-0.73
5. % Submergence	18-23	20-23
6. Vacuum - in. Hg.,	5-25	18-22
7. Dry Solids Filtration Rate[a] (lbs/ft^2 filter area/24 hours)	10-300	10-25
8. Liquid Filtration Rate - gpm/ft^2	0.1-0.5	0.1-0.35
9. Temperature of Feed - °F	40-75	60-70
10. Percent Solids in Filter Cake[b]	15-27	18-24
11. Filter Cake Thickness - inches	1/32-1	1/16-1/4
12. Flocculant Dosage[c] - lbs/ton solids (Atlasep 1A1)	0-9.9	0
13. Filter Aid Used - lbs/hr (Celite 545)	---	50-65
14. Knife Advance Rate - mils/rev.	0.6-7.2	4-5
15. Filter Cloth Used - Eimco No. PO-801 HF - polyethylene monofilament 1 x 1 plain weave - 112 x 48 thread count per sq. in. - 11 oz. per sq. yd. - air flow = 30 cfm/ft^2.		

NOTES:

[a]These rates are for mine drainage sludge only, and do not include weight of the Celite filter aid. Weight percent of Celite in the filter cake ranges from 10 to 40%.

[b]Filter cake is a combination of mine drainage sludge and Celite filter aid. Solids in Celite 545 filter medium wet cake run about 43%.

[c]A small amount of flocculant was present in most filter feeds, having been added to the thickener at dosages of 1 to 3 ppm of neutralized water.

Source: EPA 670/2-73-093

Bench Scale Filtration Testing – Sufficiently large volumes of thickener sludge were not available to make continuous filtration tests on all types of neutralized sludges produced. In providing the desired additional, supplementary data, a bench scale vacuum filtration test station was operated. The 0.1 ft^2 Eimco test leaf was covered with the same filter cloth as the large filter and was provided with a 1" deep brass guard ring which permitted the deposition of about ½" of Celite 545 on the filter cloth. Excess Celite was removed to provide a precoat surface level with the guard ring thus permitting the filter cake to be removed.

The filtration cycle was varied. The filter cake was carefully lifted from the filter medium with minimum medium contamination. The wet cake was weighed immediately, dried in an oven at 105°C overnight, and then weighed to evaluate its moisture content. The filter medium remaining on the dried sludge cake was brushed off. This separation can be made easily, permitting the weighing of the sludge directly and computing of the actual filtration rate.

Summary of Filtration Test Data – The thickener-settled sludges were subject to the filter test leaf, utilizing the previously described procedure and the following filter time cycle.

Submergence	16 seconds	21%
Drying time	50 seconds	64%
Discharge time	12 seconds	15%
Total cycle time	78 seconds	100%

Cycles per hour—46.1
Corresponding filter drum speed—0.77 rpm

The resulting filter test leaf data are presented in Tables 3.20 and 3.21. Only limestone and the aged lime containing $CaCO_3$ showed a major difference in the solids filtration rates or chemical composition. This confirms evidence from many sources that calcium carbonate produces sludges with superior settling and filtration characteristics.

TABLE 3.20: MINE WATER NEUTRALIZED SLUDGE SOLIDS FILTRATION RATES

Mine water source	% Solids in filter feed slurry	Fresh hydrated lime $Ca(OH)_2$	Old hydrated lime[a] $Ca(OH)_2$	Pebble dolomitic lime $CaO \cdot MgO$	Dolomitic hydrated lime $Ca(OH)_2 \cdot MgO$
		Sludge filtration rates - lbs dry solids/ft²/24 hrs			
Proctor No. 1	1.62	26.4[b]			
		27.8			
Proctor No. 1	2.53			54.0	
Proctor No. 2	2.70	62.7			53.2
		72.0			53.8
Proctor No. 2	7.19		131.0	63.0	
			137.0	64.0	
Proctor No. 2 (bio)	1.48		39.6		
			50.3		
Proctor No. 2 (bio)	3.26		82.5		
			81.4		
Proctor No. 2 (bio)	7.18		331.0		
Bennett's Branch	1.96	33.1			
		38.6[c]			
Bennett's Branch	2.72	40.1			
		42.9			
Tyler Run	1.58			15.6	
				21.2	
Tyler Run	5.03			128.5	
				125.3	

[a]In this case, hydrated lime stored in the plant lime storage silo for a period of 3 months, during which time it converted to about 50% $CaCO_3$.
[b]The double values reported for most tests represent duplicate determinations made on each sludge slurry.
[c]No precoat material employed.

Source: EPA 670/2-73-093

Filter feed slurries containing less than 2% solids consistently produce the lowest solids filtration rates (in the range of 15 to 30 lb/ft²/24 hr). However, at 5% solids or higher, filtration rates of 125 to 331 lb/ft²/24 hr were obtained, which makes vacuum filtration a reasonable cost unit operation.

TABLE 3.21: VACUUM FILTER LEAF TESTS OF SLUDGE FROM LIMESTONE NEUTRALIZED CMD

Temperature = 70°F

Filter Aid: Celite 545

Sludge Source: Settling Lagoon

Flocculant Used: ATLASEP 1A1 2-3ppm in original neutralized slurry

Water Source: Proctor No. 2 Biochemically Oxidized

Test No.	Vacuum inches Hg	Cycles per hr.	Filtration Rates: Solids lbs/ft²/24 hrs	Filtration Rates: Liquid GPM/ft²	% Moisture in Filter Cake	Filter Aid Condition	Filter Cake Thickness inches
Waters neutralized with Centre County, Pa. limestone (Neidigh Brothers Quarry). Free Solids: 13% by weight							
1	24	46	381	0.16	66.0	Normal Precoat	3/16
2	24	46	451	0.18	65.6	Normal Precoat	3/16
3	16	46	352	0.12	67.0	Normal Precoat	3/16
4	16	46	331	0.12	68.8	Normal Precoat	3/16
5	24	46	394	0.15	65.3	Normal Plus 0.1% Body Feed	3/16
6	24	46	372	0.14	65.6	Normal	3/16
7	24	46	340	0.14	71.0	None	3/16
8	24	23	297	0.21	71.0	None-Topload	1/4
9	24	46	796	0.42	81.0	None-Topload	3/4
10	24	46	995	0.34	78.6	None-Topload	3/4
Waters neutralized with York County, Pa. limestone (Thomasville Limestone Company). Feed Solids: 27.2% by weight.							
11	24	46	990	0.16	55.5	Normal Precoat	3/16
12	24	46	1,200	0.16	51.5	Normal Precoat	3/16

Source: EPA 670/2-73-093

Most of the sludge from limestone neutralization resulted from a flowsheet involving biochemical oxidation and the rotary limestone reactor, but some had been reacted with pulverized stone in the Densator. Such sludges have rapid settling rates, high solids content in thickener underflows, and excellent filtration characteristics. The filter leaf data given in Table 3.21 clearly show the excellent filtration rates attainable with these slurries.

The replicate tests using normal precoat vacuum filtration at vacuum levels between 16" and 24" Hg yielded greater than 350 lb/ft²/24 hr. Rates in excess of 1,000 lb/ft²/24 hr were measured with the higher solids slurry from Thomasville stone. The moisture content of 66% is a great improvement over the more typical 80% observed with hydroxide slurries. Test 7 suggests precoat media may not be necessary. With slurries of this nature, filtration becomes a realistic low-cost method for handling mine water solids.

Filter leaf tests 8, 9, and 10 were made utilizing top loading with a more open cloth and no filter aid was used. Even higher filtration rates were observed by this technique as shown in Table 3.21. In this procedure, the vacuum primarily dries the cake and is not needed to load the media. The gravity force will aid rather than hinder the operation. In making these tests a sleeve was added to the 0.1 ft^2 test leaf to permit a deep cake to accumulate. A three- to four-fold increase in solids filtration rate over submerged-type loading resulted. These tests suggest a top-loading type filter would be most effective.

With flocculated slurries, some individual flocs were so large and cohesive they appeared filterable directly on a filter cloth without the use of precoat. Several tests were made and confirmed this possibility. The results indicate that strongly flocculated pulps do not require filter aid and produce filtrates with good clarity. The elimination of filter aid would remove the highest operating filtration cost factor. A body feed of filter medium (1% of Celite 545) greatly improved the filtration of 1.96% solids slurry.

Pilot Plant Filtration Testing – Filtration testing with the Eimco drum filter was carried out with slurries pumped from the settling lagoon or from the thickener underflow. In many cases, insufficient sludge at constant solids content was available to permit smooth, steady-state operation. The plant data are presented in Table 3.22. The range of liquid and solids filtration rates are similar to those established in the test leaf filter study. Under normal, full-scale operation, filtration rates greater than 100 lb/ft^2/24 hr should be possible with a properly designed flowsheet.

Manpower Requirements in Filtration – The frequent attention required by the cake cutting knife mechanism is a major labor consideration. Frequent adjustment of the knife was necessary to maintain a clean surface (thus maximum solids capacity) due to changes in compaction of the 3" to 4" starting precoat layer. Although continuous operation with a uniform feed minimized this variable, it appears cost estimates should include a significant allowance for operating labor. Taking all phases of this operation into consideration (e.g., flocculant preparation, precoat slurry preparation and application, knife-cutting adjustment, and filter cake disposal supervision), a manpower allowance of about ⅓ man per shift appears realistic.

Precoat Requirements – Plant experience indicated a precoat usage rate of 50 pounds per hour for the 113 ft^2 drum, or 0.4 lb/ft^2/hr. Continuous, uniform operation may reduce this requirement to 0.3 lb/ft^2/hr.

A loss of 100 to 200 lb of precoat may result from residual suspension in the filter tank after precoating. Provision should permit its recovery in the precoat slurry feed tank for use in the next precoating cycle. Another precoat loss can result at the completion of a precoat loading cycle. When a precoat layer of about ¾" remains, further removal with the cutting knife could result in damage to the filter cloth. It is difficult to add another layer evenly directly on this residual layer. It is preferable to drop this residual layer into the filter tank and provide for its inclusion with the new precoat slurry.

TABLE 3.22: SUMMARY OF TESTS ON PILOT PLANT ROTARY VACUUM FILTER

Vacuum: 22 inches Hg   Temperature: 65° F

Filter Aid: 3-4 inch layer Celite 545   Drum Speed: 46 RPH

Test No.	Sludge Source[b]	% Solids in Filter Feed	% Moisture in Filter Cake	Filtration Rates Solids lbs/ft²/24 hrs	Filtration Rates Liquid GPM/ft²
1	Lagoon	4-6[c]	83.7	6.2[a]	0.07
2	Lagoon	4-6	83.0	7.4	0.06
3	Lagoon	4-6	81.5	11.9	0.19
4	Lagoon	4-6	86.7	13.6	0.19
5	Lagoon	4-6	83.3	17.5	0.12
6	Lagoon	4-6	88.9	17.2	0.10
7	Densator Underflow	1.6	71.2	35.2	-
8	Densator Underflow	1.5	72.5	21.0	0.34
9	Lagoon		60.2	10.4	0.30
10	Densator Underflow		45.7	42.0	-
11	Densator Underflow	4.5	63.8	98.5	0.17
12	Lagoon		81.4	21.5	0.11
13	Lagoon		80.0	10.4	0.54
14	Lagoon		80.0	5.2	0.24
15	Densator	3.7	81.7	68.3	0.24

Notes:

[a]These solids rates include the weight of the filter aid, which will constitute 10 to 40% of the weight over the range of 6-100 lbs/ft²/24 hrs.

[b]Feed slurries in all tests were mixtures from several types of mine water treated with partially carbonated hydrated lime.

[c]Although the % solids in these feed slurries were high, the filtration rates were poor. This sludge came from compartment two of the settling lagoon where a very fluffy, slow settling component of the original sludge settled.

Source: EPA 670/2-73-093

Filter Cake Handling — The filter cake prepared with the plant filter had a moisture content of about 80%. This cake fell freely from the cutting knife on to a moving rubber conveyor belt for discharge. No difficulties were experienced with the filter cake sticking to the belt. In subfreezing weather, the filter cake tended to freeze quickly creating handling problems in truck transportation.

Filtration Using Flocculants — The effect of flocculants on the filtration characteristics of several sludges were made in bench scale and plant tests. At flocculant levels of 20 to 40 ppm (2 to 3 lb/ton dry solids), an extensive agglomerating effect caused a very rapid separation of liquid and solids in the treated slurry. Formation of these ball-like flocs should produce much higher filtration rates.

There were higher liquid rates but no improvement in solids filtration rates. The low solids filtration rate occurred when large portions of the filter cake fell from the filter medium surface as it emerged from the liquid feed bed. In continuous practice with the high liquid filtration rate, the solids in the filter bed would soon reach an unmanageably high level. The flocculants created a slippery, noncohesive sludge flow. The cohesiveness decreases with higher flocculant dosage.

Plant Tests with Strongly Flocculated Feed Slurries — Several tests were made on the full-size precoat filter using strongly flocculated pulps. The runs at each dosage level were short, and operating difficulties prevented tests at the highest drum speeds. The results were in agreement with those from the bench scale test leaf. The most significant observations made were:

(1) Strongly flocculated filter feeds do not respond as well as weakly flocculated slurries with a conventional precoat filter, even when pulp suspension is maintained.

(2) At high flocculant dosages, the solids settle very fast to the bottom of the filter tank, providing little contact of the drum with the solids. In a separate operation, slurries could be dewatered in this manner to attain a heavier and more homogeneous pulp for filter feed slurry.

(3) In contrast to the test leaf results, high liquid filtration rates were not attained while using strongly flocculated slurries.

Correlation of Bench Scale and Plant Operations with Design Criteria — In general, the test leaf results could be extrapolated directly to predict performance of the full scale vacuum filter. The major filter design criteria for a low cost vacuum filter operation, which were developed and confirmed in both bench and plant, are summarized in Table 3.23.

TABLE 3.23: OPTIMUM VACUUM FILTER DESIGN PARAMETERS

1. Optimum solids content of filter feed - 3% solids minimum.
2. Liquid filtration rates - 0.35 gpm/ft^2 - maximum (70°F)
   0.15 0.20 average (70°F).
3. Solids filtration rates - 15 to 25 lbs/ft^2/24 hrs. @ 1% solids feed.
   120 lbs/ft^2/24 hrs. @ 4% solids feed.
   (all values for 70°F)

   A graph relating the entire range of filtration rates vs. solids content in the filter feed is presented in Figure 3.12
4. Filter drum speed - 0.8 rpm minimum.
5. Drum submergence - 25%.
6. Precoat consumption - 40-50 lbs/hr for 113 ft^2 filter area.
   0.4 lbs/hr/ft^2 filter area.
   40% of dry sludge weight @ 24 lbs/ft^2/24 hrs.
   8% of dry sludge weight @ 120 lbs/ft^2/24 hrs.
7. Knife advance - 4 mils/drum revolution.
8. Moisture in filter cake - 76 to 82%.

   Cakes of this moisture content could be handled satisfactorily on a conveyor belt.
9. Specific gravity of filter feed - 1.03 (@ 4% solids).
10. Specific gravity of filter cake - 1.13 (71 lbs/ft^3).
11. Type of filter cloth - EIMCO P.O. - 801 HF or equivalent.
12. Flocculant requirement - 2 to 4 ppm added to Densator feed.
    (will vary with flocculant)
13. Filter aid type - Celite 545 or equivalent.

Source: EPA 670/2-73-093

FIGURE 3.12: SLUDGE PRODUCTION RATES AND FILTER AREA REQUIREMENTS

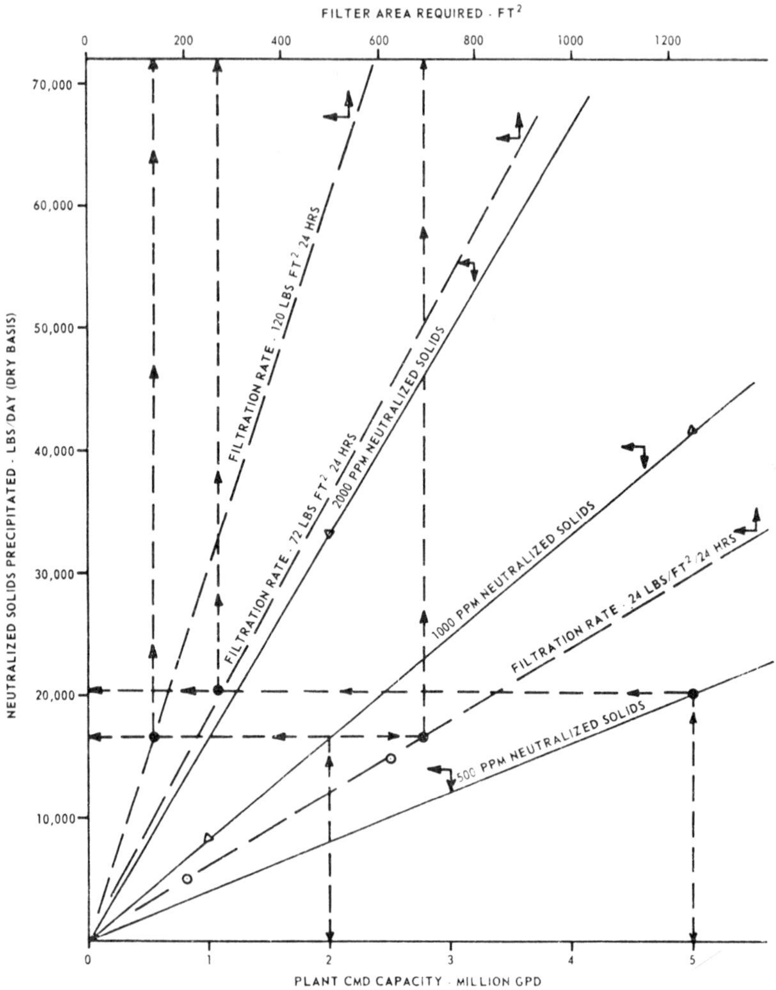

Source: EPA 670/2-73-093

These design parameters should produce satisfactory, economical filter operation if a precoat filter is to be considered. However, the preliminary test work without filter aid with strongly flocculated pulps suggested more study seeking to eliminate the use of filter aid and its high labor requirements.

In sizing a precoat filter, the capability to store sludge and the frequency and extent of filter operation requirements must be related to sludge volume and filtration rates. Once precoated, a drum filter must be kept under vacuum with continual moisture conditions to maintain the filter medium on the drum until it is all consumed. The power requirements for maintaining the vacuum represent a major portion of filtration costs, comparable only to the costs of precoat material.

Centrifuge Tests: Performance sludge dewatering experience was attained with a Dorr-Oliver Merco Bowl Centrifuge (Model Z-1-L). It has a maximum bowl diameter of 9" and an inside length of 24". This is a horizontal, continuous flow machine. The solids settle and compact against the walls under centrifugal force (up to 5,200 gravities). They are conveyed from the liquid pool for further drying under centrifugal force. The clarified centrate moves countercurrent for pump discharge at 30 psi.

The sludge dried (primarily of limestone reaction origin) was pumped from the bottom of the settling lagoon to a storage feed tank and maintained by slow stirring as a uniform feed to the centrifuge. Prior experience assisted selection of operating conditions and maintenance of reasonable clarity in the centrate. At higher flows the turbidity of the centrate increased markedly. Should this clarity be a problem the relatively small volume could be recycled through the process thickener or settling lagoon. The centrifuge cake had satisfactory handling characteristics although it showed a tendency to flow and was more subject to a loss of physical integrity than the filter cake.

The device functions satisfactorily and holds great advantage in ease of operation with a minimum of manpower requirements. As a high speed mechanical device, it is subject to high maintenance and power costs. As with other secondary dewatering devices, its response improves with feed solids content.

Sludge Drying Bed: The sludge drying bed is another approach for secondary dewatering of settled sludge to improve handling characteristics and reduce volume. Water is removed from the solids by decantation from the surface of a newly-filled bed, by percolation through the bottom of the bed, and by evaporation from the surface.

Sludge Drainage Rates—Bench Scale Drainometer Determination — A small-scale test unit was developed. A drainometer was constructed from a 100 gallon tank. The bottom of the drainometer was filled with ½" x 1" limestone to a depth of 4" upon which was placed a 6" layer of coarse sand. This simulates an ideal drainage bed medium with the larger stone providing a drainage bed while the sand would trap the sludge particles. Before loading with sludge, the sand bed was backwashed with water to remove any entrapped particles. The clean bed had a drainage rate with water of about 2 gpm/ft^2 which is comparable for a conventional sand filter. With this design, it was possible to load the drainometer to a sludge depth of about 2'.

In each test, the drainometer was carefully filled with slurry to avoid disturbing the sand bed. The slurry was allowed to drain for a period of about 48 hours, at which time the top of the cake was void of free-standing water and cracks were beginning to form from the decreasing volume of the hydrated particles. The initial drainage rates were measured at hourly intervals and subsequently at longer intervals up to 24 hours total elapsed time. Final rates, which were essentially constant, were measured at 8 hour intervals.

In two tests, a second and third layer of sludge was added to the dewatered layer to simulate the cyclical drying bed loading that was practiced in the full scale bed. The drainometer seemed to simulate the full scale bed, including the cracking phenomenon on the surface of the dewatered cake. Even the cake, which drained only to 6% solids, cracked as it further air-dried and reduced in volume. The final drained sludges had sufficient integrity to be handled by a mechanical bucket.

Each of the sludges tested was produced in the plant equipment as with the filter leaf tests. The results of sludge handling characteristics for filtration and drying bed may be compared for the same sludge. About 100 gallons of settled sludge was withdrawn from the thickener and stored for several days prior to transfer into the drainometer. During this 48 to 96 hour quiescent period under typical conditions experienced in a settling lagoon, none of the sludges (about 2' deep) compressed to a density greater than 2.8% solids. This indicated that settling techniques (such as would occur in a settling lagoon) only slowly produce any significant change in terminal sludge density.

The variations in the drainage rates obtained for lime neutralized sludge from three different waters were limited to the initial part of the cycle and approached nearly identical rates after about 20 hours drainage time. This common rate was about 0.05 gph/ft^2 at one foot hydrostatic head. The permeability rate of water through 20" of 12% solids sludge was found to be 0.06 gph/ft^2 at one foot hydrostatic head. The plot of drainage rates for CMD sludges are similar to those reported for sewage sludge, but were initially more rapid and approached the limiting values more quickly. The drying bed size is influenced by both influent solids content and drainage rates. The major conclusions from these tests are:

(1) The drainage rate for mine water sludges produced by lime neutralization attains a stable level after 24 hours of about 0.05 gph/ft^2. The drainage rate for limestone treatment sludges is nearly double (0.10 gph/ft^2). Both values are for one foot of hydrostatic head.

(2) The water permeability rate of consolidated sludge on a drainage bed bottom is about 0.02 gph/ft^2 at one foot of hydrostatic head.

(3) Multiple layers of sludge (with intermittent drying) maintain a drainage rate of 0.03 to 0.05 gph/ft^2.

(4) Drainage rates through a cracked dry bed were high initially (about 0.20 gph), but gradually decreased to the 0.05 gph/ft^2 level as the cracks filled with new sludge.

(5) Solids content of dewatered sludge after several days of draining ranged from 12 to 15% for hydroxide-produced sludges, and 20 to 30% for limestone-neutralized sludges.

(6) Shortly after drainage began, the underflow liquids were clear.

(7) Penetration of sludge into the sand bed was less than 1/16".

(8) The drainometer used could not determine the effect of hydraulic heads greater than 12" on drainage rate. However, the sewage sludge data indicate that drainage rates are directly proportional to hydraulic head, hence faster drainage rates may be attainable by operating a drying bed at a constant operating head of several feet.

Pilot Plant Parameters – The range of operating variables are indicated in Table 3.24. The test program permitted only limited production of several types of sludges which were introduced into the drying basin. The laboratory drainometer data provided a comparison of response for different sludge types.

TABLE 3.24: OPERATING PARAMETERS LIMITS OF DRYING BED

		Maximum range	Average working range
1.	Slurry working volume - ft^3/section	4,700	---
	Slurry working volume - gallons/section	35,000	---
	Drainage area/section - ft^2	---	1,270
2.	Slurry feed rates - GPM	0-127	25-70
3.	% solids in feed slurry	2-15	2-10
4.	Drainage rates - GPM/section	1-50	1-4
	GPH/ft^2/bottom area	0.03-0.05	0.03-0.06
5.	% solids in dewatered sludge	11-100	13-15
6.	Drainage time for 3 ft depth of feed slurry - days	3-10	5-7
7.	Working sludge depth - ft	0-3	1-2
8.	Weather protection - center section provided with plastic canopy.		
9.	Sludge removal facilities - access for front end loader driven onto drainage bed surface.		
10.	Sludge removal rates (with front end loader) - 2 foot depth of sludge in one section removed in three hours. Equipment: 1 front end loader - 2 cu. yd.; 2 trucks - 6 ton and 9 ton; 3 men as equipment operators		
11.	Anti-freeze provisions - none.		
12.	Dried solids disposal system - truck haulage to adjacent landfill area.		

Source: EPA 670/2-73-093

Sludge Drying Bed Operational Observations – The sludge was introduced into a single compartment of the bed by gravity flow or by pump from the settling lagoon or the thickener. Most sludges tended to settle rapidly when contacting the bottom with the supernatant water selectively moving faster over the previously dewatered sludge. This clear water percolated rapidly to the underbed and drain providing a very high initial dewatering rate.

This rate varied with the depth and moisture content of the previous sludge layer. A dried sludge layer yielded high initial drainage rates. During the evaporative months, the sludge dried to less than 10% moisture. Under these conditions, the gel-like sludge structure had been broken and rapid percolation followed. This type response favored a slow continuous introduction of the sludge, with longer drying periods.

The optimum sludge depth to be transferred in a given interval was difficult to quantify:

(1) A transfer of sludge to the drying bed creating a fresh depth greater than 6" leads to conditions favorable for decantation (the most rapid and effective dewatering procedure) and produces a greater hydraulic head which is conducive to higher percolation rates.

(2) The longer interval possible for one filling cycle in one compartment permits longer percolation and evaporative intervals for the other compartments.

(3) Evaporation of water from the sludge decreases rapidly with sludge depth after the first 6". Thus, excessive sludge depths tend to minimize and limit final sludge moisture content.

(4) Dry, nongelled sludge permits rapid percolation rates.

The following qualitative conclusions are indicated regarding the optimum sludge depth:

(1) The desirable depth of sludge in the drying bed is a reciprocal function of influent sludge density; i.e., a low density feed sludge may be allowed to accumulate to greater depths since the potential for decantation is greater, the effect of hydraulic head on the drainage rate greater, and yet the final sludge depth will be the least.

(2) The dewatered sludge depth desirable is proportional to the seasonal evaporative level; i.e., during periods of high temperature, low humidity, and high wind velocities, the sludge depth may be increased.

(3) Some minimum moisture content of dewatered sludge should be attained before a second layer of sludge is introduced.

Decantation of water in the drying basin was practiced to a limited degree by attaching a hose to a board floated on the water surface. By initiating a siphon flow, the water was transferred to the underdrain air vent. A small pump, also floated, would increase the transfer rate. The supporting board served as an automatic level limitation to prevent the discharge of sludge. Under any conditions, constant manual attention was necessary.

Percolation through the bottom of the bed proceeded rapidly during sludge introduction and slowed as the settled sludge became drainage rate controlling. After completing manual decantation, the drainage continued with a decrease in volume and bed depth until no free standing water remained. This time interval seldom exceeded 48 hours regardless of sludge depth introduced. Drainage continued at these slower rates, and surface evaporation began.

The effect of evaporation first became evident in the shallow sludge layers at the periphery of the bed. The continual decrease in sludge volume increased the growth of shrinkage cracks and continued to the final stage with about 20" of dewatered sludge containing 15 to 30% solids. Further decrease in moisture content proceeded very slowly and the effect of sludge depth upon drying became obvious.

It was during the dewatering stages preceding the final stage that a plastic bed cover was most effective. Physically, the cover was quite stable and no special problems were experienced. The thin plastic is subject to loss of strength from ultraviolet radiation during the spring and summer months. The weakness develops at any crease or fold and extends upon pressures as from high winds. Replacement appears necessary at six month intervals. Higher temperatures developed under the canopy were evidenced year round.

The excellent ability of the sand-surfaced bed to retain sludge was evidenced by a very clear underflow and by direct observations of the bed. When the dewatered sludge reached depths of 24" to 36" having at least 15% solids, it resisted flow and could be hauled by dump truck without leakage. Rubber-tired equipment did not have sufficient traction to manipulate in the sludge bed, unless the bed was allowed to achieve much higher solids content. Equipment with treads performed well although the 15° access slope seemed excessive when covered with sludge. The use of heavy equipment further disrupted the bed.

The weight of the sludge removal equipment demonstrated an inadequacy of low-cost red dog as a porous bed support. This material did not have sufficient physical stability and was compressed during sludge removal reducing bed drainage rates. The compressed material did protect the clay drain tiles from breakage. Beds probably should utilize physically-stable gravel (such as grade 3A) for the bottom 12" followed by a second foot of 1" stone to support the sand filter layer.

Quantitative Drying Basin Data — With the basin containing about 6" of dewatered sludge, a sludge settled from limestone treatment of Proctor No. 2 waters was drained by gravity from compartment No. 1 of the settling lagoon to drying basin No. 2. A composite sample contained 8.8% solids by weight. Over 13,000 gallons of sludge were transferred in four hours raising the bed depth about 15". This represented a rapid transfer rate, which exceeded sludge production levels. In 20 hours, the bed had dropped nearly 13", more than 0.6"/hr.

After draining and evaporating ten days, an additional transfer of similar sludge was made to the same compartment which was allowed to dewater for 72 hours when a third transfer of similar sludge was made. The dewatering response of the drying basin under these conditions is detailed in Table 3.25.

The residual volume and solids content of the dewatered sludge in the drying basin was quite different from that indicated by a sludge settling test. In the drying basin the residual volume was 24.6%, in contrast to the 63.1% shown by the settling test.

In further drying operations, following five days of dewatering as described in Table 3.25, 10,005 gallons of similar sludge (10.6% solids) were transferred to drying bed compartment No. 2 which showed a drainage rate of 0.10 gph/ft^2 in the next 40 hours and a volume reduction of 65% in the next four days.

Nearly simultaneously with the above dewatering operation, about 14,000 gallons of the same sludge were transferred to the adjacent drying bed No. 3 which gave a drainage rate of 0.15 gph/ft^2 in the first 40 hours.

TABLE 3.25: SLUDGE DRYING BASIN EVALUATION

Initial bed conditions:	Drying Basin No. 2 - Covered Depth of Sludge in Bed - Approximately 12 in. Solids Content (Surface) - 15.8% by wt. Drainage period since previous transfer - 72 hours Weather - Temperature - 59-84°F Overcast, humid with rain on fourth day of dewatering
Sludge transfer:	Origin - sludge from settling lagoon, developed from limestone neutralization of Proctor No. 2 water after biochemical oxidation. Pumped from settling lagoon compartment no. 1 to sludge drying basin no. 2. Suspended Solids - 9.5% by wt. (composite) Transfer Rate (average) - 73 gpm for 170 minutes Total Volume Sludge - 12,396 gallons. 11,775 gal. water, 10,315 lb dry solids (using estimated density - 1.05) Bed Level Increase - 14.25 inches
Drainage rate:	Averaged 0.19 g/hr/ft^2 over first 40 hours (see Figure 58 for details)
Sludge drying period:	5 days (121 hours), 97% of total volume reduction in 40 hours
Final bed condition:	Sludge Bed Depth Increase - 3.50 inches Solids Content (surface) - 44.8% by wt. Final Volume Sludge Added - 3,045 gallons ($407\ ft^3$) (Using estimated density - 1.10) - 1,850 gal water, 12,522 lb dry solids
Water rejection:	84%
Operating time:	1 man hour during initial sludge transfer
Sludge dry bed capacity[a]: (3 beds - 36 inches of sludge)	80 days 94,000 gallons dewatered sludge 176 tons dry sludge - 390 tons dewatered sludge 12,500 ft^3 = 463 yd^3 587,000 gallons settled sludge 59,000,000 gallons coal drainage (47% excess capacity of plant design)
Cost to remove dewatered sludge:	\$231.50 (\$0.50/yd^3) 0.31 cents/1,000 gal.

[a]Based on addition of 4 inches dewatered sludge to alternate beds every 3 days, thus a nine day drying cycle. Proctor No. 2 with limestone, 1% settled sludge volume, 350 gpm.

Source: EPA 670/2-73-093

Response of Frozen Sludge – During periods of subfreezing temperature, the sludge in the drying basin and the dewatered sludge placed for disposal as landfill will freeze and retard dewatering. As the water in the sludge changes physical state, the sludge structure is modified which leads upon thawing to major reductions in moisture content and volume. Although freezing of the dewatering sludge requires planning in basin design and operation to provide adequate storage volume, the overall response to sludge freezing is favorable. To quantify this response, a sludge freeze-thaw test was made.

A cubic foot of dewatered sludge was cut from the drying basin. The physical instability of the sample led to a bulk volume of 2.05 ft^3 upon transfer to a wire mesh freezing basket. The sample was allowed to thaw at about 20°C, while observing the changes in weight and water drained from the sample. The water ceased to drain from the sludge after 143 hours when the original 1 ft^3 volume had reduced to 0.73 ft^3 and the material was transferred to a tray to prevent sludge loss. At that time the total weight loss was 54.4% of which 52.2% was from drainage water, and the balance from evaporation. After an additional 28 days drying, the moisture content was less than 6% and the final volume was 0.25 ft^3 or one-fourth the original dewatered sludge volume. Freezing assists in final dewatering and drying.

Sludge Drying Bed Design Criteria – The data obtained from drainometer tests and actual drainage bed operation was used to determine the design parameters presented in Table 3.26.

TABLE 3.26: SLUDGE DRYING BED DESIGN PARAMETERS

Bed Drainage Rate Capacity[a]	0.5 GPH/ft^2 minimum
Percentage of Solids in Feed Slurry	High as possible
Percentage of Solids in Dewatered Sludge	15-30%
Final Sludge Depth[d]	3 ft
Bone-Dry-Solids Capacity	
(a) for hydroxide slurries	8-12 lbs/ft^3
(b) for limestone slurries	13-20 lbs/ft^3
Specific Gravity, Feed Slurry	1.03-1.04
Specific Gravity, Dewatered Sludge	1.07-1.13
Bulk Density, Dewatered Sludge	67 to 71 lbs/ft^3
Bed Drying Time(after influent cut-off)	10 days minimum
Total Bed Depth[b]	6 ft minimum
Bed Holding Capacity	30 day production min.
Average Bed Drainage Rate Through Sludge	0.03 to 0.05 GPH/ft^2
Number of Beds Required	2 - minimum
Influent Feed Distribution	Even distributed over bed area.
Bed Length	300 ft maximum
Bed Width	See Note c
Evaporation Losses[e]	Varies with weather

Notes:

[a]The bed should contain at least 4 inches of clean coarse sand laid on 8 to 12 inches of 1/2 by 1-inch stone supported by 12-inches 1/4 by 3-inch stone with at least two perforated under-drain tiles running the length of the bed. The 12-foot spacing utilized at Hollywood was adequate.

[b]In freezing climates, extra bed depth should be available to provide sufficient bed volume to store 2 to 4 months sludge production during the winter months when drainage rates may decrease drastically.

[c]Bed width should be chosen to be compatible with the type of sludge removal system to be used. A dragline appears desirable for a large bed, hence the bed width should not exceed twice the casting radius of the dragline.

[d]Greater sludge depths may be used, but "drying" times may be excessive for the sludge to attain handling consistency for belt or truck haulage.

[e]Evaporation will aid the final drying of the top 6 to 12 inches of sludge.

Source: EPA 670/2-73-093

The drying bed size and number of beds must be established with a selected operating cycle time. Bed size selection must also be dictated by the requirements of the sludge removal system. It appears bed lengths should be limited to 300' and widths to about 100'. For large bed area requirements use multiple ponds.

Sludge Removal Facilities – Since sludge removal requires most labor, efficient and low-cost methods of sludge handling should be employed. The sludge must be removed by a system that will not unduly disrupt the sand filter bottom. An option may be the use of a special dragline. The bed size and location should be compatible with a particular dragline configuration.

Most plants would require at least two drying beds for continuous operation. In order for two or more beds to be serviced by a single dragline, a radial pie-segment arrangement has been suggested. This system provides for the servicing of two drying beds by a dragline tower located at the center of the quadrant sector and the stacking of the dried sludge around the periphery of the sludge bed area. This arrangement should permit rapid, low-cost removal of the sludge, with dragline power equipment provided at intervals as needed. The underbed drainage lines are located for servicing using a common sump for control discharge.

Disposal of Dewatered Sludge: The dewatered sludge attained adequate consistency for truck transportation without leakage, whether dewatered by the drying basin, filter or centrifuge. In all cases only the minimum volume and weight had to be transported. The relative efficiency of handling is better when removing sludge from the drying beds since the truck was used constantly only for the day or two required. In contrast, the truck had to be available continuously in handling sludge from the filter or centrifuge or a temporary storage developed which required multiple handling.

The dewatered sludge was trucked to an abandoned "borrow pit" (several acres) adjacent to the treatment facility which would have capacity for years of plant operation at design capacity. The drying and volume reduction of the dewatered sludge continued slowly in the disposal area by percolation and evaporation whether in warm or subfreezing weather. The volume reduction as a result of freezing was about 75%. The volume reduction from drying of a dewatered limestone sludge was about 80%. After the dumped sludge reached a moisture content near that of the surrounding soil, it was graded, compacted, and seeded with common grasses, which germinated and grew. The seeding prevents possible windage of the dried sludge.

Producing High Density Sludge by Recycle

Densator Operation: The full-sized densification unit was not amenable to short time evaluation. The test described covered about 100 operating days necessitating a modification of the planned project program. Unfortunately, plant layout precluded simultaneous evaluation of other processes while the Densator was in operation, although some experience with biochemical Fe(II) oxidation was carried out concurrently. The biochemical effluent, as indicated in the flowsheet (Figure 3.13), formed a portion of the Densator feedwater. This concurrent study undesirably decreased the Fe(II)/Fe (III) ratio, complicating the analysis and control.

Accordingly, the system was operated only with Proctor No. 2 water which was the most severely mineralized. The experience gained, including the laboratory and pilot tests, indicated the sludge recycle concept holds its greatest advantage when applied to highly mineralized drainage waters which characteristically form large volumes of low density (0.2 to 1.0%) settled sludges. Lightly mineralized waters, forming slow settling floc of negligible solids content, are less amenable to the process. Such waters tend to have low iron and aluminum content and normally produce less than 1 lb solids/1,000 gal.

FIGURE 3.13: FLOWSHEET FOR PRODUCING HIGH DENSITY SLUDGE

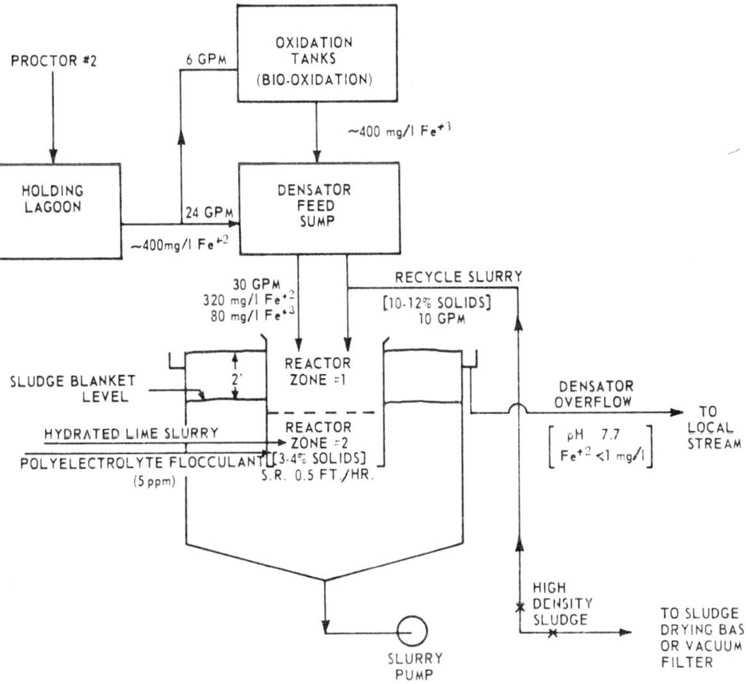

Source: EPA 670/2-73-093

The sludge recycle rate sought to maintain approximately 3.0% solids in the primary reaction zone of the Densator. These solids were composed of a mixture of freshly-precipitated and recycled sludge. The Proctor No. 2 waters alone yielded about 1,500 ppm suspended solids upon neutralization. To maintain the control level of solids in the primary reaction zone, as was recommended by Infilco, an optimum ratio of recycled to precipitated solids was about 20:1. The feed water flow rate to the unit was established in conjunction with the other variables, primarily by the settling rate of the sludge blanket.

The high density sludge concept functions from a highly ferrous system and alkali addition to maintain a pH level higher than conventional neutralization processes. A 4:1 Fe(II)/Fe(III) ratio was maintained during most of the test, although this ratio and total iron levels were decreased by rain dilution near the end of the test. A decrease in solid content of the underflow resulted from these concentration changes. No provision is made for iron oxidation in the system. This represents a capital and operating cost savings. Air introduced into the secondary reaction compartment interferes with sludge flocculation and settling rate.

Hydrated lime slurry was pumped into the secondary reaction zone in the Densator to maintain a final pH of about 7.7. At this pH level, the iron concentration in the overflow was less than 1 mg/l. A flow rate to the Densator in the range of 50 to 130 gpm permitted proper continuous operation, with the lower rates offering better overall response. The flow rate was modulated with the settling characteristics of the prevailing sludge settling rate. The use of a weakly anionic flocculant was desirable, if not necessary.

The dosage was in the range of 4 to 5 ppm based upon influent volume. It appears that the flocculant dosages should be determined from the solids loading in the secondary reaction zone and expressed in terms of pounds of flocculant per ton of dry solids rather than on a raw water volume basis. Under these conditions, high density sludges ranging from 10 to 16 wt % solids were produced consistently. The initial increase in sludge density was quite slow. The initially-formed sludges settled with typically low solids contents.

Accordingly large volumes of settled sludges were recycled seeking to attain the desired 3% solids level in the reaction sections. Over 50 operating days were expended to meet the 3% criteria. As sludge density builds in the unit during this period, the normal sludge free settling rate is minimized by entering early stages of compression. It appears start-up should utilize low raw water and recycle flow rates, ignoring solids levels in the reaction sections. This would provide minimum rise velocities and permit greater sludge compression to develop in the sludge collection well.

No sludge density improvement was indicated by premixing the lime slurry with the recycle sludge before reaction with the raw water. Individual underflow samples showed solids content to nearly the 30% level. The samples were taken from a pumped stream. The solids content values cited represent recycle sludge and not that from additional settling. Recycle sludge densities varied widely during adjacent samplings.

This sampling difference was associated with the position of the bottom rake, the depth of the sludge compression zone, and sludge flowability. The recycle sludge showed no tendency to rat-hole, as do gelled hydroxide sludges. Its viscosity reduces flow rates. The recycled sludge was viscose enough to support a stirring rod upright. These sludges were ferrous in character (dark green).

No direct evaluation was made of the pollution potential of these ferrous recycle sludges after blowdown. They are highly alkaline in character and oxidized very rapidly at the sludge-air interface when transferred to the sludge drying basin. This highly alkaline nature should be adequate to neutralize any acid formed during oxidation of the Fe(II) content.

Water effluent quality was excellent during most of the test except when the control pH was allowed to drop below 7.5. Even under these conditions the overflow maintained an iron content consistently less than 1 mg/l.

A summary of the operating parameters is given in Table 3.27. A wide fluctuation in operating variables was experienced as the parameters were modified during preliminary operation of the system. It was difficult to maintain steady-state flow conditions because of the slow settling rates of the high level sludge blanket in the Densator. This sludge, which was well into the compression zone range, had an average settling rate of less than 0.5 ft/hr, which made it difficult to operate at continuous water flow rates much greater than 50 gpm. During these operations the level of the sludge blanket varied up to about 10" of the overflow. The input flow rate was the variable utilized to control the sludge blanket level.

TABLE 3.27: DENSATOR RECYCLE TEST—OPERATING PARAMETERS

	Maximum	Minimum	Optimum
Raw Water			
pH	3.4	2.9	
Fe^{+2} - mg/l	530	100	
Total Acidity - mg/l	1,700	1,500	
Sulfate - mg/l	2,500	2,300	
Temperature - °C	18	8	
Densator Effluent			
pH	6.3	9.1	7.7
Fe - mg/l	91	<1	<1
Densator Feed			
Total Feed Flow Rate	160	50	40
Precipitable Solid (PS) - ppm	1,400	400	
Recycled Solids (RS) - ppm	20,000	60,000	30,000
Ratio RS:PS	14	150	30
Total % Solids in Primary Zone	2	6	3-4
Flocculant Addition - ppm	5	0	5-10
Recycle Slurry Flow Rate - gpm	60	10	10
Raw Water Feed Flow Rate - gpm	140	15	30
Ratio Fe^{+2}/Fe^{+3}	4:1		
Settling Rate - ft/hr[a]	2.3	0.3	1.0
Densator Underflow			
% Solids (average operation)	20	2	10-12
Densator Area required for optimum conditions - ft^2/ton PPTD solids/24 hrs - 660			

[a]At solids concentrations of greater than 4%, the slurries are already at compression zone conditions; hence, free settling rates are not obtained.

Source: EPA 670/2-73-093

The recycle sludge produced during the tests had the desired properties sought. Besides high density (thus significant sludge volume reduction), the sludge had near ideal dewatering characteristics. Upon transfer to the sludge drying basin, the recycle sludge drained immediately, and never maintained any free-standing supernatant. It also provided the best filtration characteristics of any sludge tested. With a filter feed suspended solids level between 10 and 17% solids by weight, a filter cake in excess of ½" could be maintained without blinding or buildup of filter vacuum. The cake cut cleanly from the drum with the very minimum of precoat. The filter cake had about 24.5% solids with excellent handling characteristics. The solids filtration rate averaged about 312 lb dry solids per square foot per 24 hours.

Simple material balance calculations readily indicate the impracticality of the recycle concept when applied to weakly mineralized waters. The sludge recycle load to maintain 3 to 4% solids in the reaction zone would be unrealistic. The recycle system is susceptible to wide fluctuations of performance as feed water characteristics change due to rain or events in the originating mine.

Filtration of High Density Recycle Sludge: In the start-up operations of the vacuum filter, the heavy (>15% solids) recycle sludge was selected anticipating good filtration properties. The response was excellent, in fact the best observed, attaining filter cakes with thickness in excess of ¾". The cake was dry and readily handleable. The filtration rates were so high that the total Densator storage supply of recycle sludge was consumed in the qualitative start-up procedures.

Properietary Sludge Recycling Process: In the process of *A. Cywin; U.S. Patent 3,617,559; November 2, 1971; assigned to the U.S. Secretary of the Interior* precipitated sludge containing substantial quantities of iron oxides is recycled back to the mixing and/or aeration steps.

Referring to Figure 3.14, a dilute acid waste stream **1**, containing dissolved iron in the ferrous state, is passed into mixer-contactor **2** where at least partial neutralization occurs by reaction with a finely divided limestone-neutralizing agent introduced via conduit **3**. The mixer may be of any conventional type which provides thorough mixing and contact between the acid waste and the limestone-neutralizing agent. Dissolved carbon dioxide suppresses the pH level attainable in the mixer to a maximum of about 5 to 6. At this pH level, substantial quantities of ferric iron, but little if any ferrous iron are precipitated.

FIGURE 3.14: SLUDGE RECYCLING PROCESS

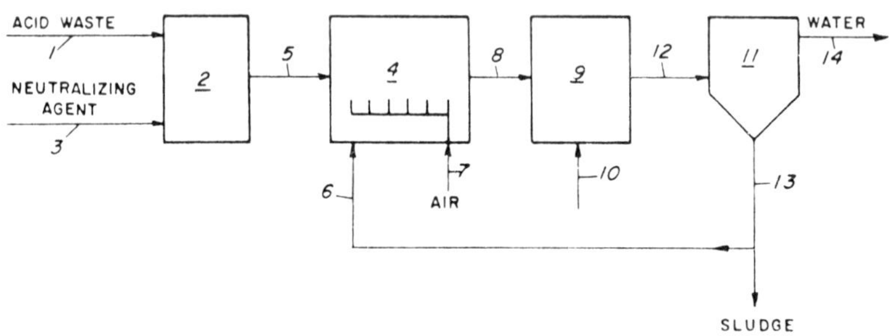

Source: U.S. Patent 3,617,559

The reaction mix is then passed to aerating vessel **4** by means of conduit **5** where the mix is contacted with substantial quantities of recycled iron oxide containing sludge from a later clarification step. Recycle sludge, which also contains some unreacted limestone, is introduced into the aerating vessel by means of line **6**. Air is introduced into the vessel by means of sparging lines **7**.

The amount of sludge recycled may vary from 5 to 95% of total sludge production but it is preferred that recycle be in the range of about 25 to 75% of the total. In a preferred embodiment sludge recycle and aeration detention times are adjusted to produce a waste sludge composition at the time of precipitation comprising a mixture of ferrous and ferric oxide corresponding as closely as possible to the ferrous-ferric ratio found in magnetite, or about 1:2. By producing a mixed valence, iron oxide sludge, a number of significant process advantages are achieved. Such a sludge displays superior settling and dewatering characteristics and is of greater density than either a ferrous hydroxide or ferric hydroxide sludge.

Less neutralizing agent is required to produce an iron-free-clarified effluent. This follows because during oxidation of ferrous to ferric iron, one equivalent of acidity is produced for each equivalent of iron oxidized. Normal neutralizing processes oxidize substantially all of the ferrous iron to the ferric state. Recycling of substantial quantities of sludge also reduces the conditioning time required to achieve superior sludge characteristics and may increase the capacity of each unit operation of the process. In addition, sludge recycle permits better control of the oxidation rate in the aeration step.

From aerator **4**, a stream of neutralized effluent, containing substantial amounts of precipitated solids, is passed via line **8** to flocculation tank **9**. There, under the influence of gentle agitation, the suspended sludge solids are agglomerated into larger particles. It is sometimes advantageous to add small quantities of a chemical coagulating aid, such as a liquid polyelectrolyte, to the floc tank via feed conduit **10**.

From flocculation tank **9**, the suspension of mixed hydrous oxides is passed to clarifying means **11** by way of line **12**. Here, a concentrated sludge suspension **13** and a clarified wastewater stream **14**, suitable for direct stream disposal in most instances, are recovered. The clarifying means may comprise a conventional clarifying vessel, a filter or centrifuge, or may even comprise a settling pond. In any case, sludge suspension **13** is divided into two portions, one of which is recycled back to an early stage of the process via line **6** as has been previously described, and the other fraction is discarded from the process as waste. The waste sludge fraction may be dewatered prior to disposal as by filtration or centrifugation.

In another embodiment, sludge recycle line **6** is introduced into acid waste stream **1** rather than into the aeration vessel. A portion of recycle stream **6** may also be passed directly into mixer-contactor **2**.

IRON OXIDATION

The material in this chapter was excerpted from EPA-430/9-73-011; EPA-670/2-73-093; PB 208 820; PB 229 649; PB 232 764; U.S. Patents 3,617,562 and 3,823,081 (see Bibliography, p 390).

The exposure of iron sulfides to the atmosphere and moisture causes them to oxidize to ferrous sulfate. These salts then dissolve into ground or surface waters forming mine drainage. If there is an overabundance of these salts and little alkalinity available in the water, the mine drainage will be acid. Iron present in the mine drainage is as serious a pollutant as the acidity. The iron compounds coat the bottom of streams, leaving them uninhabitable for aquatic life.

As mine drainage is formed, iron is first present in the ferrous form. This form of iron is very soluble but will precipitate as a hydroxide as the water becomes alkaline. Minimum solubility occurs in a pH range of 9.3 to 12.0.

As the water becomes alkaline, ferrous iron will oxidize to the ferric form. This oxidation is dependent on the pH of the water, and is very slow at pH's less than 4.0, slow in the range of 4.0 to 6.0, moderate in the 6.0 to 8.0 range, and proceeds quickly above this point. Ferric iron is much less soluble than ferrous iron, and will precipitate as the hydroxide at a pH of 5.0 with minimum solubility at a pH of 8.0.

In a mine drainage treatment system, such as any of the chemical neutralization processes, it is most advantageous to oxidize any ferrous iron present to the ferric form so it can be removed as an insoluble hydroxide at near-neutral pH's. A number of methods are available to accomplish this oxidation process, and these are discussed in the following sections.

BIOCHEMICAL OXIDATION OF FE(II) IN LIMESTONE NEUTRALIZATION

The use of limestone to neutralize coal mine drainage has several limitations. Limestone has a slower reaction rate which is controlled by its limited solubility. The limestone particles are subject to surface coating with reaction products.

Further, the maximum pH which can be developed in the aqueous system with limestone cannot exceed about 8.3 due to several chemical equilibria, including the presence of soluble carbon dioxide. In fact, only under extremely vigorous agitation and with the use of very finely divided limestone particles can the pH of the treated effluent rise above 7.0. Although this situation prevents over-treatment of CMD, the lower pH levels place restrictions upon oxidation of ferrous iron, which normally follows neutralization and is necessary to complete the treatment.

Experience indicates a practical limitation of the use of this desirable reagent to waters containing less than about 100 mg per liter Fe(II). This arbitrary value appears to be related to the oxygen solubility attainable. Waters with somewhat higher levels of Fe(II) may be treated satisfactorily if the limestone neutralization reaction is followed by lengthy air oxidation periods in the presence of large excesses of pulverized limestone. During this oxidation period the limestone must be maintained in suspension.

To permit the effective use of limestone neutralization in ferrous waters above the 100 mg per liter level, it becomes desirable to oxidize Fe(II) in an acid system. Oxidation by air (oxygen) under these lower pH conditions is impractically slow. Stronger chemical oxidants such as peroxides, permanganate, and chlorine may be used but are deemed too expensive, at least to the degree that their use would increase the costs to the level which would preclude the advantages of limestone.

It is generally accepted that autotrophic or chemolithotrophic bacteria, such as *Ferrobacillus ferrooxidans* and *Ferrobacillus thiooxidans* play a role in the development of mineralized mine drainage from pyrites. It is also known that these bacteria will oxidize soluble ferrous iron in an acid solution. This response by the bacteria may be considered a catalysis of the reaction between atmospheric oxygen and Fe(II) ions, which is thermodynamically favored, but apparently kinetically hindered.

The Hollywood facility (see Description of Pilot Plant, page 68) was designed to permit the biochemical oxidation of Fe(II). In addition to the vat type of oxidation vessel, a trickling filter type of unit operation was included at Hollywood to carry out this biochemical oxidation. Bench-scale studies established the feasibility of this latter concept, including its combination with a rotary kiln-type limestone reactor.

This trickling filter type of unit operation, termed a surface biochemical reactor, has several inherent advantages: [1] No energy is required to control the water distribution or to introduce oxygen and carbon dioxide into the reacting water. The pressure supplying the water to the reactor, whether achieved by pump or gravity pressure head, should be adequate to distribute the water. Porosity through the media bed permits distribution of oxygen and carbon dioxide from the air. [2] The reactor provides reaction surface area which may be controlled by selection of the media particle configuration. [3] Capital cost of the unit is low.

Designs less complex and costly than the Hollywood reactor may be feasible based on this experience. Glover (1) cited as a disadvantage of the percolating filter concept a tendency to build up insoluble iron oxidation products. Although such deposition occurs, it is the high surface area associated with the deposition which supports the bacteria and provides interphase transfer of nutrients. The cementitious nature of the deposit which builds on the surface of the media in the reactor creates a reaction area responsible for superior oxidation rates. The buildup in the surface reactor at the Hollywood facility did not reach a level, in over one year's operation, detrimental to flow through the filter. Continual deposit growth will lead to channelling and a reduction in efficiency. Preliminary tests indicate several approaches to remove the sludge: partial solubilization and water scrubbing in situ, or after drying, the deposit will spall from the media surface with minor vibration. Chemical control to prevent Fe(III) hydrolysis is possible but does not appear to be feasible as a solution to the deposition problem.

The two controlling factors in this biochemical process include the growth rate of the bacteria and the rate at which the bacterial system oxidizes the ferrous ion. In a continuous treatment system, the retention time in the reactor is the critical design factor.

Parameters which establish the bacterial growth rate include: inoculation level, temperature, Fe(II) concentration, bacterial nutritional requirements (as oxygen, carbon dioxide, phosphate, calcium, nitrogen, and trace elements), agitation, and the unique characteristics of a given bacterial strain, including tolerable population levels. These parameters determine the required bacterial population. Evidence suggests minimal cell counts of 5×10^6 cells/ml are necessary, whereas higher levels are more desirable.

Cell levels in excess of 10^7 cells/ml have been maintained in continuous flow in systems, whereas counts approaching 10^9 cells/ml have been achieved in laboratory cultures. Glover reports the activated sludge serves to maintain an appropriate concentration of bacteria, a conclusion not wholly supported by the Hollywood studies.

The bacterial iron oxidation rate is dependent upon maintenance of a viable cell population (growth rate, cell viability time, system cell loss), temperature, Fe(II) concentration, pH, and at least one inadequately-defined parameter which is related to liquid-solid interface surface areas.

The last-named factor may be involved in Glover's activated sludge, and may explain the high level of oxidation activity at the surface of natural stream yellowboy. Glover reports the bacterial sludge to have a low cell count, which was confirmed in the Hollywood studies. It is the liquid-solid interface area which is believed to be responsible for the much higher oxidation rate found in the surface biochemical reactor in contrast to those reported in vat-type systems.

Associated with the effect of pH upon the biochemical oxidation of iron is another related equilibrium: the hydrolysis of the ferric ionic species (presumably as some sulfate complex) which results in the insoluble hydrous ferric oxide deposit. When the Fe(II) oxidation is carried out chemically with oxygen at higher pH levels, this hydrolysis proceeds very rapidly, reducing the soluble iron species to less than 1 mg per liter. As the oxidation is carried out at low pH levels (pH 2.9 to 3.0), the hydrolysis reaction is slow and the resulting Fe(III) species tend to remain in solution. Under acid conditions, the biochemical oxidation system effluent remains relatively free of suspended iron oxide, but does deepen its amber-like color as the concentration of the ferric ion species increases. The sludge deposition appears to grow from nucleation sites rather than direct precipitation containing ferric ion. A slight increase in pH (by 0.1 to 0.3 unit from addition of lime slurry) of the water slowly develops a colloidal haze but the formation of distinct particles is retarded.

Higher pH levels from minor lime additions to waters are apparently detrimental to the bacteria metabolism. Other bacterial species may prefer higher pH levels. Glover's activated sludge (2) may play a role in this hydrolysis reaction, perhaps beyond that as a carrier of bacterial cells. The recycled activated sludge may serve as a physical growth nucleus favorable to this reaction. This phenomenon may be a key to the further enhancement of this biochemical oxidation process and may be the undefined parameter which is related to surface area. In the Hollywood studies, the merit of separating and recycling the sludge formed after biochemical iron oxidation could not be supported, and was discontinued.

Chemical analysis of the sludge developed by biochemical oxidation shows it to be characterized by high iron and sulfate concentrations as well as low aluminum concentrations, when contrasted to sludge from alkali neutralization. Apparently between 15 and 16% of the iron present is associated with the sulfate. The high sulfate content suggests another advantage of biochemical iron oxidation in achieving some sulfate concentration reduction. The potential to enhance this sulfate reduction is of great interest.

The iron oxidizing bacteria were named *Ferrobacillus ferrooxidans* by Leathen et al (3) but are regarded by research workers as closely related to members of the Thiobacillus genus since many of them can oxidize sulfide compounds as well as iron. These bacteria are variously referred to as *Ferrobacillus ferrooxidans, Ferrobacillus thiooxidans* and *Thiobacillus ferrooxidans*. In waters containing Fe(II) which is associated with sulfide, there no doubt exist several strains of these bacteria that vary slightly in appearance and in their biochemical activities.

Members of this group of microorganisms are non-spore-forming, gram-negative rods varying from 1.0 to 2.0 microns in length and having polar flagella. The organisms can be observed in vivo with a phase contrast microscope and appear to be straight or slightly curved rods, actively motile and varying in diameter as well as length.

The organisms derive their basic metabolic carbon, hydrogen, and oxygen from carbon dioxide, water, and air. They grow slowly compared to common heterotrophic gram-negative rods, having generation times of about five or six hours at their optimum temperature of 25° to 28°C. Temperatures higher than 35°C are harmful to most natural isolates. Glover reports 40°C to be immediately fatal. The organisms grow at colder temperatures, although their growth rate decreases with temperature. Laboratory experiments with several strains have indicated that at 10°C, the bacteria require from 24 to 30 hours to double their numbers. They achieve a low cell weight (1 gram, dry weight) from the much larger amounts of iron (180 grams).

A growth lag period of 10 to 20 days was reported by Glover. High levels of inoculant encourage initial growth. If the initial plant system has cell levels of the order of 10^2 cells/ml, a much longer incubation period can be anticipated. The permissible pH range appears to be between 1.5 and 4.0, with an optimum near 3.0. The pH levels which might be specified in this biological process illustrate the complex nature of the reactions which occur. There are at least two optima which should be satisfied: the pH condition which relates to the growth levels of the bacteria, and that pH condition which relates to the Eh level most favorable to the Fe(II) oxidation. The latter level increases with rising pH, thus developing conditions more conducive to chemical oxidation. Efforts to enhance the oxidation rates by maintaining higher pH levels in the biochemical oxidation tanks (by the addition of lime slurry) have not been advantageous, suggesting that the metabolic levels of the bacteria decrease in these higher pH ranges. This observation confirms a distinct role of the life process in this biochemical oxidation. Further study of the mechanisms involved is strongly indicated.

Under continuous flow conditions, Glover reported Fe(II) oxidation rates of nearly 11 mg per liter per hour when maintaining an equilibrium concentration of 50 mg per liter of Fe(II). Rates nearly three times greater were reported in ten-day-old laboratory cultures. Air was introduced by sparger into Glover's oxidation reactors. In subsequent, larger-scale studies, inoculation of the water and slow development of an activated sludge, the rate increased to 120 mg per liter per hour while maintaining a temperature of 15° to 20°C. The suspended solids were maintained near the 1% level.

At near-freezing temperatures, the rate dropped to 12 mg per liter per hour. The activated sludge was recycled and became very tacky and cohesive. He employed retention times of one hour at 20°C and up to five hours at a temperature near 0°C. Huddleston et al (5) reported much higher rates in laboratory cultures with synthetic solutions but were unable to sustain them in continuous flow with actual mine water (at 20° to 26.5°C) in a small pilot plant. These authors found approximately 0.1 volume of air per volume of water to be optimum. They also utilized about 20 mg per liter each of ammonia and phosphoric acid as nutritional supplements.

DATA FROM THE HOLLYWOOD FACILITY

The bacteria employed were isolated from samples of Hollywood waters collected in sterile containers, and, in one instance (Strain Z), from a Central Pennsylvania coal refuse product derived from Lower Kittanning seam coal.

Experience indicated the advantage of starting the biochemical oxidation systems from high-cell-count, pure, laboratory cultures rather than seeking to increase the population from a low-cell-count, natural water or attempt to introduce a natural culture from the surface of the yellowboy deposit in a stream. A series of batch steps is desirable in which a minimum of 10% inoculant is utilized in each step.

Typical batch oxidation rates are illustrated in Figure 4.1 while the conditions utilized are detailed in Table 4.1. In stage A (Figure 4.1), a 150-gallon tub was employed as the reactor with the matured growth being transferred to the sludge sump for stage B and subsequently to one of the concrete oxidation tanks (see Figures 3.1 and 3.2) for stage C. In proceeding to stage D, the third stage product was equally distributed between the two oxidation tanks which were then filled with raw water. A final Fe(II) concentration of about 10 mg per liter was reached within 24 hours. Little Fe(III) is hydrolyzed and precipitated from solution during these batch oxidations. With the maturation of the culture representing two 10,000-gallon stages, the batch operation was converted to a dynamic multistage system by the continuous introduction of raw water.

FIGURE 4.1: FE(II) OXIDATION RATES WITH BATCH INOCULATION OF MINE DRAINAGE

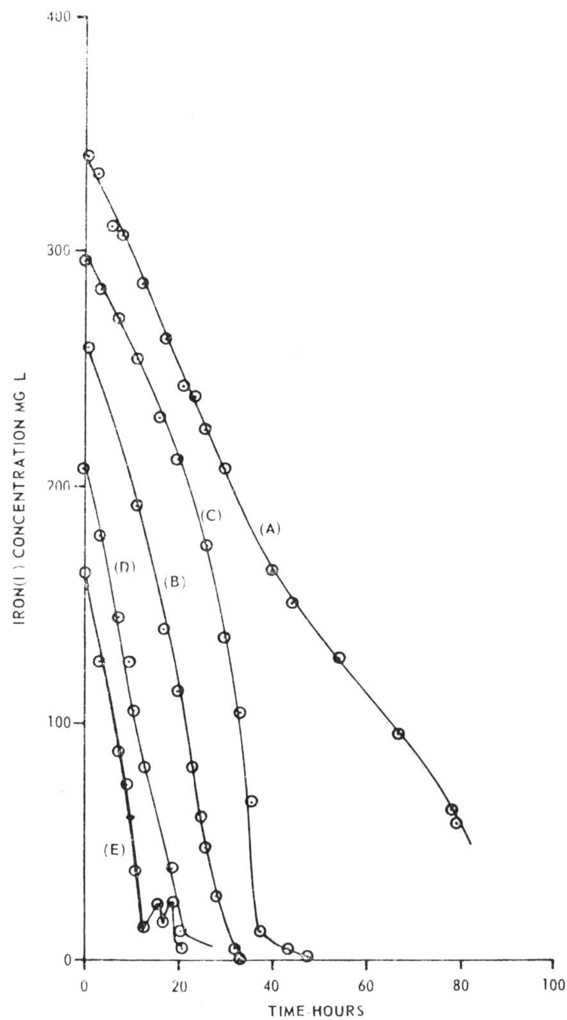

Source: EPA-670/2-73-093

TABLE 4.1: DETAILED CONDITION OF BATCH INOCULATION

Stage	Reactor Volume, gallons	Inoculation Level, volume %	Overall Oxidation Rate, mg Fe/l/hr
A	150	3.2	3.6
B	1,257	10.8	7.9
C	9,930	12.4	7.6
D-1	9,930	50.0	9.8
D-2	9,930	50.0	11.5

		Stage A	Stage C
Nutrient supplement:	NH_4^+	1.6 mg/l	1.0 mg/l
(as $NH_4H_2PO_4$)	$PO_4^{\equiv}$	3.4 mg/l	5.1 mg/l

Aeration: Air sparger in stages A and B at approximately 1/10 vol air/min/vol water
Surface aeration in stages C-D-E

Agitation: Slow mechanical stirrer by surface aerator
24 rpm in stages C and D

Water temperature: 61°F

Culture: Strain Z, 93.5 hours old

Raw Water Analyses:	
Fe(II):	340 - 358 mg/l
Fe (total):	410 - 420 mg/l
Sulfate:	1,950 mg/l
pH:	3.0 - 3.5
Acidity:	1,350 - 1,500 mg/l

Source: EPA-670/2-73-093

In addition to the water loading rate to the surface reactor (>0.16 gpm/ft^2 surface), the retention time in the reactor is critical. The retention time of the surface reactor was found to range between 4.7 and 5.3 minutes in the flow range considered. This parameter will not deviate extensively with flow changes until a flooding condition would develop which probably would not permit sufficient oxidation time. This retention value was attained by batch introduction of a concentrated tracer (sulfuric acid and sodium sulfate were both helpfully used) to the surface of the media and the conductivity of the effluent measured as a function of time. The first increase in conductivity was observed after one minute. The response of the tracer had passed in about 12 minutes.

The overall oxidation performance of the reactor indicated the extent of Fe(II) reacted varied between 60 and 94%. In comparing deep vat and surface reactor biochemical Fe(II) oxidation, three pragmatic conclusions are made which favor the surface-type reactor:

[1] The surface reactor is least sensitive to disruptive influences, i.e., change of Fe(II) concentration, metabolic inhibitants, flow rates, etc.

[2] The surface reactor permits much higher reaction rates than deep vat systems under continuous flow operation conditions: at least 6,000 mg per liter per hour versus 50 mg per liter per hour.

[3] The surface reactor has no power requirements to control water flow, introduce oxygen and carbon dioxide.

MICROBIOLOGY OF ACID MINE WATER TREATMENT IN PACKED BED COLUMNS

The research discussed in this section was undertaken in an effort to identify the agents responsible for the transformation of ferrous iron to the ferric state during application of acid mine water to columns of trickling filter medium located in the bed of a full-scale trickling filter. The surface contact system was examined with regard to:

[1] Efficiency of ferrous iron transformation at two depths of trickling filter medium. It is stressed that the phenomenon of ferrous iron oxidation, not total iron removal, was the item of interest although it was obvious from the condition of filter media during experiments that some iron removal had actually taken place.
[2] Presence of microorganisms and their contributions to the oxidation of ferrous iron.
[3] Microbial population densities in active iron oxidizing systems
[4] Rates of ferrous iron transformation.

Materials and Methods

Ferrobacillus ferrooxidans (ATCC 13661) and bacterial strain Z, a gob pile isolate, were used in developing isolation and enumeration techniques for application to field studies. Field studies were carried out at the acid mine water treatment facility of The Pennsylvania State University at Hollywood, Pennsylvania.

Experimental Apparatus: The full-scale trickling filter was 35 feet in diameter and 5 feet in depth. During the winter and spring of 1973, a wooden A frame covered with a plastic sheet protected the filter bed from the cooling effect of air currents. During severe weather, a propane heating unit was available to heat the ambient air inside the covered filter. The filter bed consisted of argillite stone, ranging in size from 0.5 to 4 inches in diameter. The effective surface area was 45 ft^2/ft^3 of filter medium with approximately 46% void space.

To facilitate a study of the iron oxidizing activities in the filter, three cores, three feet in depth, were hewn in the filter bed and prepared to receive removable baskets of filter medium. Each core was cased with a plastic sewer pipe to prevent filter bed cave-in. Cores were located in an arc of the filter bed so that individual baskets of filter medium would receive influent mine water from the same dosing spouts on the rotary distributor of the trickling filter. Baskets were cylindrical (length, 12 inches; inside diameter, 11.5 inches; outside diameter, 12 inches) and were fabricated from clear, 0.25-inch methyl methacrylate sheets (Plexiglas). The floor of each basket was perforated with 0.75-inch diameter holes to allow passage of percolating mine water.

Prior to each experimental run, certain baskets were filled with either argillite stone or 1.5-inch polypropylene Flexirings having a surface area of approximately 40 ft^2/ft^3 and 96% void space. The baskets of filter medium and an effluent sample collection basket were inserted into the cased cores of the trickling filter bed. Four long-term studies were made with the trickling filter and are designated runs 1, 2, 3, and 4. Run 1 consisted of an analysis of ferrous iron oxidation during passage of mine water through one foot of argillite filter medium. The remaining three runs involved a comparative study of ferrous iron oxidation in mine water percolated through baskets which contained polypropylene filter medium and which were stacked in cores to provide zero, one, and two-foot depths (columns) of the synthetic filter medium. The column of baskets containing no (zero) filter medium served as a control on ferrous iron oxidation.

Mine Water: Influent mine water used in all experimental runs was Proctor No. 2 abandoned coal mine drainage. A typical chemical composition of the mine water is given in Table 4.2.

TABLE 4.2: CHEMICAL COMPOSITION OF PROCTOR NO. 2 MINE WATER

Chemical Characteristic	Value
pH	2.83
Acidity as $CaCO_3$	1,230 mg/l
Iron, total	341 mg/l
Iron, ferrous	245 mg/l
Iron, ferric	96 mg/l
Aluminum	81 mg/l
Calcium	161 mg/l
Magnesium	62 mg/l
Manganese	7 mg/l
Sulfate	1,777 mg/l

Source: PB 229 649

It should be noted that home (individual) wastewater treatment systems were present in the area of field study and these were suspected to contribute soluble organic matter to the mine water regime. During run 1 and most of run 2, mine water was pumped to a distribution basin from which it flowed by gravity to the trickling filter. During runs 3 and 4, mine water was pumped directly from the source to the filter since it was discovered that appreciable ferrous iron oxidation already had taken place during detention of mine water in the distribution basin. It was learned, also, that approximately one year in advance of the current project, mass cultures of axenic iron oxidizing bacteria were inoculated into the distribution basin to provide seed for biological mine water treatment units under study at that time.

Field Studies with Trickling Filter Media

A summary of the conditions which characterized the four field studies (runs) is given in Table 4.3. Ferrous iron concentrations of influent mine water varied markedly depending upon the extent of dilution by rainfall and the degree to which ferrous iron oxidation took place in the holding basin. The latter factor was largely overcome in later runs by pumping the influent mine drainage directly to the trickling filter from the source.

Argillite Stone Filter Medium: Run 1 — Argillite stone is a calcium silicate presumed to be relatively inert to chemical action and weathering. It is these properties which are important in the consideration of a trickling filter medium. Percolation of influent mine water through a single foot of argillite stone contained in a plastic basket produced a median efficiency of ferrous iron oxidation of 65% during a 41-day equilibrium period which was reached in two weeks following startup. At an hydraulic loading of 1.76 gpd/ft^2 of effective filter medium surface area, 41 mg ferrous iron/ft^2/hr was oxidized at median treatment efficiency when influent mine water contained 225 mg per liter of ferrous iron.

TABLE 4.3: EXPERIMENTAL CONDITIONS OF FIELD STUDIES

Run No.	Period	Filter Medium Type	Depth of Filter Medium Dosed	Typical Hydraulic Loading (gpd/ft^2)	Influent Temperature (°C)	Mine Water Properties pH	Mine Water Properties Ferrous Iron (mg/l)*
1	7/72-9/72 (8 weeks)	Argillite stone	1'	1.76	7.2±0.6	2.65±0.15	85-225
2	10/72-2/73 (15 weeks)	Polypropylene Flexirings	1' and 2'	3.29	8.9±2.28	2.71±0.29	17-221
3	2/73-5/73 (15 weeks)	Polypropylene Flexirings	1' and 2'	2.43	11.0±0.6	2.61±0.25	5-206
4	6/73-8/73 (8 weeks)	Polypropylene Flexirings	1' and 2'	1.29	12.7±0.55	2.67±0.03	159-232

*Ferrous iron concentrations fluctuated in the influent mine waters, presumably owing to dilution by precipitation and oxidative action in holding basin.

Source: PB 229 649

Yellow-orange deposits were clearly evident on the stone after only one week of mine water dosage. The full-scale trickling filter, already in operation for one year, produced ferrous iron oxidation efficiencies which ranged from 64 to 91%. A water suspension of solids scraped from the surface of selected stones was found to contain 1.3 x 10^6 iron oxidizing bacteria per ml as determined by the microtiter plate-MPN technique. During this run, influent mine water was received from the previously inoculated holding basin.

It was discovered, upon inspection of the trickling filter bed, that excessive fragmentation of stones occurred at a depth of two to three feet. Disintegrated stones filled void spaces between other intact stones obstructing the downward flow of mine water. This condition affected a large area of the trickling filter which resulted in ponding and short-circuiting of mine water in the filter bed. It was decided that argillite stone was not suitable as a trickling filter medium for use in the treatment of mine water.

Consequently, all succeeding experiments were conducted using polypropylene filter medium. The synthetic medium was especially advantageous in that individual units were (a) uniform in size, shape, and surface area; (b) could be more easily manipulated to remove adherent solids; and (c) was not visibly reactive with or affected by the chemical properties of the mine water.

Polypropylene Filter Medium: Run 2 — Ferrous iron oxidation efficiencies of up to 67 and 93% were obtained after mine water had been percolated for 77 days through one and two feet of synthetic filter medium, respectively. At these peak iron oxidation efficiencies and a hydraulic loading of 3.29 gpd/ft^2 of effective filter medium surface area, 77 and 30 mg ferrous iron/ft^2/hr were oxidized in the first and second-foot sections of filter medium, respectively, when the influent mine water contained 221 mg per liter ferrous iron. Unlike the rapid achievement of equilibrium ferrous iron oxidation experienced with argillite stone, a steadily increasing efficiency of ferrous iron transformation over a long duration was observed in this run. However, during the first six weeks of run 2 and especially on weekends, influent ferrous iron concentrations were generally low.

It was apparent that significant ferrous iron oxidation was taking place during periods of long retention of mine water in the holding basin. To relieve this condition, influent mine water was pumped directly to the filter from the source. In spite of this alteration in mine water feeding procedure, heavy rainfall during weeks 7, 8, and 9 of run 2 maintained the condition of a dilute ferrous iron concentration in the influent. It is believed that these unfavorable circumstances were responsible for the restricted attainment of equilibrium ferrous iron oxidation during run 2.

During the course of ferrous iron oxidation with synthetic filter medium, individual units became coated with iron deposits. As a first test of the hypothesis that ferrous iron oxidation on the filter medium was responsive to the activities of iron oxidizing bacteria, baskets of filter medium with red-orange deposits were removed from cores on the eighty-third day of the run and soaked in a solution of the disinfectant, Zephiran Chloride. After rinsing the disinfected baskets of filter medium, containers were returned to the trickling filter bed and percolation of mine water was resumed.

It was found that the efficiencies of ferrous iron oxidation in passage of mine water through one and two feet of synthetic filter medium dropped to levels of 1.9 and 3.6% respectively. During the three weeks which followed disinfection of filter medium, ferrous iron concentration in the influent mine water ranged from 125 to 212 mg per liter and the ferrous iron oxidation activity was rapidly reestablished in the baskets of filter medium. In fact, a degree of ferrous iron oxidation was achieved in the three-week post disinfection period that had required ten weeks to accomplish before disinfection.

Several possible explanations may account for this unusual occurrence: (a) influent ferrous iron concentrations which became available to the iron oxidizing bacteria were higher during postdisinfection due to elimination of the distribution basin from the flow scheme and, as such, may have stimulated rapid growth of the bacteria on the filter medium, (b) disinfectant action may have been bacteriostatic rather than bacteriocidal thus permitting the iron oxidizing bacteria on the filter medium to recover rapidly and reestablish a high degree of ferrous iron oxidation, and (c) solids already present on the disinfected filter medium may offer effective attachment sites for viable iron oxidizing bacteria present in percolating mine waters. In any case, presumptive evidence of microbial mediation in the oxidation of ferrous iron was established in run 2.

Run 3 – In contrast to run 2, a plateau (equilibrium) in the efficiency of ferrous iron oxidation developed and was reached after approximately 28 days of mine water percolation through polypropylene filter medium. During the equilibrium period, median ferrous iron oxidation efficiencies of 58 and 83% were obtained with one and two feet of synthetic filter medium, respectively. At these median efficiencies and a hydraulic loading of 2.43 gpd/ft^2 of effective filter medium surface area, 46 and 20 mg ferrous iron/ft^2/hr was oxidized in the first and second foot sections of filter medium, respectively, when the influent mine water contained 206 mg per liter ferrous iron. During the first four to five weeks of run 3, both iron oxidizing and heterotrophic microorganisms maintained a trend toward increased numbers on the filter medium in each of the top baskets of two separate columns. Interestingly, viable heterotrophic microorganisms began to diminish in number toward the end of the run whereas iron oxidizing bacteria remained relatively stationary.

The trend in production of filter medium solids roughly paralleled the growth of iron oxidizing bacteria, heterotrophic microorganisms, and the efficiency of ferrous iron oxidation until the latter stage of the run when solids accumulated rapidly while microbial populations and ferrous iron oxidation remained stationary.

Using that portion of the data which relates to the apparent equilibrium period of ferrous iron oxidation, median values were calculated for microbial densities of filter medium solids sampled from the surface of those baskets in each of the two columns which directly received mine water. The median numbers of iron oxidizing bacteria and heterotrophic microorganisms were 3.82×10^4 and 4.63×10^3 cells per mg of filter medium solids, respectively (surface basket of one column) and 1.49×10^4 and 3.2×10^3 cells per mg of filter medium solids, respectively (surface basket of other column). Notwithstanding possible inadequacy of the enumeration technique and culture medium, heterotrophic microorganisms were consistently lower in number in filter medium solids than were iron oxidizing bacteria.

Treatment of a single basket of microbiologically active filter medium with the disinfectant, as described earlier, produced the effect of curtailing ferrous iron oxidation in a manner similar to that which was observed during run 2. In this run, ferrous iron oxidation efficiency dropped to 5% following disinfection of the filter medium and the iron oxidizing bacteria were inactivated. Care was taken to assume that solids attached to the filter medium were not lost during the course of disinfection.

Whereas iron oxidizing bacteria were seen to be adversely affected by the disinfectant, heterotrophic microorganisms were not. The control basket of filter medium, which underwent all procedures of disinfection save the substitution of tap water for the disinfectant, showed no decrease in any of the parameters previously described. Peak ferrous iron oxidation was restored in the disinfected filter medium in three weeks following treatment with the Zephiran Chloride disinfectant.

Percolation of mine water through synthetic filter medium had little effect on pH. In comparing the pH of influent and effluent mine waters during run 3, it was discovered that effluent waters generally had slightly higher pH values than influent waters; however, the widest difference in pH between the two waters was only 0.16 unit. The slight, although inconsequential, increase in the influent pH during the early stage of the run was unexplained.

Run 4 — Equilibrium ferrous iron oxidation was reached after an initial 15 days of mine water percolation through columns of filter medium. Median ferrous iron oxidation efficiencies of 68 and 89% were achieved during the equilibrium period in passage of mine water through one and two feet of synthetic filter medium, respectively. At these median efficiency values and a hydraulic loading of 1.29 gpd/ft^2 of effective filter medium surface area, 32 and 10 mg ferrous iron/ft^2/hr was oxidized in the first and second foot sections of filter medium, respectively, when the influent mine water contained 232 mg per liter ferrous iron.

The iron oxidation data obtained in run 3 and run 4 were used to produce a plot of median percent ferrous iron remaining in the mine water effluent versus the depth of filter medium dosed (Figure 4.2). The linear relationship observed indicates that the oxidation of ferrous iron approximated a first order decay scheme apparently unaffected by changes in hydraulic loading.

FIGURE 4.2: EFFICIENCY OF FERROUS IRON REMOVAL FROM ACID MINE WATER

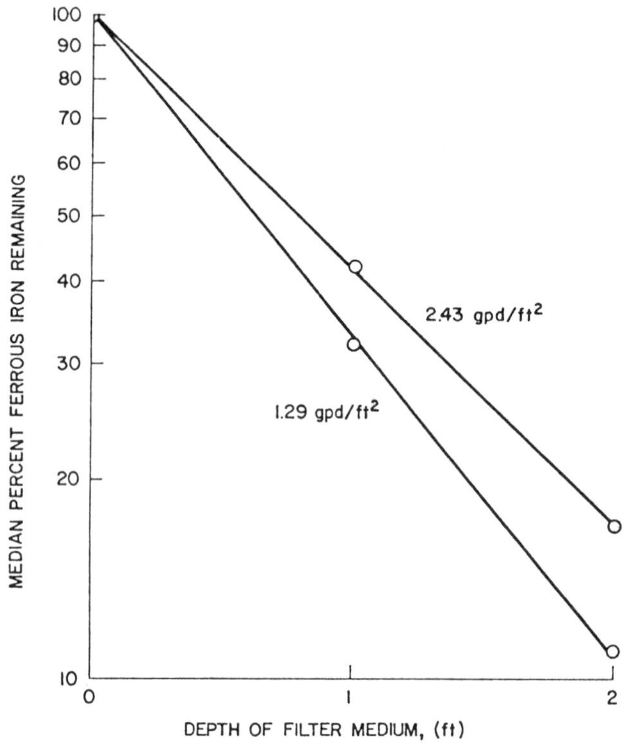

Source: PB 229 649

In run 4, two baskets of filter medium located in separate columns in the trickling filter bed were the source of filter medium solids to be analyzed for microbial content. One of the baskets was located at the top of a column and directly received influent mine water. The other basket was situated beneath the top basket of a second column and received the effluent mine water from the basket of filter medium which it physically supported. The microbial enumeration data obtained during the ferrous iron oxidation equilibrium period were used to calculate median densities of microorganisms in filter medium solids. During the equilibrium period, iron oxidizing bacteria and heterotrophic microorganisms in concentrations of 4.16×10^3 and 6.0×10^2 cells per mg of filter medium solids, respectively, were present in the solids sampled from the basket of filter medium directly dosed with mine water. Iron oxidizing bacteria and heterotrophic microorganisms in concentrations of 4.39×10^3 and 4.03×10^3 cells per mg of filter medium solids, respectively, were present in solids sampled from the subsurface basket of filter medium which received effluent from the uppermost basket in the column.

It is obvious from the results of runs 3 and 4 that the native microorganisms in acid mine water influent are sufficient to inoculate the filter medium and provide for the eventual establishment of dynamic microbial populations. During these runs, fresh filter medium was dosed only with mine water pumped from the source and the previously inoculated holding basin was eliminated from the flow scheme.

Although prior inoculation of a biological acid mine water treatment unit with iron oxidizing bacteria may serve to expedite attainment of maximum ferrous iron oxidation it does not seem necessary to do so in order to experience biological ferrous iron oxidation at a desirable rate eventually. It may even prove futile and expensive to attempt adaptation of laboratory-trained cultures to the mine water environment.

The high volatile matter content of filter medium solids was considered suggestive of the presence of organic matter although it was recognized that a serious loss of water of hydrate from the iron compounds would tend to liberalize the estimate of volatile solids. The organic nature of the filter medium solids was realized better when it was discovered that acid cleaned (iron-free) solids, following purification and hydrolysis, yielded positive results in tests for the presence of reducing substances including glucose.

Purity of the hydrolyzed matrix initially dissolved in sodium pyrophosphate was questionable since peaks were observed in the ultraviolet spectrum, possibly indicative of nucleic acid and protein contaminants. Specimens of the iron-free matter were found by microscopic examination to comprise dense populations of microorganisms, mostly bacteria embedded in a gelatinous matrix. Solid TGBYM culture medium, streaked with samples of filter medium solids, supported growth of several kinds of heterotrophic microorganisms including fungi, yeasts, and bacteria.

The bacterial isolates consisted of 6 gram-negative rods, 3 gram-negative cocci and a gram-positive bacillus. The bacteria were not characterized. No attempt was made in this study to isolate iron oxidizing bacteria from solids concentrate; however, these bacteria were always observed microscopically to be present in specimens obtained from microtiter plate cultures of serial dilutions of the filter medium solids.

The percentage of volatile matter in filter medium solids was found to decrease in the later stages of the run apparently reflecting the effect of a substantial accumulation of mineral matter, presumably ferric hydrates, without corresponding increases in microbial cells or insoluble organic matter (gelatinous residue). Possibly, increased levels of inorganic substances influenced the decline of viable heterotrophic microorganisms which was especially noticeable during run 3. In addition, generally lower median microbial densities in filter median solids were observed in run 4 relative to run 3, possibly reflecting the influence of a larger fraction of inert (ferric hydrates) matter in the total solids.

Iron Oxidizing Bacteria in Influent and Effluent

Microtiter plate-MPN (most probable number) analyses were conducted on samples of influent and effluent mine water during the course of run 3 and run 4. A direct comparison of bacterial concentrations of influent and effluent mine waters sampled during run 4 revealed consistently higher numbers of the bacteria in the effluents especially during the preequilibrium phase of the run. This finding is interpreted to mean that some of the iron oxidizing bacteria which developed on the polypropylene units were picked up by the percolating as it went downward through the filter medium.

Newly Isolated Iron Oxidizing Bacteria: Four iron oxidizing bacteria were isolated from Proctor No. 2 mine water and briefly characterized along with *F. ferrooxidans* ATCC 13661 and bacterial strain Z. Especially interesting was the discovery that certain of the newly isolated bacteria differed in appearance from the characteristic straight cells of *T. ferrooxidans*, that is, cells were distinctly curved, and in some cases slightly larger in diameter. In spite of these differences, all strains utilized ferrous iron and elemental sulfur as energy sources although none grew on thiosulfate including the *F. Ferrooxidans* ATCC 13661 strain.

Discussion

Although additional details are still being sought, it is generally agreed that both oxidative and hydrolytic chemical processes are operative in the chemistry of acid mine water formation. The acidophilic, chemolithotrophic iron and sulfur oxidizing bacteria have been repeatedly implicated as the mediators of biogeochemical transformations involving the reduced iron and sulfur compounds which ultimately give rise to the acidic end products which contribute to pollutional characteristics of acid mine drainage.

Considerably more is known about the physiology and nutrition of these bacteria than about their distribution and numbers in the environment. The problem of adequately assessing the activities and magnitude of populations of iron oxidizing thiobacilli in the natural environment has not been solved. Several workers have attempted to devise enumeration techniques in the laboratory using axenic cultures of *Thiobacillus ferrooxidans* but little information is available on the applicability of the methods to evaluations of bacteria in natural samples.

In this study, efforts were made to develop nutritionally sound culture media and enumeration procedures which ultimately might be used to establish a relationship between ferrous iron oxidation and iron oxidizing bacteria existing in the surface contact mine water treatment system. However, experience with laboratory cultures of *T. ferrooxidans* has led to the belief that these organisms cannot be reliably cultivated for quantification on solid culture media either by direct spread plating or concentration on membrane filters. Nevertheless, it is possible to recover strains of the bacteria from natural samples in colonial form and one should avoid dilution-dispersion techniques as a means of acquiring isolates for pure culture studies. The ease of obtaining the organisms is largely determined by the composition of the culture medium and the nature of the solidifying agent. Factors such as nutrient and waste product diffusion, organic matter, and moisture are probably critical to the successful cultivation of iron oxidizing bacteria on solid culture media. That these conditions are apt to be of lesser significance in liquid culture media may account for the better recovery of viable cells in samples evaluated by multiple tube-MPN procedures. However, MPN determinations are statistical evaluations of the bacterial numbers which probably existed in a sample in order to have produced a specific result (growth at various dilutions) and the values are subject to very large confidence limits.

This study has successfully demonstrated that ferrous iron oxidation will proceed during the percolation of acid mine water through a porous medium. The iron oxidation appears to proceed by a first order decay scheme analagous to that described for the typical first stage BOD decomposition. Furthermore, iron oxidizing bacteria are the responsible catalytic agents as evidenced by their growth and activity on the filter medium concomitant with mine water ferrous iron oxidation and the fact that treatment of active filter medium with Zephiran Chloride significantly reduced both ferrous iron oxidation in mine water and the numbers of viable iron oxidizing bacteria in filter medium solids.

Conversely, heterotrophic microorganisms were not affected by the disinfectant treatment. Interestingly, it required only about one half the time to reestablish ferrous iron oxidation during post disinfection as was needed to achieve a similar level of activity in fresh, uncolonized filter medium.

Both chemolithotrophic and heterotrophic microorganisms were present in filter medium solids and it would be of interest to learn what kinds of interaction exist between the two populations. Heterotrophic microorganisms have been observed in acid mine waters on several occasions and sometimes bacteria occur in massive aggregates or acid streamers. Although gram-positive bacilli have been isolated, gram-negative bacteria are predominant in acid mine waters. Apparently, organic matter is often present at high levels in mine waters and it is reasonable to expect acid tolerant heterotrophs to thrive there. In fact, some of the heterotrophic microorganisms in the acid mine water environment must have a wide range of pH tolerance since the pH of the culture medium used in this work to enumerate these organisms was within one pH unit of neutrality.

Perhaps recovery of heterotrophs could have been improved by the use of a culture medium of low pH. Unfortunately, it is not possible to adjust the pH of a protein-based medium to extreme acid pH values without effecting denaturation of proteins. Since acidophilic, iron oxidizing bacteria normally grow in dispersed cell form and have not been found to produce organic, extracellular, polymeric substances which might aid attachment of cells to clean solid surfaces, it may be that the slime-forming heterotrophs function to prepare objects for colonization by the chemolithotrophs. The filter medium solids referred to in this study consisted of flakelike, gelatinous substances coated with the ferric hydrates for which *T. ferrooxidans* is believed to have a strong affinity.

The nature of the filter medium solids is suggestive of much heterotrophic activity taking place in direct proximity to ferrous iron oxidation; hence, organic molecules must be present. Organic matter has been reported to be inhibitory to biological ferrous iron oxidation. In the microbial assemblage of filter medium solids, heterotrophic microorganisms may benefit the iron oxidizing bacteria by utilizing potentially toxic organic molecules naturally present in the mine water or produced and secreted by the chemolithotrophs. On the other hand, heterotrophic microorganisms may adversely affect optimum activity of the iron oxidizing bacteria by producing organic substances or competing for available oxygen. In the aerobic biological treatment of organic wastewaters, it is the vigorous utilization of oxygen by heterotrophic organisms that prohibits chemolithotrophic nitrification from taking place until the carbonaceous oxygen demand of the wastewater is considerably reduced.

However, it is doubtful that the organic carbon present in mine water exerts an oxygen demand at all comparable to that in wastewater. Temperature is probably a more important factor affecting ferrous iron oxidation in that the temperature of mine water is typically in the mid-psycrophilic range at least 5° to 10°C lower than the mean temperature of domestic wastewater.

The optimum temperature for growth of *T. ferrooxidans* and maximum ferrous iron oxidation is stated to be 28°C and 37°C, respectively. One might speculate on the existence of psycrophilic iron oxidizing bacteria in the mine water environment. The distinct morphological variation seen among isolates of iron oxidizing bacteria obtained in this study is suggestive of the existence of different strains of these organisms in mine water. It is presumed that all of the new isolates could be appropriately designated *T. ferrooxidans* although none were able to grow on sodium thiosulfate under the conditions tested.

The residence time of mine water in the five-foot-deep trickling filter described in this report has been estimated to range from 4.7 to 5.3 minutes. Assuming a residence time of two minutes for a two-foot column of polypropylene filter medium and an equilibrium transformation of 200 mg ferrous iron per liter to the ferric state, it would be possible to achieve an oxidation of 6,000 mg of ferrous iron per liter per hour. This ferrous iron oxidation rate is in excess of one hundred-fold that reported for a suspended growth process by Glover (2) and Whitesell et al (6).

Singer and Stumm (7) summarized the chemistry of acid mine water according to three chemical reactions:

$$[1]\ 2FeS_2 + 7O_2 + 2H_2O \longrightarrow 2FeSO_4 + 2H_2SO_4$$

$$[2]\ 4FeSO_4 + 2H_2SO_4 + O_2 \longrightarrow 2Fe_2(SO_4)_3 + 2H_2O$$

$$[3]\ Fe_2(SO_4)_3 + 6H_2O \longrightarrow 2Fe(OH)_3 + 3H_2SO_4$$

Whereas it is apparent that iron oxidizing bacteria play a critical role in the oxidation of ferrous iron to the ferric state (reaction 2), a question arises concerning the formation of the insoluble ferric compounds which develop and accumulate at the surfaces of filter medium units. Reaction 3 proceeds to the formation of insoluble ferric hydroxide only at pH values near 5.0. The pH of the mine water is generally between 2.8 and 3.0. Possibly, bacteria on the filter medium modify the pH of their microenvironment, enabling precipitates to form. It is likely that the ferric iron substances are not of the composition described as $Fe(OH)_3$.

It was observed both during the latter stages of runs 3 and 4 that filter medium solids sharply increased even though populations of iron oxidizing bacteria were stationary. Since early in both runs solids production paralleled reasonably well the development of iron oxidizing bacteria it is difficult to explain the unexpected rise in solids. Sampling error is not ruled out.

The conventional trickling filter is a costly and permanent installation which requires maintenance and supervision. In some cases, especially regarding biological treatment of abandoned mine drainage, it would be difficult to justify the expense involved. Nevertheless, the trickling filter offers advantages over suspended growth systems in that

[1] continuous recycling of iron oxidizing bacteria is not required to sustain ferrous iron oxidation in the influent mine water,
[2] no mechanical or forced aeration is needed for oxygenation and mixing of bacteria with influent mine water, and
[3] supplementation of the influent mine water with nutrients (nitrogen and phosphorus) does not appear necessary.

At the hydraulic loadings experienced in this study, it would not be necessary to construct greater than a two or three-foot-deep bed in order to achieve a high degree of equilibrium ferrous iron oxidation and the effluent would be readily amenable to acid neutralization with limestone, thus avoiding the use of the more expensive hydrated lime. The installation of a cover over the trickling filter bed would constitute a simple and effective measure to prevent chilling air currents from excessively depressing the desired rate of ferrous iron oxidation.

Synthetic (polypropylene) filter medium was found as effective in supporting ferrous iron oxidation as argillite stone. An examination of the trickling filter bed at a depth of two or three feet revealed that the stone was severely fractured. Interestingly, clogging of the void space in the bed at this depth was due as much to disintegrated stone as to accumulated ferric hydrates. Interrupted mine water flow through columns of synthetic filter medium was never experienced. The capital cost of synthetic medium, although much in excess of stone, should not be grounds for rejecting out-of-hand its use for mine water treatment. Polypropylene medium appears to weather well and the necessary size and depth of the trickling filter may not require its use in large quantities.

The maximum treatment efficiency of a surface contact system requires that all of the available surface area be wetted by the wastewater. Such is not the case with the trickling filter. Other devices may be more effective in this regard and the rotating biological contactor is under study for its potential in biological treatment of acid mine water.

It is expected that the biological treatment of ferrous iron in acid mine water can be closely evaluated using this treatment device. If the rotating biological contactor can be applied successfully to the treatment of ferrous iron in acid mine water, its usefulness will be enhanced by the fact that it is portable and can be transported to the site of active mining operations and removed when the worked mine is effectively sealed.

Conclusions

[1] Both natural (argillite stone) and synthetic (polypropylene units)

trickling filter media are capable of supporting the oxidation of ferrous iron in the treatment of acid mine water by the surface contact principle.

[2] Ferrous iron oxidation under the conditions of surface contact mine water treatment is microbially mediated.

[3] Ferruginous solids formed on filter media during ferrous iron oxidation contain gelatinous organic matter with embedded bacteria.

[4] Both chemolithotrophic iron oxidizing bacteria and heterotrophic microorganisms are present in the filter medium solids.

[5] Ferrous iron oxidation efficiencies ranging from 87 to 96% may be achieved in percolation of acid mine water through a two-foot column of polypropylene trickling filter medium at a hydraulic loading of 1.3 gpd/ft^2 of effective medium surface area.

[6] Spread plating and membrane filter cultivation techniques are not reliable for enumerating iron oxidizing bacteria although these may be satisfactory for isolating the organisms from natural samples.

[7] Argillite stone is technically inadequate as a filter medium for acid mine water treatment, apparently due to its inability to withstand the chemical properties of the mine water.

ELECTROCHEMICAL OXIDATION–TYCO LABORATORIES, INC.

Basic Electrochemistry

An acid mine water, containing on the average 500 mg per liter of Fe^{2+} and 1,000 mg per liter H^+ and variable amounts of aluminum, calcium and manganese, is an electrolytically conducting solution. As such, it is capable of electrolysis without resort to additional salt additives. The electrochemical reaction $Fe^{2+} \rightarrow Fe^{3+} + e^-$ would take place at an inert (nonconsumable) anode. Concurrently, on an inert cathode, the reaction $H^+ + e^- \rightarrow \frac{1}{2}H_2$ would take place, resulting in the generation of ½ mol of hydrogen gas for every mol of iron that is oxidized. The study discussed in this section describes the technical and economic evaluation of an AMD treatment process based on these reactions.

Upon the electrolysis of acid mine drainage water, the following electrochemical processes are possible for the major soluble ionic species:

(A) At the negative electrode (the cathode):

[1] $2H^+ + 2e^- \rightleftharpoons H_2$ E = -0.35 V

[2] $Fe^{2+} + 2e^- \rightleftharpoons Fe$ E = -0.79 V

[3] $Fe^{3+} + e^- \rightleftharpoons Fe^{2+}$ E = +0.52 V

(B) At the positive electrode (the anode):

[3a] $Fe^{2+} \rightleftharpoons Fe^{3+} + e^-$ E = +0.52 V

[4] $2H_2O \rightleftharpoons O_2 + 4H^+ + 4e^-$ E = +0.88 V

[5] $2H_2O \rightleftharpoons H_2O_2 + 2H^+ + 2e^-$ E = +1.41 V

[6] $2HSO_4^- \rightleftharpoons S_2O_8^= + 2H^+ + 2e^-$ E = +1.8 V

[7] $2SO_4^= \rightleftharpoons S_2O_8^= + 2e^-$ E = +1.7 V

Electrical potentials are thermodynamic values versus a saturated calomel reference electrode (SCE) for concentrations of 0.02 N. The sign (+ or -) results from use of the standard reduction potential convention for half-cell reactions. The desired anodic process is reaction [3a], the desired cathodic process reaction [1]. All other processes are parasitic for development of a viable AMD treatment. Reactions 5 through 7 are disregarded; they occur at potentials more positive than those readily accessible in dilute aqueous electrolytes. Current (reaction rate)-potential dependencies of the remaining reactions were found experimentally.

Laboratory Studies

AMD water varies in Fe^{2+} concentration and pH; 0.02 M in Fe^{2+} and in H_2SO_4 was taken as representative. In the initial studies, the experimental conditions were simplified by using pure $FeSO_4$ and pure H_2SO_4. Subsequently, the results were confined in synthetic AMD water produced in the laboratory by draining water through waste coal.

Normally this effluent solution from the waste coal was cloudy yellowish brown in color and varied in total soluble iron concentration from $\sim 4 \times 10^{-3}$ M to $\sim 2 \times 10^{-2}$ M (both Fe^{2+} and Fe^{3+}). The solution was generally pH 2.7. When these water samples were left for extended periods of time, the Fe^{2+} concentration usually decreased and Fe^{3+} increased, presumably from oxidation of the former by dissolved O_2. (The participation of bacteria in this process was not studied.) Thus, over a 5-day period, initial concentrations of 1.7×10^{-2} M Fe^{2+} and 1.2×10^{-2} M Fe^{3+} changed to 1.4×10^{-2} M Fe^{2+} and 1.3×10^{-2} M Fe^{3+}. Some iron was apparently lost through precipitation during the period.

In most experiments, this synthetic AMD was first centrifuged for 10 minutes before use, in order to remove solid residues. This treatment caused only a small drop in concentration of both Fe species; the solution was then clear. However, a yellowish brown residue settled out in the electrochemical cell over a period of days. Similarly when the solutions were passed through a 0.45-μ Millipore filter, the cloudiness was removed but again a light yellow residue settled out in the deaerated cell over several days. This latter treatment also slightly reduced the concentration of both Fe species in solution. The values for both Fe^{2+} and also Fe^{3+} in these solutions are somewhat greater than might normally be found in AMD in streams, i.e., after natural dilution. In some experiments, therefore, this synthetic AMD was diluted with 0.02 M H_2SO_4 before use.

Carbon was selected as the anode since it is inexpensive, can be obtained in tubular form, is very resistant to chemical attack, and has a relatively wide potential range over which it is electrochemically stable (i.e., does not evolve O_2, H_2, or dissolve). As cathode, 316-stainless steel rather than pure iron was chosen, since stainless steel is more resistant to corrosion in dilute acids.

A rotating disc electrode setup was used. The equipment consists of a sturdy stand on which a 1/15-hp Bodine motor and a precision ball bearing for the 0.025-

inch rotating shaft were mounted. The motor speed was controlled by a Minarik speed control. The coupling of the motor and the electrode shaft was accomplished by a nonslip belt. The rotation rate of the electrode was continuously monitored by frequency modulation resulting from magnetic coupling of an electromagnet with an iron gear mounted on the rotating shaft. This signal was amplified and displayed on a frequency counter. Electrical contact to the disc electrode was accomplished through a mercury pool in the top of the rotating shaft.

The electrochemical cell was made of a 50-mm O-ring joint with a 30-mm thick Teflon cover. The Teflon cover contained tapered holes to accommodate a liquid seal in the center with a 29/40 standard tapered joint surrounded by four 10/30 joints for gas inlet (N_2) and outlet as well as the counter- (platinum) and reference electrodes (calomel). A Teflon bell was fixed to the rotating shaft. Mercury was used as a sealant liquid to prevent O_2 from contacting the solutions. The main feature of the electrode is the vitreous carbon anode which is press-fitted into hot Teflon and contacts the upper mercury pool via a simple spring. A rotating disc electrode was constructed inhouse from 316 stainless steel rod.

A series of experiments was conducted using the above-described cell which leads to the following conclusions.

[1] The anodic oxidation of Fe^{2+} to Fe^{3+} does take place on a carbon electrode at a mass transport limited rate; the oxidation region is about 0.80 V in extent; the potential plateau for this process is shifted to more positive values by increasing the Fe^{3+} concentration; this shift ($<$0.2 V) is not sufficient to cause problems in an operational system.

[2] Excessive oxidation (high positive potentials) at the carbon electrode will inhibit somewhat the Fe^{2+} oxidation; this will not cause a problem in a practical reactor, since such high positive potentials will not be reached.

[3] Hydrogen evolution occurs on a polished 316 stainless steel cathode (in deaerated 0.01, 0.02 M H_2SO_4) at potentials more negative than -0.5 V (SCE); these currents are not diffusion-limited but are controlled by a slower electrochemical kinetic step.

[4] The cathode in 0.02 M H_2SO_4 is passivated by an oxide film at potentials more positive than -0.04 V; this film is removed only at potentials more negative than -0.90 V, the normal operating region of the cathode.

[5] The diffusion-limited back reduction of Fe^{3+} to Fe^{2+} occurs at potentials more negative than -0.15 V.

[6] Soluble ferric iron is reduced to the metal on a 316 stainless steel cathode in 0.02 M H_2SO_4 at potentials $\leqslant$-1.0 V; large currents from H_2 evolution can be supported before iron deposition occurs.

[7] A highly selective membrane of polyolefin-based material is available which will prevent the penetration of Fe^{3+} species while still passing Fe^{2+}; the membrane has acceptable ohmic and structural characteristics.

[8] Smooth carbon electrodes show a slow degradation as the result of poisoning by soluble species; the degradation is enhanced in AMD water prepared from waste coal; the degradation has not been detected with porous carbon electrodes.

Pilot Plant Studies

The economics of any large-scale electrolytic process is dependent upon the capital costs of the electrolysis cells. Obviously, for a given set of electrode materials, the smaller the required electrode area, the less expensive will be the system. For the total process being considered here, i.e.,

$$Fe^{2+} + H^+ + \text{electrical power} \longrightarrow Fe^{3+} + \tfrac{1}{2}H_2$$

the overall area of the electrodes is determined by the rate of diffusion of Fe^{2+} to the anode-electrolyte interface. To a first approximation, this diffusion rate is proportional to the concentration of Fe^{2+} in the bulk electrolyte and to the diffusion coefficient of the hydrated Fe^{2+} ions. The rate is inversely related to the thickness of liquid boundary adjacent to the electrode surface.

The diffusion coefficient is an inherent property of the AMD solution and is not amenable to external control. The concentration of Fe^{2+} is fixed by the specific AMD water being treated and by the desired conversion factor. Thus the only variable available for experimental manipulation is the distance Fe^{2+} must diffuse in order to react on the electrode surface. (In this treatment, mass transport mechanisms such as convection are considered as techniques for reducing the diffusion boundary layer and are essentially specified by the reactor configuration.)

Three reactor configurations (simple parallel electrodes, fluidized bed electrodes, and packed bed electrodes) were evaluated in the context of minimizing this diffusion boundary layer and hence the eventual size of an electrolytic cell. Specifically, the investigation of each configuration involved the reduction of experimentally measured reactor performance to an analytical equation. Then, choosing a typical AMD concentration and an acceptable conversion factor, these equations were used to compare the reactor types via a minimum geometric anode area. The derivation of the equations for each reactor configuration and the computer programs developed are given in PB 208 820.

All the experimental engineering investigations were performed in a small pilot plant constructed entirely of glass and 316 stainless steel. The electrolytic reactor was formed inside a 4-inch glass pipe lined with a ½-inch thick graphite anode; a smaller column was available for preliminary studies of fluidization phenomena. The system was designed so that variations in reactor length and electrode geometry could be made without excessive downtime.

At 100 amp (the current capability of the power supply) the maximum possible AMD flow (0.02 M Fe^{2+}) for 100% conversion is 48.5 gal/hr. A high-speed Teflon gear pump, together with a variable bypass stream, provided a smooth, nonpulsating flow from a 200-gal polyethylene tank. Flow through the reactors was controlled and measured by valved rotameters below the columns. All experiments were performed using a chemical AMD solution containing either 0.01 or 0.02 M Fe^{2+} and 0.02 M H^+. The stock solution was prepared and stored in the

200-gal tank incorporated in the system. The principal system variables were flow rate and current. Iron (total and Fe^{2+}) was determined via standard techniques. A periodic check was made on the stock solution to determine if any Fe^{2+} oxidation (from dissolved air) was occurring spontaneously; no Fe^{3+} was found.

For all runs, the Fe^{2+} conversion was determined after the effluent had reached a constant concentration. Although it was desired to measure the rate of hydrogen production by means of a wet test meter, it was found that the gas production rate (at these low currents) was too far below the range of the instrument to permit reliable readings.

Annular Flow Reactor: The basic annular flow reactor contained a 3-inch diameter carbon anode and a 2.875-inch stainless steel cathode. The effective column length was 4 feet. The cathode was wrapped with a single layer (0.010 inch) of porous polyethylene felt in order to prevent shorting to the close-spaced anode. Table 4.4 presents the data obtained for a typical test run. The flow rate was 1.7 gal/hr. The current was varied stepwise and the indicated final Fe^{2+} concentration was measured. Cell voltages are not presented since they represent mostly iR losses in the current leads and are thus specific to this single experimental setup.

TABLE 4.4: CONVERSION DATA FOR FLOW OF 1.7 GALLONS/HOUR

Current, Amperes	Final Fe^{2+} Concentrations, M	Conversion, %	Current Efficiency, %
0.5	0.0174	12.5	100
0.75	0.0162	19	100
1.00	0.0153	24	93
2.00	0.0136	31	60
3.00	0.0130	35	49

Source: PB 208 820

The increased ferrous oxidation at the higher currents (1 ampere and above) results from the oxidation of Fe^{2+} by oxygen evolved at the carbon anode. Note that, although higher Fe^{2+} conversions are now possible, the process is relatively inefficient.

These data were then compared to the results predicted by applying a suitable mass transfer equation to the experimental reactor configuration. A correspondence in results would establish the validity of the mass transfer equation as used for the system under study.

A computer program was used to provide analytical confirmation of the experimental data obtained from the pilot plant. The computed concentration profile for a flow rate of 1.7 gal/hr predicted a final concentration of 0.0162 M which agreed with the maximum observed conversion at 100% current efficiency. Further confirmation of the applicability of the analytic equations used was obtained by extending the data to include a wide range of flow rates. These results are shown in Table 4.5. Here again, the agreement between the experimental and analytical values is excellent.

TABLE 4.5: FINAL Fe^{2+} CONCENTRATIONS AS A FUNCTION OF FLOW RATE

	Fe^{2+} Concentration, M		
		Predicted	
Flow, gal/hr	Observed	No Back Reduction	Back Reduction
5	0.0184	0.018	0.018
2.5	0.0173	0.0168	0.0170
1.7	0.016	0.0160	0.0162
1.0	0.016	0.015	0.0154
0.5	0.0132	0.012	0.013
0.26	0.010	0.0092	0.0093
0.0	0.007	0.00	0.0071

Source: PB 208 820

The effect of the competing back reaction (the reduction of Fe^{3+} on the cathode) is apparent. The predicted maximum conversions indicate that this back reduction of Fe^{3+} does not become of major significance until the total conversion reaches 50%. Since conversions in excess of 90% are desired, it is apparent that a means of suppressing the competing back reduction is required. As is shown in Table 4.6, where the computer predicted concentration profiles for a flow of 0.1 gal/hr are compared, conversions of 75% or more can be obtained in relatively short reactors operated at low flow rates once the ferric to ferrous back reduction is suppressed.

Consideration of several alternative approaches to eliminate this reactor inefficiency led to an evaluation of ion selective membrane materials. A polyolefin-based material produced by RAI Industries was found to fulfill the basic requirements of stability in dilute sulfuric acid solutions, low ohmic resistance (i.e., permeability to solvated protons and anions), reasonably low cost, and a low permeability for hydrated ferric ions.

TABLE 4.6: CONCENTRATION PROFILES FOR FLOW OF 0.1 GALLON/HOUR

Distance, cm	No Back Reduction Concentration, M	Back Reduction Concentration, M
10	0.0154	0.0157
20	0.0130	0.0137
30	0.0113	0.0124
40	0.0099	0.0114
50	0.0088	0.0107
60	0.0079	0.0102
70	0.0072	0.0097
80	0.0065	0.0093
90	0.0059	0.0090
100	0.0054	0.0088
110	0.0050	0.0086
120	0.0045	0.0084

Source: PB 208 820

Fluidized Beds: In a fluidized bed, the continuous agitation of glass beads suspended in the interelectrode space by the AMD flow reduces the thickness of the diffusion boundary layer and thus increases the limiting current density. A further advantage of a fluidized system lies in elimination of the hydrodynamic boundary layer which is present in a simple flow system.

The experimental configuration was similar to that described above. The carbon anode was 2 feet long and had an internal diameter of 3 inches. The stainless steel cathodes were 0.84 and 1.90 inches in diameter. Since wall effects on fluidization velocity are not important, these variations in annular size permit the fluidization of a wider range of bead sizes than could be possible if the reactor geometry were fixed. All experiments shown were performed with commercial grade glass beads. The particle size given thus represents an average value.

Note that the bead size and the liquid velocity cannot be independently varied, since each size bead requires a specific liquid velocity for fluidization. The fluid velocity at 65% bed voidage ranges from 0.05 cm/sec for 35 μ beads to 2 cm per second for 375 μ beads. A graph of fluid velocity (at a measured 65% bed voidage) versus particle diameter shows that the observed velocity varies as the average particle size to the 1.54 power.

Table 4.7 presents the experimental data for the treatment of chemical AMD (1,000 mg/l Fe^{2+}, 2,000 mg/l H^+). The actual current passed through the reactor was consistent with the theoretical current calculated on the basis of the observed ferrous to ferric conversion. For each experimental point, the reactor was operated at constant current until the effluent Fe^{2+} concentration reached a steady value.

The quantity of glass beads in the bed was adjusted so that at the required bed height (2 feet) bed voidage would be 65%. This value has been shown to result in optimum mass transfer rates. Note that in the case of the larger (and hence more mobile) beads, conversions are now approximately a factor of 20 times greater than a simple annular reactor of equal anode area.

TABLE 4.7: FLUIDIZED BED DATA

Flow Rate, gal/hr	Bead Size, μ	Current, amp	Final Fe^{2+} Concentration, M	Conversion, percent
34.3	375	9	0.017*	12
20	200	5	0.0176**	12
1.2	35	1	0.0112**	44

*(initial 0.0193)
**(initial 0.02)

Source: PB 208 820

Since in these experiments no membrane was used around the cathode, the back reduction of ferric ion must be considered in evaluating the effectiveness of the apparatus. In these experiments, conversions approaching the desired 95 to 99% were not obtained. According to the subsequent analysis, this was primarily because the reactor length was too short for the flow rates being employed.

It was shown subsequently that in the case of the 35 μ beads, a column 320 cm long would have been needed to produce a Fe^{2+} conversion above 95%. The alternative of using bead sizes below 35 μ was not feasible due to entrainment of the very fine particles.

Packed Bed Reactor: In this configuration, the entire anode space was filled with porous, conductive particles in direct contact with the electrode wall (the feeder electrode). The intent was to extend the electroactive surface into the bulk of solution thus minimizing the average diffusion required by the ferrous ion. The exact distance over which the electrode potential was effective remained to be determined. It is to be expected that the total exposed electroactive surface would be greater than in either of the previous configurations considered.

The anode bed was formed from 4 x 10 mesh carbon granules within a 3-inch-diameter carbon cylinder. Bed heights of 1 foot and 2.5 feet were used. The stainless steel cathodes (perforated in order to allow the generated hydrogen bubbles to leave the reactor without obstructing electrode area) were centrally located along the length of the carbon bed. Cathode diameters of 0.83 and 1.90 inches were used in order to investigate the effect of bed width. For the data shown, the cathodes were wrapped in a single layer of a ferric impermeable membrane.

For each experimental point, the column was operated at constant flow rate and current until a steady effluent concentration was obtained. The Fe^{2+} stock solution (0.01 M Fe^{2+}, 0.02 M H_2SO_4) was stored in a 200-gal polyethylene tank. The time for each point varied from 5 hr at 1 gal/hr to less than 30 minutes at 20 gal/hr. Iron concentrations were determined via standard titration techniques.

Figure 4.3 illustrates the basic operating data obtained on packed bed reactors. The results are plotted in terms of the observed ferrous to ferric conversions versus the applied current (expressed as a percentage of the theoretical current required for complete conversion). Data are given for flow rates of 1, 2, 5, 10, 15, and 20 gal/hr. The data in Figure 4.3a were obtained at a bed length of 1 foot and a bed width of 1.58 inches. Figure 4.3b represents runs carried out at the same electrode spacing but with a bed length of 2.5 feet. The effect of changes in bed width from 1.58 inches to 0.55 inch is shown in Figure 4.3c for a bed length of 1 foot.

Below a critical mass transfer limitation for each flow rate, the steady state concentrations fall on the 100% efficiency line. When this point is exceeded, most of the excess applied current is consumed in the anodic generation of oxygen. Some of this oxygen produced will oxidize additional Fe^{2+}, but this process is relatively inefficient as can be deduced from the low slope of the operating lines deviating from the 100% efficiency area. Table 4.8 compares the mass transfer limitations observed with the 1.08 and 0.55-inch bed width. These data were obtained from Figures 4.3a and 4.3c.

Since slightly higher conversions are exhibited by the bed which contains less carbon, it must be presumed that much of the carbon contained in the wider bed is electrochemically inactive. The width of this active carbon zone determines the maximum electrode separation and hence the total area of the stainless steel required in the final reactor.

FIGURE 4.3: PACKED BED DATA

a.

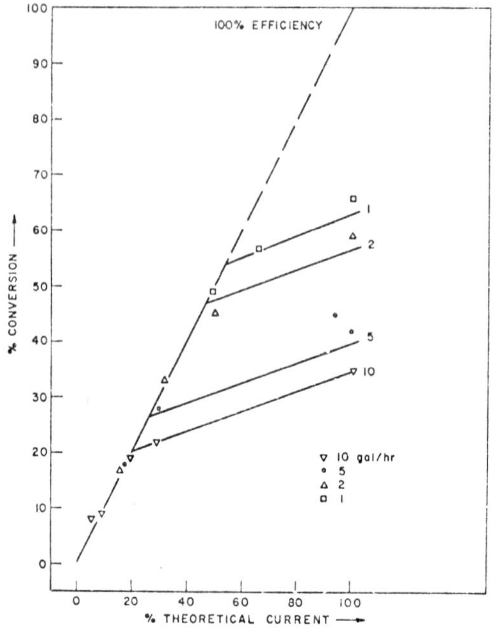

b.

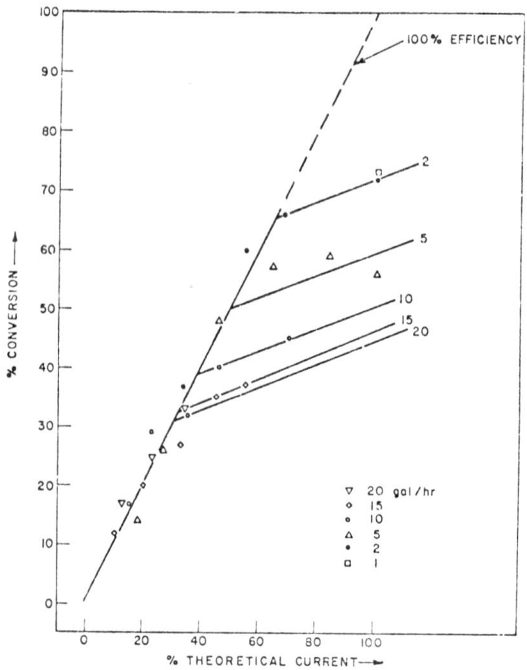

(continued)

FIGURE 4.3: (continued)

c.

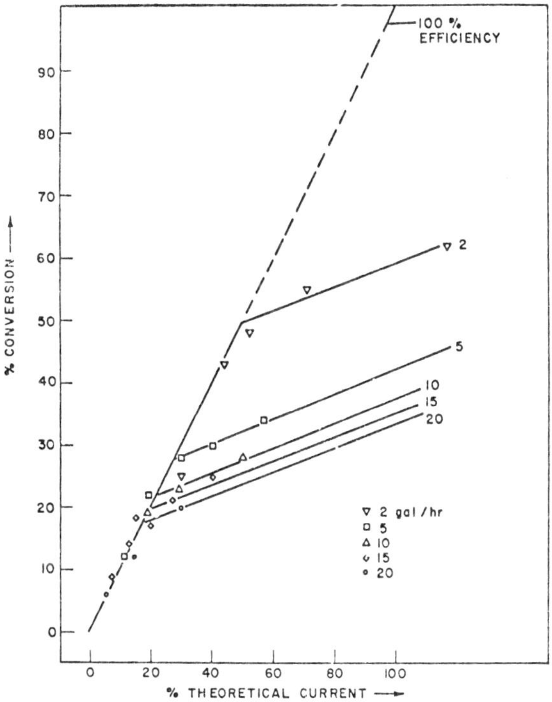

TABLE 4.8: EFFECT ON COLUMN PERFORMANCE OF CHANGES IN BED WIDTH

Flow Rate, gal/hr	Observed Conversion, % 0.55-Inch Width	1.08-Inch Width
2	50	47
5	28	27
10	22	20
15	20	-
20	18	-

Source: PB 208 820

At a given AMD flow rate, the actual velocity in the bed containing less carbon is higher because of the reduced flow area. Therefore the higher conversions shown in Table 4.8 are to be expected as a result of increased local turbulence in the fluid surrounding each carbon particle. The next size parameter is bed length. Table 4.9 compares the data obtained from the 1-foot (Figure 4.3a), and 2.5-foot (Figure 4.3b) columns. The data represent the maximum conversion obtained at 100% current efficiency, i.e., the mass transport limitation of the columns at each flow rate.

TABLE 4.9: EFFECT OF BED LENGTH ON COLUMN PERFORMANCE

Flow Rate, gal/hr	- - - Observed Conversions, %- - - 1-Foot Bed	2.5-Foot Bed	Exponent
2	47	66	2
5	27	50	2.5
10	20	39	2.5
15	-	33	-
20	-	31	-

Source: PB 208 820

If, as expected, the diffusion limiting current for ferrous oxidation is proportional to the ferrous concentration, the effect of bed length on the overall conversion should be given simply by: $(1 - c)^{\ell} = F$ where c = conversion in a column of unit length, ℓ = number of unit lengths in the column, and F = final conversion observed. Computed values (obtained by plotting the observed conversions versus bed length on log-log coordinates) for the exponent ℓ were given in Table 4.9. Since the values determined experimentally are close to 2.5 (the ratio of the column lengths) the above equation should be applicable in predicting the performance of columns of arbitrary length.

When the mass transport limited current was plotted against flow rate for the region where the overall conversion was relatively constant using data from Tables 4.8 and 4.9, a slope of 0.67 was indicated. For liquid flow through packed beds, standard correlations predict a slope of 0.69. The excellent agreement obtained allows justifiable scaleup to beds operating at higher superficial velocities.

Comparison of Reactor Types: A summary of the pilot-plant data for each of the three reactor configurations considered is shown in Table 4.10. All the reactors had the same basic inside diameter of 3 inches. Since the data points chosen are at currents where significant conversion via oxygen production was not occurring, the currents passed represent the absolute amount of Fe^{2+} that was removed.

TABLE 4.10: COMPARISON OF REACTOR CONFIGURATIONS

Reactor Type	Initial Fe^{2+} Concentration, M	Flow Rate, gal/hr	Conversion, percent	Reactor Length, feet	Current Passed, amp
Flowthrough	0.02	5	8	4	1
Fluidized bed	0.02	34	12	2	9
Packed bed	0.01	10	20	1	2

Source: PB 208 820

It is difficult to compare the three systems directly from these data since the concentrations, flow rates, and reactor sizes are different. However, it is evident that the simple flowthrough reactor cannot compare with the other configurations. In comparing the fluidized bed with the packed bed, it must be noted that the fluidized bed current includes a factor of 2 for the increased Fe^{2+} concentration, a factor of 1.5 ($3.4^{0.4}$) for the increased flow rate, as well as factor of 2 for the reactor length.

A more effective comparison of the three systems can be given in terms of the anode electrode area needed to handle a practical fluid flow. For this purpose, treatment to 95% conversion of the following AMD water was considered: 1,000 mg/l of ferrous iron, 0.01 M H_2SO_4 at 6,000 gal/hr. The electrode area was computed from the experimentally based extrapolation equations. These comparison data are given in Table 4.11.

TABLE 4.11: COMPARISON OF ELECTRODE SYSTEMS VIA ANODE SIZE

System	Area, ft^2
Parallel plate	50×10^3
Fluidized bed	6.2×10^3
	3.25×10^3
Packed bed	$1.74 \times 10^{3*}$

*Area of stainless steel cathode.

Source: PB 208 820

A few more details regarding Table 4.11 are desirable. The value for the parallel plate system is based on a minimum partial electrode spacing of 0.005 inch since the closer the electrodes, the more efficient the reactor. It is assumed that a ferric iron impermeable membrane can be used; it is also assumed that there will be no problem in feeding solution into the electrolyte gap. Any departure from these assumptions will increase the size of the reactor.

The value of 3.25×10^3 square feet for the fluidized bed is based on being able to fluidize 1-mm glass beads. The bead size is felt to be the upper practical limit. At 65% bed voidage, the Reynolds number is 100 and the coefficient of performance is 0.6.

The design point for the packed bed reactor was the observed conversion of 17.5% at 20 gal/hr. The electrode area figure in Table 4.11 is for a reactor consisting of two series-connected vertical tanks, each 8 feet high by 3 feet. Each tank is partitioned into 35 flow channels, 1-inch wide and treating 170 gal/hr of AMD. Clearly then, of the three systems, the packed bed reactor requires the smallest amount of electrode area and should therefore be the least expensive.

Conclusions: Evaluation of the fluidized bed, packed bed, and annular flow electrochemical reactors lead to the following conclusions: Of the three the packed bed electrode gave rise to the smallest and hence most economical configuration. An evaluation of process economics, using a packed bed reactor and limestone to precipitate the iron indicated total treatment costs in the range of \$0.11 to \$0.72/1,000 gal, compared to alternate process costs of \$0.20 to \$2.00/1,000 gal. Additional cost savings are possible from the sale of electrolytic hydrogen. Because of transportation and collection costs, these savings would be realized when handling large volumes of highly polluted AMD water. A detailed cost analysis is given in PB 208 820 which is dated 1972. As compared with alternative oxidation methods, the electrochemical process is free of the safety hazards associated with radioisotope-induced oxidation and should as well be free of the temperature sensitivity associated with biological oxidation methods.

ELECTROCHEMICAL OXIDATION–ECOTROL, INC.

The Ecotrol, Inc. study covers the use of electrolytic cells containing a semiconductive bed of particles between the electrodes. The bed particles, while not reacting chemically with the contaminants in solution, behave as small intermediate electrodes in the electric field between the major electrodes. That is, ends of the particles facing the major cathode exhibit anodic characteristics, and ends facing the major anode exhibit cathodic characteristics. The particles themselves become bipolar. The result is a cell with a multiplicity of anodic and cathodic sites.

The cell can also be modified to provide a bed with predominantly anodic or cathodic character. When the solution to be treated is passed through this medium, the distance contaminant ions have to migrate to any one of the numerous electrode sites is drastically reduced. This feature overcomes the excessive residence time required by conventional cells in the treatment of dilute waste water. Since the particles are conductive, and because of their proximity in the cell, abnormally high voltages are not needed to oxidize or reduce trace quantities of dissolved impurities.

While initial development of the Ecotrol cell was directed towards applications in the metal finishing industry, exploratory tests with synthetic AMD containing up to 500 ppm ferrous iron showed that the iron could be oxidized at an acceptable rate in the cell. In addition, it was found that in AMD with a pH of 3 or higher, the ferrous iron was converted to a relatively dense red solid that appeared to be readily filterable. The laboratory work was followed by a field study on the treatment of pH 2 and pH 5 AMD. The field study was conducted at the Delaware Pumping Station, Wilkes Barre, Pennsylvania.

Phase I–Laboratory Studies

The purpose of this part of the project was to examine the effects of cell configuration, electrode bed material, power, AMD composition, and temperature on the degree of treatment in bench-scale cells in order to gain some idea of the range of operational capabilities which had to be incorporated in the design of the Phase II system. While consideration was given to wide variations in AMD composition, some emphasis was placed on those conditions which were to be encountered at the mine site chosen for the field demonstration.

Laboratory Cells and Associated Equipment: Cells were designed in two different configurations. The Ecotrol Electrochemical System in its basic form consists of a series of cells having rectangular plane-parallel anodes and cathodes with the semiconductive bed of particles dispersed in between. From earlier studies cylindrical cells with concentric anodes and cathodes were found to give more favorable results than the rectangular type in certain applications. Although the cylindrical configuration was chosen for AMD study several Phase I studies were conducted using the rectangular cell.

It should be noted that no attempt was made to effect 100% conversion of Fe(II) or reduction in total iron (Fe_T) concentration during the Phase I studies, as the cells used were too small to provide the required residence time. The main objective was to cause enough change in the concentration of these species by various cell and operating adjustments so that the effects of these adjustments could be clearly distinguished and the data applied to the design of the Phase II system.

Screening Runs with Rectangular Cell: Initial experiments consisted of evaluating the effects of power and the addition of a PVC screen diaphragm to the cell. The results of these initial runs indicated that for efficient Fe(II) conversion the Ecotrol cell should be modified with the inclusion of a diaphragm against the major cathode to increase the anodic character of the bed and that means must be provided for periodic purging of the bed to remove nonconductive solids precipitated during the treatment of AMD at pH 3 or higher.

Evaluation of Electrode Bed Media: The materials listed in Table 4.12 were evaluated as static beds in a rectangular cell.

TABLE 4.12: EVALUATION OF ELECTRODE BED MATERIALS

Bed	Volts	Amps	Percent Reduction in Concentration of Fe(II)	Fe_T
Norit RB-II	6.4	3.0	32	18
Graphite 9H	16.0	3.0	23	11
Pittsburgh BPL	8.8	3.0	18	12
Witco 337	10.9	3.0	17	12
Graphite 6B	8.0	3.0	17	8
Nuchar WV-W	9.7	3.0	15	10
Columbia LCK	11.4	3.0	13	9
Anthracite	26.0	0.25	10	10
Witco 256	8.4	3.0	10	6
Bituminous	25.0	0.20	6	5

Source: PB 232 764

The results of the evaluation are also shown in Table 4.12. The applied potential, current and percentage figures are averages of values measured on an hourly basis after the cell attained equilibrium in regard to Fe(II) and Fe_T removal. The pH values of the effluents were generally the same as the feed (3.0) or slightly less (2.8). Solids were generated in every run, the most voluminous being observed when anthracite and bituminous coals were used.

The bed materials are listed in the order of decreasing effectiveness for AMD treatment. Apparently the high surface area and porous structure of activated carbon are not essential properties for an effective electrode bed material in view of the results with the smooth, nonporous particles of Graphite 9H. It was impossible to pass a current greater than 0.25 amp through the anthracite and bituminous beds even at an applied potential of 25 volts. Consequently, the degree of Fe(II) oxidation was poor for these systems.

From a consideration of the data in Table 4.12 and costs of the materials, it was concluded that, of the solids tested, Norit RB-II would be the best electrode media for demonstration of the treatment of AMD.

Effects of AMD pH: The effects of varying the pH of synthetic AMD solutions containing Fe(II) in essentially two different initial concentrations was determined. Runs were conducted at pH 2.0, 3.0, 4.0 and 5.3. Fe(II) concentrations were about 500 ppm and 50 to 60 ppm. A rectangular cell with 10.2-cm spacing of the major electrodes (12.7-cm wide by 17.8-cm high) and static anode bed was used. The flow rate was 0.15 liter per minute and running time generally about 12 hours during which samples were taken from the cell exit port and analyzed on an hourly basis.

It was concluded that the effect of pH on the Fe(II) conversion lies mainly in the extent of ferric hydroxide formation. Conductivity differences can be compensated for by appropriate adjustment of the applied potential. While a decrease in oxidation efficiency with increased pH was observed with the high ferrous AMD, the pH effect was negligible for the 50 to 60 ppm solutions since much less solid was deposited during the pH 5 run by virtue of the low iron concentration.

Variation of Fe(II) and Fe(III) Concentrations: The same cell and operating conditions as used for the work on pH effects were applied in a study of Fe(III)/Fe(II) variations. For the first three runs, the concentration of Fe(II) was varied in pH 2 water containing very little Fe(III). To examine the effect of the Fe(III)/Fe(II) ratio, the same pH and Fe(II) concentrations were used for the next three runs. However, Fe_T was kept as close as possible to 500 ppm by the addition of Fe(III) in $Fe_2(SO_4)_3 \cdot xH_2O$.

The last three runs were performed to observe the effect of Fe(II) concentration in AMD with pH 5. Attempts were not made to maintain a Fe_T value of 500 ppm in this series, as the excess Fe(III) would merely have precipitated when added to the pH 5 feeds.

Figure 4.4 contains plots of effluent Fe(II) concentrations versus the corresponding feed values. A comparison of the curves for the pH 2 solutions with and without initial Fe(III) addition indicates some degree of back reduction of Fe(III) to Fe(II) in the AMD with high Fe(III)/Fe(II) ratio. Although the electrode bed was made predominantly anodic, the probability for Fe(III) reduction at the surface of the major cathode would be higher in solutions with a high (Fe(III)/Fe(II) ratio than in the high Fe(II) water. Some chemical reduction of Fe(III) by hydrogen generated at the cathode could also occur.

Although a lower conversion of Fe(II) takes place in the high ferrous AMD at pH 5 than at pH 2, the precipitation of Fe(III) from the high pH solution would result in less back reduction to Fe(II). Also, a lower concentration of hydrogen would be produced in the pH 5 water. The difference in results between the high and low pH AMD therefore, approaches zero at low Fe(II) feed concentrations.

Attempts to fit the data to simple first and second order rate equations were unsuccessful. However, the plots in Fiugre 4.4 can be used as working curves to determine the minimum length of cell needed to effect a desired degree of

Fe(II) oxidation. Consider, for example, the pH 2 solution with 500 ppm Fe(II) and no Fe(III). According to the corresponding curve in Figure 4.4, as this solution is passed through the unit cell used in this case, the effluent will have a Fe(II) concentration of 305 ppm. The concentration of the effluent that would be obtained by flowing the 305 ppm solution through another unit length of cell can then be approximated by reading from the curve the effluent value corresponding to a 305 ppm feed. This procedure is continued until the required cell length is determined.

FIGURE 4.4: EFFLUENT CONCENTRATION vs FEED CONCENTRATION OF Fe(II)

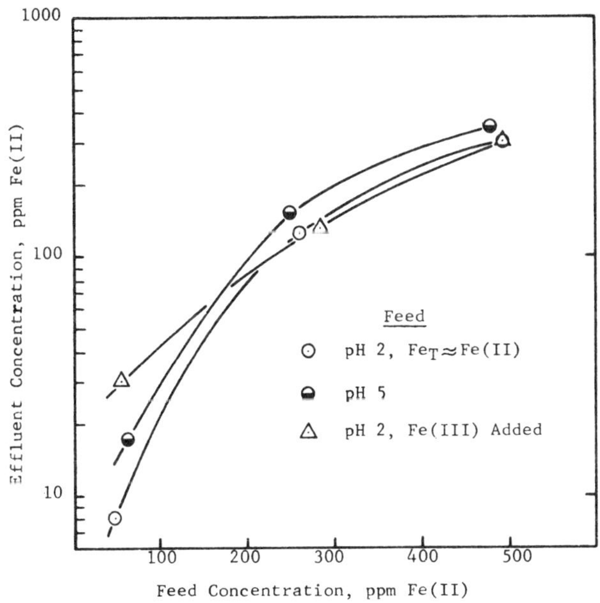

Source: PB 232 764

Treatment of AMD Containing Other Metals Besides Iron: A series of runs were conducted with synthetic AMD containing aluminum, calcium, magnesium and manganese. It was of interest to determine the effect of electrolysis on these metals and the effect of these metals on the degree of Fe(II) oxidation. Cells and operating conditions were the same as used for the pH and iron concentration studies.

As for the effect of electrolytic treatment on Mn(II) in the electrode bed cell, only a small percent reduction in this metal's concentration was observed. In regard to an effect of Mn(II) on the conversion of Fe(II), the data show a reduction of about 25% in the fraction of initial Fe(II) oxidized in the presence of

50 or 90 ppm Mn(II). These results present an anomaly as one might expect the opposite effect when known redox potentials are considered; i.e., oxidation of Mn(II) to MnO_4^- which in turn would oxidize Fe(II). No explanation can be offered at this time.

It was concluded from this series of runs that, in view of the small deleterious effect of manganese on the Fe(II) conversion, the treatment of actual AMD containing 50 ppm or more of this metal would require a somewhat longer residence time than drainage waters having little or no manganese. The results of the runs with calcium, magnesium and aluminum indicate essentially no effect on the degree of Fe(II) oxidation and a small percentage removal of these metal ions in the high pH AMD.

Effect of AMD Temperature: All of the tests previously described were conducted with the feed solution at room temperature. In anticipation of the lower temperatures which would be encountered at the field site, runs were carried out with feeds kept at 1.1° and 8.8°C (deep well temperature). Synthetic AMD with a pH of 5 and Fe(II) concentration of approximately 400 ppm was treated in the same cell as used for pH and iron concentration studies. At a constant cell current, the degree of AMD treatment was approximately the same in each case.

The only difference among the three runs is the voltage which had to be applied to maintain a current of 3 amps. As was expected, the necessary applied voltage increased with decreasing temperature. A plot of the cell voltage versus temperature of the feed yielded an average temperature coefficient of -0.12 volt per degree which indicated that temperature variations to be encountered in the field could be compensated for by small adjustments in applied voltage.

Solids Generated by Electrolysis: Throughout the course of the laboratory work, it was repeatedly observed that during electrolysis of synthetic AMD at a pH of 3 or greater, a small fraction of the total solids generated in the cell consisted of what appeared to be a dark reddish-brown crystalline material that settled at a rapid rate compared to the usual gelatinous $Fe(OH)_3$ precipitate. Some effort was directed in Phase I to a closer examination of the properties of this solid and of the sludge formed after lime neutralization of the electrolyzed mine water. Characteristics of the electrolysis solids were compared with those observed for AMD treated by aeration.

The solids generated in the cell settled more rapidly than those formed during aeration of the AMD. Although the difference was not as great for the lime-treated waters, the settling rate of the solid in the electrolyzed effluent still appeared to be higher than that for the aerated AMD.

Final sludge volume and density data for the sludges present before lime addition concur with the difference in settling rates. The opposite is true for the solids after lime addition. However, the beneficial effect of the small amount of crystalline solid in the effluent right after electrolysis might have been diminished by dilution with excess $Fe(OH)_3$ formed during the neutralization. Also, it is possible that upon standing over a long period in neutral or high pH water, the crystalline solid formed by electrolysis undergoes transition to an amorphous state.

Cell Evaluation and Sizing: Rectangular Cells — A series of runs were conducted to evaluate the degree of Fe(II) conversion at various flow rates in a cell with constrained bed and with the bed free to expand at relatively high feed velocities.

An increase in efficiency derived from the operation with a slightly expanded bed was exemplified by comparing results from this system with data obtained using a compact bed. Although a larger bed volume and higher potential were applied in the former case, approximately the same percent conversion of Fe(II) was attained at one-third the current drawn by the compact bed and at about 60 times the mass loading.

Circular Cells — For the evaluation of the cylindrical configuration, measurements were made first with a small cell containing a static packed bed of Norit carbon. Because of the difference in surface areas of the concentric major electrodes, the main purpose of the runs with the small cell was to determine any difference in power characteristics and degree of AMD treatment between the configuration having inner cathode-outer anode and vice versa. Runs were also carried out with a PVC diaphragm placed against the cathode wall.

During electrolysis in the small circular cells without diaphragm, a comparison of the current-voltage values indicates that the cell behaves to some degree like a diode due to the difference in surface areas of the inner and outer electrodes, i.e., the configuration with the small anode conducts better than that with the small cathode. At the same cell current, the degree of Fe(II) oxidation was less in the cell with the inner anode. In addition, a pronounced odor of hydrogen sulfide was generated from this cell which was not noticed during electrolysis with the small cathode configuration nor during treatment in any of the rectangular cells.

With the screens placed against the cathodes, the same diode effect as described above was observed. The potential drop across the cells with diaphragm is higher than that without diaphragm, since the screen prevents the semiconductive bed from coming in direct contact with the cathode. Thus the current is lower in the diaphragm cell at a given applied voltage. Introduction of the diaphragm to the small cathode cell caused only a slight increase in Fe(II) removal.

With the small anode cell, however, the presence of the screen doubled the percent reduction in Fe(II) concentration. In addition, no sulfide odor was detectable, as was the case without the screen. On the basis of these results, further laboratory studies with circular cells were performed using the inner anode-outlet cathode configuration with diaphragm against the cathode.

As was done with the rectangular cells, a series of runs were conducted to evaluate the degree of Fe(II) conversion at relatively high flow rates in large cylindrical cells containing beds without constraints. Bed expansion was more uniform than in the rectangular cells.

Electrochemical measurements were made with one cell consisting of an outer stainless steel cathode and inner coated titanium anode. A graphite rod was also substituted for the titanium to compare results. A second cell with outer steel cathode and inner graphite anode was also used. Runs were made with AMD at pH 2 and 5. All runs except the first one with pH 2 AMD were conducted with beds free to expand. It appeared that maximum oxidation occurred at a feed flow rate of about 7.2 l/min. At this velocity, the bed was barely expanded. A comparison of these results with the data for the first run indicates that conversion was higher in the compact bed. However, it should be noted that 2.5 times the current was drawn in the latter case. There was no difference in the results with graphite and coated titanium anodes. The first three runs with pH 5 water showed essentially no change in the amount of Fe(II) oxidized as the flow was increased from 6.8 to 9.5 l/min.

Choice of Configuration, Sizing and Materials for Phase II Pilot Cells: It was difficult to make a choice of configurations based solely on the raw data obtained during the Phase I study. For example, for the treatment of pH 5 AMD at approximately the same flow rate in rectangular and circular cells of equal length and with approximately the same major electrode spacing, the percent reduction in Fe(II) concentration in the rectangular cell was twice that observed for the circular cell despite the higher current drawn by the latter and longer residence time in this higher volume cell.

When the same AMD was pumped through the same two cells at a flow rate of 6.8 l/min the difference in performance was not as great as in the first case. The percent reduction in Fe(II) concentration by the circular cell was 3.7 times that in the rectangular system. However, the circular cell drew 3.9 times the current. A consideration of residence time would still let the rectangular cell appear to be somewhat better, but the exact degree of improvement cannot be estimated. For runs with the pH 2 AMD, the reverse was found to be true. The degree of Fe(II) oxidation was slightly higher in the circular cell.

However, although this cell drew 3.2 times the current in the rectangular cell, a consideration of the higher mass loading in the circular system and the higher residence time in the rectangular cell would make the circular configuration with compact bed appear to be superior for treating low pH AMD.

After examining the raw data, calculations were made for scale-up to pilot size (18.9 l/min or 5 gal/min average, 500 ppm initial iron, 10 ppm or lower final iron) to compare dimensions and gain some idea of the design of the Phase II system depending on cell configuration. The total residence time or length of cell required for the desired treatment was determined from working curves such as those in Figure 4.4. Figure 4.5 shows a comparable working curve plotted from the data on treatment of pH 5 AMD with different initial iron concentrations in the large circular cell.

In those cases where the data were not sufficient to permit plotting of actual working curves, an average percent removal of Fe(II) per unit length cell based on available numbers was taken as a guiding factor. After the required total length of cell was determined for each configuration and flow rate employed in the laboratory studies, the dimensions required for a 18.9 l/min (5 gal/min) loading were calculated.

Based on the results of such calculations and a consideration of factors such as availability of components with the desired size, ease of construction, and timing, the circular cell was found to be the preferred configuration. For sizing purposes, the laboratory data on treatment of pH 5 AMD was used, since the rate of Fe(II) oxidation was lower in this solution than in the pH 2 water. From calculations based on the working curve in Figure 4.5 and information on the diameter of commercially available major electrode materials, a basic unit cell was designed which is 45.7 cm (18 inches) long and has a 10.2-cm (4-inch) diameter inner anode enclosed by an outer cathode cylinder with 25.4-cm (10-inch) i.d. (7.6 cm or 3-inch spacing of major electrodes).

FIGURE 4.5: EFFLUENT CONCENTRATION vs FEED CONCENTRATION OF Fe(II)–LARGE CIRCULAR CELL

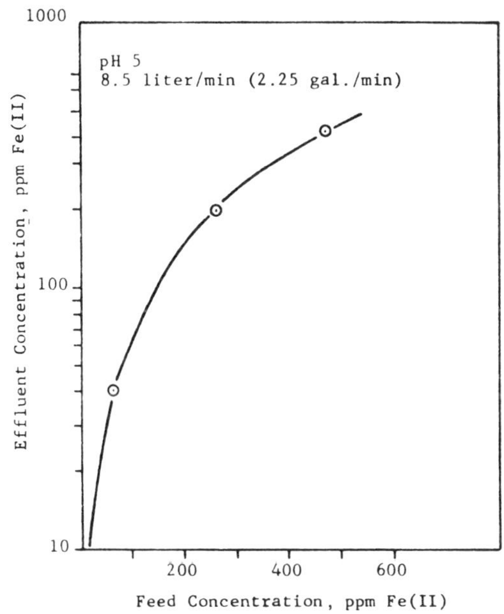

Source: PB 232 764

The total required length of cell was broken down into 45.7-cm (18-inch) segments to provide flexibility in testing and operation. Also, it was thought that advantage could be taken of a known multiple entry effect, whereby a greater net conversion of Fe(II) could occur if the AMD is passed through a number of unit cells than if the solution were passed through a single cell with a length equal to the total of the unit cells. The unit cells possessed means for treating the AMD on a compact and expanded bed and for adding the PVC-screen diaphragm.

As for the choice of material for the major electrodes, it was decided that graphite would be used for both cathode and anode. Since no data were obtained which suggested that coated titanium would yield superior results, graphite was the immediate anode candidate in view of the lower cost of this material. Plans were made to elaborate on the previously discussed polarity reversal tests during initial systems testing in the field. Use of the all graphite cell would enable this to be accomplished. On the basis of the laboratory evaluation, Norit RB-II pellets were chosen as the electrode bed media in the pilot cells. From a comparison of the size of the Phase II cells and the voltage-current values applied to the large circular cell during the laboratory work, a DC power supply providing a maximum of 50 volts and 500 amps was considered to be ample for the field study.

Phase II—Field Studies

Description of Pilot Plant: Figure 4.6 shows a schematic diagram of the Phase II system. Raw AMD was pumped into a 568-liter stainless steel tank for mixing either with 10% sulfuric acid and/or 1.5% ferrous sulfate solution if a change in pH or composition were required. In addition to electrochemical processing, the AMD was treated by aeration in order to develop comparative economics. Aeration runs were carried out in the Operation Yellowboy Van, a mobile pilot plant built for this purpose. For evaluation of electrolytic treatment the AMD was fed to the bank of cells and the effluent subsequently treated with lime in a flash mixer housed in the Yellowboy Van.

FIGURE 4.6: PILOT PLANT FLOWSHEET AND CELL

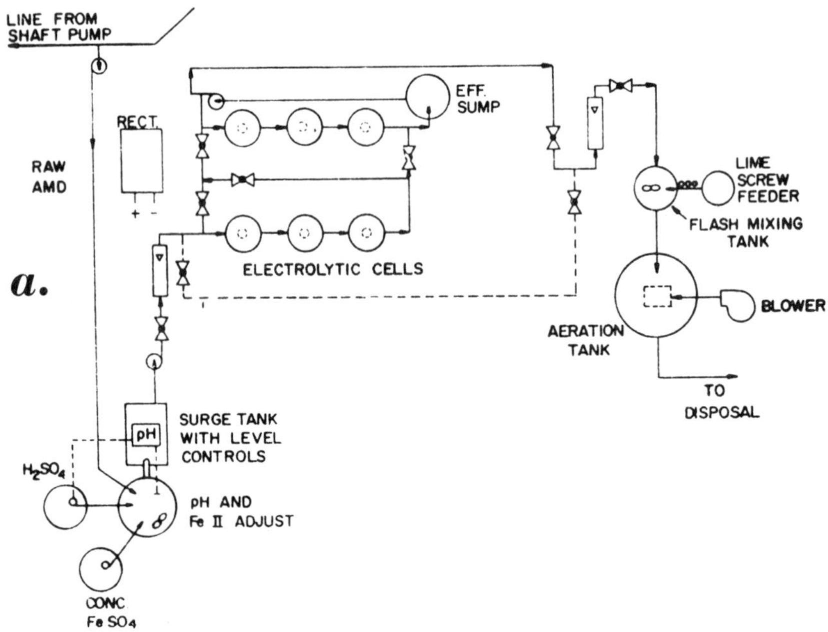

Flow Diagram of Pilot Plant

(continued)

FIGURE 4.6: (continued)

b.

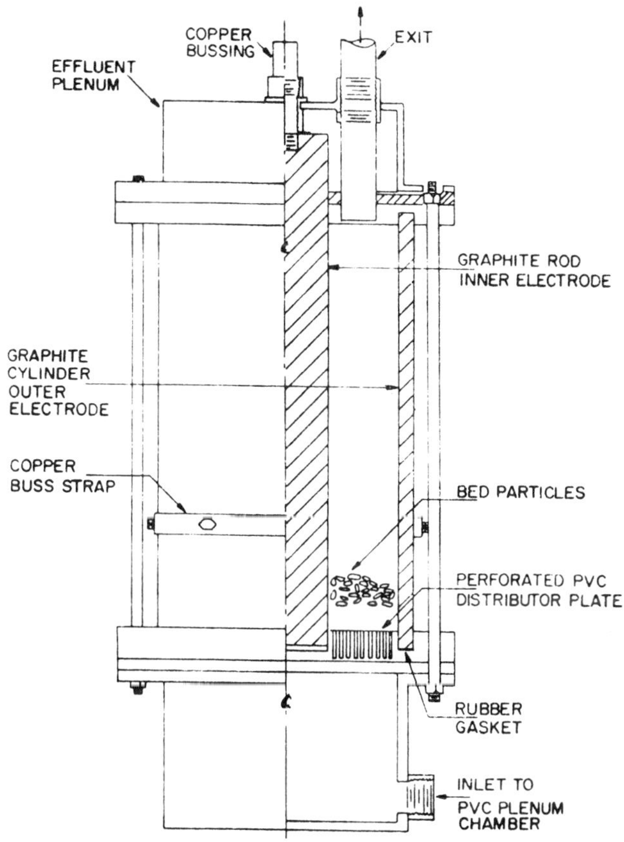

Unit Cell Used in Pilot Plant

Source: PB 232 764

During the aeration runs, AMD feed was pumped directly to the flash mixer for pH adjustment with lime prior to treatment in the aeration tank also housed in the van. Samples for analysis were taken from the overflow of the feed pH adjust tank, the cells effluent sump, and from the overflow of the aeration tank.

Figure 4.6 shows a cutaway view of a unit cell. The outer electrode consists of electrolytic graphite pipe (45.7 cm or 18 inches high, 25.4 cm or 10 inches i.d., 2.54-cm or 1-inch wall) coated on the outside surface with phenolic resin to prevent seepage. A four-inch-diameter graphite rod served as the central electrode. To provide electrical contact, a copper strap was bolted to the outer shell, and a copper rod was threaded into the central electrode. Not shown are the PVC screen diaphragm which was fitted against the cathode and a top restraining screen which was used when runs with a compact bed were conducted. For convenience in initial systems testing, six of the unit cells were connected for series hydraulic operation with the flow upward in each cell. For all runs the cells were connected electrically in parallel. Phase II runs were conducted using the AMD at its natural composition and with Fe(II) concentration adjusted to 250 ppm instead of 500 ppm as originally planned.

Initial Systems Testing – Initial runs were conducted using the 50 ppm Fe(II) mine water with pH adjusted to 2. Feed was pumped through individual cells while observations of percent Fe(II) conversion versus cell current were made. With this system a reevaluation of the effect of major electrode polarity was conducted. The resulting data confirmed observations made during Phase I that conversion is higher when the central graphite electrode is anodic and the PVC screen is placed against the cathode wall.

The next series of runs were performed to further investigate the difference between a compact and expanded bed in the circular cells. For these tests four 3-stage cells were used, two with compact beds and two with sectionalized beds with space provided for bed expansion. Feed with 250 ppm Fe(II) and pH 2 was processed at 3.8 and 18.9 l/min (zero and 1% expansion of the free beds). At the limiting current values the percent conversions of Fe(II) on the compact and loose beds were approximately equal at 3.8 l/min, a flow rate below the threshold value for bed expansion.

For the 18.9 l/min flow the expanded bed effected about 25% higher conversion than the compact system. Also, the optimum conversion at 18.9 l/min with the expanded bed was approximately equal to the maximum conversion at 3.8 l/min. The latter result further emphasized the beneficial effect of bed expansion in view of the higher mass loading with the 18.9 l/min flow.

Finally, the complete system was assembled. In view of the favorable results obtained with the previous runs, provision was made for bed expansion in each of the 18 unit cells. With all six columns in operation tests were conducted with 250 ppm Fe(II) pH 2 AMD at flow rates of 3.8 l/min, 11.4 l/min, 18.9 l/min and 26.5 l/min. The data obtained for the first three flow rates are plotted in Figure 4.7 as percent Fe(II) conversion versus the applied current, expressed as a percentage of the theoretical current required for complete conversion of 250 ppm ferrous iron.

Maximum conversion at 3.8 l/min was slightly higher than at 18.9 l/min as expected based on a consideration of contact time. However, conversion at 11.4 liters per minute was less than at 18.9 l/min. The optimum combination of residence time and degree of bed expansion most likely occurred at 18.9 l/min, the designed loading for this system, since at 26.5 l/min an Fe(II) conversion higher than 10% could not be attained regardless of the power applied.

Although conversion was maximized at 5 as per scale-up, the highest degree attained with the 250 ppm Fe(II) feed was 76%. One possible reason for the lower than expected conversion is the occurrence of back reduction of Fe(III) by hydrogen generated in the pH 2 water. In order to decrease the volume of hydrogen accumulated as the liquid flowed through the cells, vent pipes were installed in the exit line of each cell. Repetition of the 18.9 l/min run with this modification resulted in an increase in percent Fe(II) conversion to 85. This figure represents the highest degree of Fe(II) oxidation attainable by electrolysis alone during this study.

FIGURE 4.7: EFFECT OF FLOW RATE ON Fe(II) CONVERSION

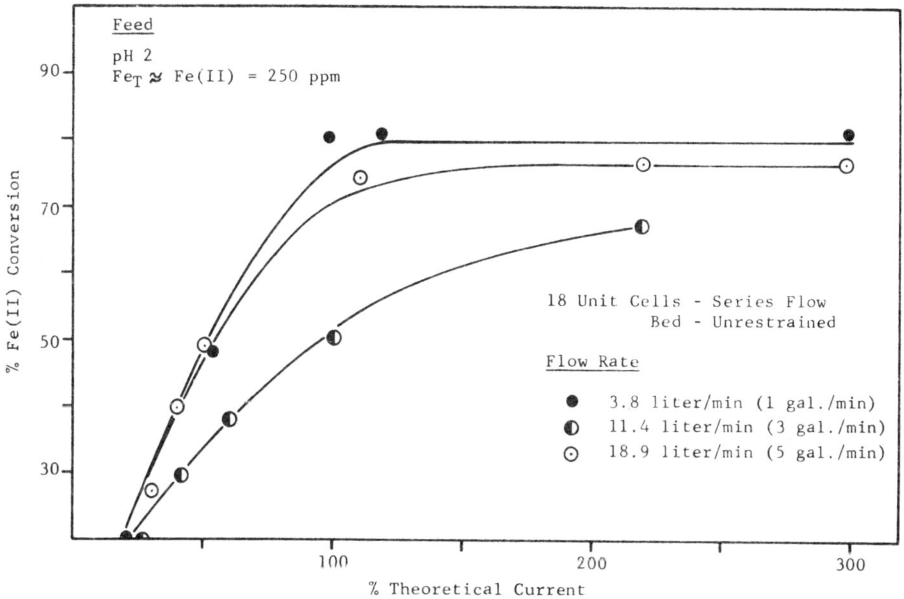

Source: PB 232 764

Final Electrolysis and Aeration Runs: In order to generate cost figures for electrochemical treatment, pH 2 and pH 5 AMD with Fe(II) concentrations of 40 and 250 ppm were processed at the rate of 18.9 l/min. Comparative 18.9 l/min aeration runs were conducted with the high ferrous AMD at pH 2 and 5. The results of electrochemical treatment followed by neutralization are shown in Table 4.13. Table 4.14 contains the data obtained by aeration after neutralization. Except for the Al, Mn, Ca, and Mg values, the numbers represent averages of many measurements taken over a forty-hour period after an equilibrium condition had been reached during electrolysis treatment, and over a twenty-hour period in the case of the aeration series.

Toward the end of each run the feed, cell effluent, neutralization and neutralization-aeration overflows were sampled for Al, Mn, Ca, and Mg analyses. All samples including the neutralization overflows were filtered through medium porosity paper before analysis. Considering first the results of electrolytic processing alone (feed and cell effluent values in Table 4.13), it is seen that for treatment of the low ferrous AMD, the cell system as designed produced an expected effluent containing less than 10 ppm Fe(II).

TABLE 4.13: AMD TREATMENT BY ELECTROLYSIS AND NEUTRALIZATION

AMD*							Cell Effluent*							% Conv.	Neutralization Overflow*						
pH	Fe(II)	Fe(III)	Al	Mn	Ca	Mg	pH	Fe(II)	Fe(III)	Al	Mn	Ca	Mg	Fe(II)	pH	Fe(II)	Fe(III)	Al	Mn	Ca	Mg
(18.9 l/min, 7.5 volts, 220 amps)															(Lime feed rate, 1.3 g/l)						
2.0	38.8	2.2	0.92	4.1	140.0	97.9	2.0	5.0	36.0	0.34	3.9	(300)	79.9	87.5	7.6	0	0	1.12	1.9	500	87.6
(18.9 l/min, 10 volts, 350 amps)															(Lime feed rate, 2.2 g/l)						
2.0	250.0	5.0	1.0	4.2	95.0	85.0	2.0	37.0	218.0	0.54	3.8	(130)	80.0	84.8	7.3	0	0	0.54	1.6	1,000	125.0
(18.9 l/min, 12.5 volts, 350 amps)															(Lime feed rate, 0.38 g/l)						
5.4	35.0	0	0.35	3.5	100.0	94.9	5.4	7.5	27.5	0.20	3.3	(110)	95.0	78.0	7.1	0	0	0.52	1.5	330	98.3
(18.9 l/min, 28 volts, 380 amps)															(Lime feed rate, 0.66 g/l)						
5.4	245.0	10.0	1.0	4.6	100.0	102.0	4.8	110.0	28.0	0.96	4.3	84	104.0	55.2	7.1	0	0	0.52	2.3	240	148.0

*All metal concentrations measured as ppm or mg/l.

TABLE 4.14: AMD TREATMENT BY NEUTRALIZATION AND AERATION

AMD*								Lime Feed	Aerator Blower	Neutralization—Aeration Overflow*						
Flow (l/min)	pH	Fe(II)	Fe(III)	Al	Mn	Ca	Mg	(g/l)	(l/min)	pH	Fe(II)	Fe(III)	Al	Mn	Ca	Mg
18.9	5.2	228	2.0	0.72	4.5	135	89	0.48	453	7.5	3.0	9.0	2.17	1.8	740	102
18.9	1.9	245	5.0	0.44	4.6	140	93	1.0	453	6.2	2.5	0	1.04	3.8	600	104

*All metal concentrations measured as ppm or mg/l.

Source: PB 232 764

The results for the high ferrous waters are short of the design value especially in the case of the pH 5 AMD. The lower conversion in the high pH system is believed to be due to the deleterious coating of ferric hydroxide on the electrode bed. The 18.9 l/min liquid flow apparently was not vigorous enough to aid in expelling the solids from the cells.

After the runs with the pH 5 waters were completed, attempts were made to clean the beds by pumping dilute sulfuric acid through the cells while periodically reversing the polarity of the major electrodes. Analysis of the effluent acid solution showed mainly ferric and very little ferrous iron in the wash. Following this treatment, pH 5 high-ferrous AMD was passed through the cells while no DC potential was applied to the electrodes.

When the effluent pH and Fe(II) concentration equaled the feed values, power was applied. Analysis of the effluent during a few hours' operation at 18.9 l/min showed a significant improvement in Fe(II) conversion over the value indicated by the data in Table 4.13. These observations suggest that for electrolytic treatment of high pH AMD, a periodic bed maintenance procedure as described above might have to be included in the operation of a large-scale plant in order to maintain the required degree of Fe(II) oxidation.

It is difficult to draw definite conclusions regarding the effect of electrolysis on aluminum and manganese in view of the low concentrations of these metals in the raw AMD. However, it appears from the analytical data that reductions in aluminum concentration of 40 to 60% might have been attained except during treatment of pH 5 high-ferrous AMD when, along with the relatively low Fe(II) conversion, the reduction in aluminum value dropped to 4%.

Manganese concentrations were reduced by 5 to 10% in accordance with the laboratory results. No explanation other than analytical error can be offered for the apparent anomaly of higher calcium concentrations in the cell effluent than in the feed. In view of the fact that lime was used for neutralization, this anomaly was not further investigated.

Natural aeration of the cell effluent during lime addition and neutralization resulted in complete oxidation of residual ferrous iron, as this species was not detected in the neutralization overflow (see Table 4.13). Also, essentially all of the iron was removed from the overflow by filtration. Manganese concentrations were further reduced during neutralization. The increase in aluminum values is believed to be due to impurities in the lime used.

Approximately 10 times the stoichiometric amount of air was fed to the mine water during the neutralization-aeration runs (Table 4.14). Both ferrous and ferric iron were detected in the filtered overflow. Manganese concentrations were detected in the filtered overflow. Manganese concentrations were reduced by 50%, while the aluminum values increased with lime addition.

A comparison of the lime feed rates for high ferrous AMD in Tables 4.13 and 4.14 shows that the lime demand for neutralization after electrolysis was higher than for aeration, more so for the pH 2 than for the pH 5 runs. In this regard, it should be noted that two different batches of lime were used for these tests. An older lot which had been stored in the Yellowboy Van from previous projects was depleted during the electrochemical runs. A new batch obtained from the

D.E.R. was used for the aeration studies. It is possible that the older lot had deteriorated on standing, thus the apparent higher lime demand of the electrolyzed water. In addition to this factor, the significantly higher lime uptake of the electrolyzed versus aerated pH 2 AMD most likely occurred as a consequence of pH adjustment of the raw mine water.

This was done essentially by pH titration. Since the required pH of 2 lies on the standard acid-base titration curve in the region where significant changes in volume of added acid have only a small effect on pH, more acid might have been used in adjusting the pH of AMD treated in the cells than added to the mine water treated by aeration, which was done on a different day.

During electrochemical treatment of the pH 5 AMD, the dark red and apparently crystalline solid observed during the laboratory runs was formed with the usual ferric hydroxide precipitate. However, the volume fraction of the desired solid generated in the actual AMD was no greater than found in the laboratory. X-ray patterns of the isolated solid again showed no evidence for a crystalline material.

Characteristics of the sludges formed after lime treatment of the electrolyzed and aerated AMD were compared. Although final volumes and percent solids (9.7) of the settled sludge were approximately the same for each case, there was a definite improvement in sludge settling rate when the AMD was electrolyzed before lime treatment.

Cost Analysis

Based on the data in Tables 4.13 and 4.14, estimates were made of the capital and operating costs for 473 l/min plants treating pH 2 AMD [95% oxidation of Fe(II)] by electrochemical oxidation and by aeration. It is realized that some degree of lime treatment must precede aeration in order to provide the appropriate pH for rapid oxidation of Fe(II). However, in view of the lime demand discrepancy between the pH 2 runs discussed earlier and since costs of lime neutralization for both treatment systems should be approximately the same, these costs were not included in the present analysis.

Further bases for scale-up are that the cells system be constructed as one piece of equipment and that power could be scaled up linearly from the 18.9 l/min data. Tables 4.15 and 4.16 show the capital cost estimates for electrochemical treatment and aeration respectively. Operating costs are compared in Table 4.17. As indicated by these figures, the electrochemical system is more expensive in fixed capital requirements and in operating costs.

The difference in operating costs between the two processes can be lessened if total treatment including neutralization is taken into account. It is known that whereas high-ferric AMD can be satisfactorily neutralized with limestone, neutralization-aeration of high-ferrous drainage requires the addition of the more expensive lime in order to attain the appropriate pH for rapid oxidation of ferrous iron. Limestone could not be used for this project because of the nature of the apparatus in the Yellowboy Van (lime screw feeder).

However, limestone is the preferred material for neutralizing the effluent from the electrochemical cells. Although hydrogen production was not studied during the present project, further cost savings for operation of a large-scale electrochemical plant might ensue from the recovery and sale of by-product hydrogen. The foregoing discussion pertains to the removal of iron from AMD. The effluents generated during total electrochemical-lime treatment of the Delaware Pump water showed a reduction in manganese levels from 4 to 2 ppm, but most of this occurred in the neutralization step.

TABLE 4.15: CAPITAL COST ESTIMATE FOR ELECTROCHEMICAL SYSTEM*

Purchased equipment		
Cells and pump	$32,500	
Installation @ 22%	7,150	
Piping @ 40%	13,000	
Instrumentation @ 5%	1,625	
		$ 54,275
Electrical		
DC power (100 KW installed)	2,600	
Buss bars installed	1,300	
Plant electrical @ 10%	3,250	
		7,150
Buildings @ 35%		11,375
Land and improvements @ 10%		3,200
Carbon bed @ 50¢/lb		5,400
Physical plant cost		**81,400**
Engineering and construction @ 25%		20,350
Direct plant cost		**101,750**
Contractor's Fee @ 8%		8,140
Contingency @ 15%		15,260
	Fixed Capital	**$125,150**

BASIS: pH 2, 250 ppm Fe(II), 95% conversion.
Cells system designed as one piece of equipment.
Linear power scale-up.
Neutralization equipment not included.

*473 l/min, 125 gal/min

Source: PB 232 764

TABLE 4.16: CAPITAL COST ESTIMATE FOR AERATION SYSTEM*

Purchased equipment		
4,000-gal capacity tank	$5,000	
Pump	1,000	
Aerator	1,000	
		$ 7,000
Installation @ 22%		1,540
Piping @ 40%		2,800
Instrumentation @ 5%		350
Electrical @ 10%		700
Building @ 35%		2,450
Land and improvements @ 10%		700

(continued)

TABLE 4.16: (continued)

Physical plant cost		**15,540**
Engineering and construction @ 25%		3,885
Direct plant cost		**19,425**
Contractor's fee @ 8%		1,555
Contingency @ 15%		2,915
	Fixed Capital	**$23,895**

BASIS: pH 2, 250 ppm Fe(II), 95% conversion.
Neutralization equipment not included.

*473 l/min, 125 gal/min

TABLE 4.17: ESTIMATES OF OPERATING COSTS FOR ELECTROCHEMICAL AND AERATION SYSTEMS*

	Electrochemical Oxidation	Aeration
Raw materials (annual bed replacement)	$ 2,000	-
Power	13,480	5,200
Maintenance and supplies @ 5% fixed capital	6,260	1,200
Labor	12,000	12,000
Indirect labor @ 40%	4,800	4,800
Total annual costs	**$38,540**	**$23,200**
¢/1,000 gal	71.4	43.0

BASIS: 95% conversion, pH 2 AMD
24-hour day, 300 days/year.
Labor, 4 men at 40% time to both systems
Neutralization costs not included.

*473 l/min, 125 gal/min

Source: PB 232 764

In conclusion, from scale-up estimates for a 473 l/min loading based on data obtained from the pilot tests, this electrochemical system does not appear to offer a more economical approach to AMD treatment than conventional neutralization-aeration.

AERATION METHODS

Ferrous iron in mine drainage can be oxidized to the ferric form in the presence of oxygen. This oxidation is pH dependent with the reaction proceeding rapidly at pH's above 8.0. In chemical neutralization systems, this pH requirement can be maintained through the addition of a suitable alkali. The oxidation of iron then becomes dependent on the availability of oxygen. The oxidation of ferrous iron occurs through the reaction shown on the following page.

$$4FeSO_4 + O_2 + 10H_2O \longrightarrow 4Fe(OH)_3 + 4H_2SO_4$$

In this reaction, the theoretical requirement for oxidation is one part of oxygen for seven parts of ferrous iron. Oxygen has a low solubility in water. For the oxidation reaction to proceed as quickly as possible, oxygen must be intermixed with the water. Vigorous aeration is the simplest method to accomplish this. It has been found that the oxidation reaction can be accomplished within a 15 to 30-minute period under the proper conditions of pH and excess oxygen.

The aeration of water to accomplish oxidation of ferrous iron is accomplished by either diffused or mechanical aeration equipment. This equipment is usually mounted in tanks with a depth of 3.05 to 4.57 meters. The efficiency of oxygen transfer in the process can be calculated from many factors involved in the design of the unit. Efficiencies of 10% for diffused aeration, and 1.5 kg of oxygen per kilowatt hour are presented as average design factors.

For a diffused aeration system, one kilogram of ferrous iron will require about 5.5 cubic meters of air to accomplish the complete oxidation. For mechanical aeration, a system containing one kilogram (2.2 pounds) per hour of ferrous iron would require a 0.11 kilowatt (0.15 horsepower) unit to accomplish this oxidation.

The capital costs involved in an aeration system consist of the aeration basin, which is usually of earthen or concrete construction, and the mechanical equipment involved; e.g., blowers, diffusers, turbine units, etc. Capital costs are available from a number of sources, but these vary considerably. Operating costs are a function of the power consumption of the equipment and the maintenance required which should be minimal. Operating costs will vary from 10 to 20% of the total plant operating cost.

OZONE OXIDATION

A study has been conducted by the Brookhaven National Laboratory of the U.S. Atomic Energy Commission on the use of ozone to oxidize ferrous iron to the ferric form. Ozone production was considered using electrical discharge, isotopic radiation and chemonuclear methods. Both on-site and central ozone production facilities were considered in preparing cost estimates for comparison to other processes. Following oxidation of ferrous iron by this method, limestone would be used for final neutralization of the acidity present in acid mine drainage.

The study concluded that ozone could be used to oxidize ferrous iron under acidic conditions to the ferric form. The process control is much simpler than with present aeration methods. The electric discharge method of ozone production gave the highest costs for on-site ozone production; however, this method is the only one for which production equipment is presently available.

Cost estimates were presented in the 1970 report of the study for the ozone-limestone system with comparison to a conventional treatment system using lime and forced-air aeration to accomplish the iron oxidation. For a 3,785-cubic-meter-per-day plant, capital costs of $350,000 for the lime-air system and $280,000 for the on-site electrical discharge ozone-limestone facilities were presented. Operating costs were comparable at 4.5 cents per cubic meter treated.

PROPRIETARY PROCESSES

Catalytic Oxidation Process

In the process of *A. Cywin and E.A. Mihok; U.S. Patent 3,617,562; Nov. 2, 1971; assigned to the U.S. Secretary of the Interior* ferrous iron-containing acid wastewaters are neutralized to form a dense, compact, and easily settleable sludge. Ferrous to ferric iron ratios are adjusted prior to neutralization by catalytic oxidation to conform approximately to that of magnetite; $1Fe^{++}$ to $2Fe^{+++}$. Neutralization of the acid waste and precipitation of a mixed valence iron oxide is accomplished using a finely divided limestone slurry as the preferred neutralizing agent. Ferrous iron is oxidized to the ferric state in a reactor which may be of any conventional type suitable for the contacting of a solid catalyst with a liquid and gas stream. Preferably the reactor comprises a contacting tower packed with catalytic material. Granular, activated carbon has been found to be desirable for use as the catalyst.

A stream of oxidizing gas, preferably air, is passed into the reactor along with the acid stream. The air and acid streams may contact the catalyst in either a cocurrent or countercurrent fashion. An oxidized, ferric iron stream is removed from the reactor and is merged with the ferrous iron-containing stream to form a combined stream having a ferric to ferrous ion molar ratio of approximately 2:1. Ratio of ferric to ferrous iron is controlled by proportioning the flow of the two streams by controlling the amount of oxidant introduced into the reactor by controlling contact time of the acid stream and oxidizing gas within the reactor or by any combination of these methods.

The combined stream is then passed into a mixer-contactor where neutralization occurs by reaction with a neutralizing agent. The neutralizing agent is an alkaline material in solution or slurry form. A preferred meutralizing agent is a slurry of very finely divided limestone having a size range such that substantially all particles will pass a 400-mesh screen and having a median particle size of less than 10 microns.

Electrolytic Conversion of AMD to Potable Water

R.W. Treharne and D.E. Wright; U.S. Patent 3,823,081; July 9, 1974; assigned to Kettering Scientific Research, Inc. provides a process of electrolytically converting acid mine water to potable drinking water. As illustrated in Figure 4.8a, the acid mine water enters a pretreatment reservoir where the iron and other metals in the acid mine water may be preoxidized, if desired. This may be done by use of the oxygen evolved in the anode compartment. As shown, the oxygen may be converted to ozone by an ozone generator, the ozone being a better oxidizer for this purpose.

However, this is not an essential step in the process since the oxygen produced or even air may be used as the oxidizer. In any event, the oxygen or ozone can be used to preoxidize the acid mine water to increase the rate of precipitation of iron hydroxide, and thus increases the efficiency of the system. The pretreatment reservoir also serves as a sludge trap to remove as much $Fe(OH)_3$ as possible and to prevent certain large contaminants, such as leaves, sticks, etc., from being pumped into the electrolysis cell. As shown in the flow diagram a pump is used to transfer the acid mine water from the pretreatment reservoir to the cathode compartment and to circulate a part of the water from the cathode compartment back through the pretreatment chamber.

FIGURE 4.8: ELECTROLYTIC CONVERSION OF AMD TO POTABLE WATER

a.

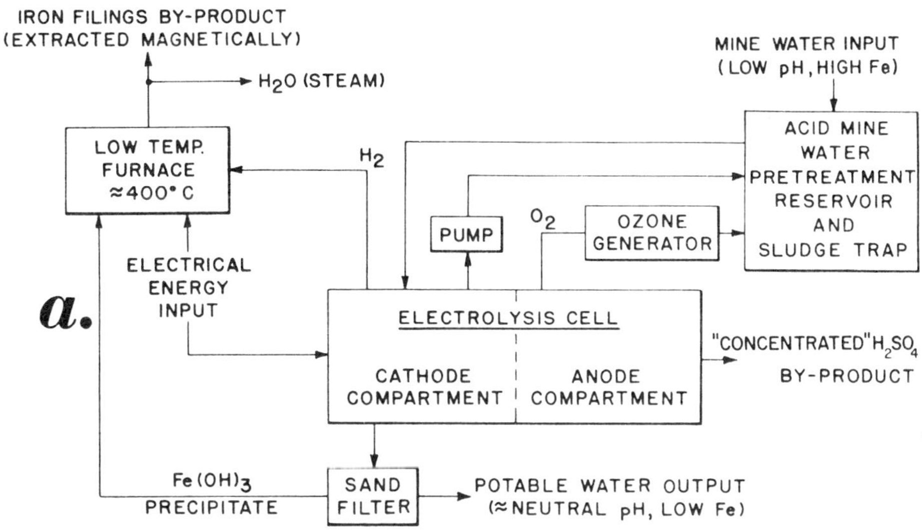

b.

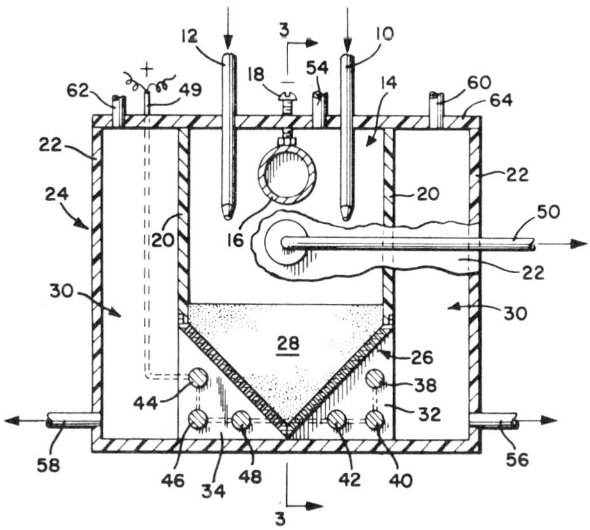

(continued)

FIGURE 4.8: (continued)

c.

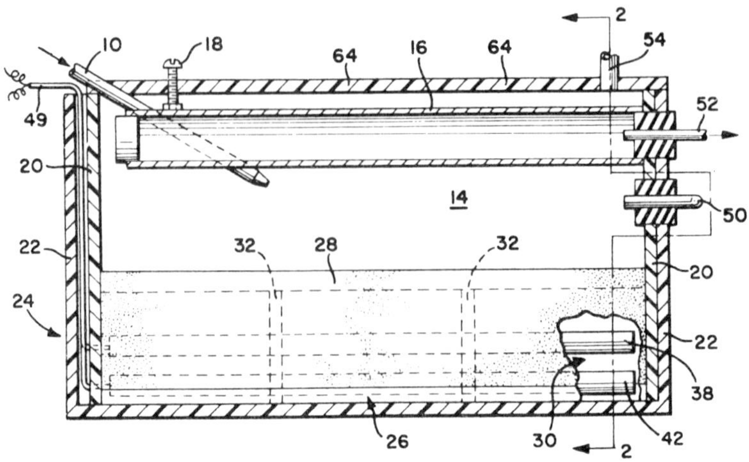

Source: U.S. Patent 3,823,081

Referring to Figures 4.8b and 4.8c, input jets **10** and **12** are used to introduce the acid mine water into cathode compartment **14**. Inside the cathode compartment is a hollow cathode **16** which is attached by means of a bolt **18** to the negative end of a power source. The cathode may be cast iron, stainless steel, carbon, or any similar type cathode material.

The cathode compartment walls **20**, as well as the walls **22** of the electrolysis cell **24**, are made of a nonconducting material such as plastic, glass, or cement. The bottom **26** of the cathode compartment is a nonconductive, porous material or material rendered porous. It may be, for example, plastic or concrete with holes in it, backed by a fiber glass cloth. The bottom supports the sand barrier **28** which separates the cathode compartment from the anode compartment **30**.

The porous support material, forming a V-shaped bottom of the cathode compartment, is itself supported by a series of triangular-shaped supports or posts **32** and **34** on each side of the V. These posts are spaced intermittently along the bottom to permit flow of the liquid within the anode compartment.

The two halves of the anode compartment are, however, partially separated (but not isolated) by the sand barrier and the porous bottom of the cathode compartment. Therefore, it has been found desirable to use two series of anodes **38**, **40**, **42** and **44**, **46** and **48**, one series on each side of the sand barrier. The anodes may be of iron, stainless steel, carbon, etc. Preferably, however, they are made of a high-silicon-content iron which does not corrode as rapidly as the above-mentioned materials. Anodes such as those sold under the trademarks Duriron and Durichlor are particularly effective. The anodes are connected via electrical line **49** to the positive end of a power source.

During electrolysis the acid mine water in the cathode compartment becomes more basic as hydrogen gas is evolved leaving behind an excess of hydroxyl ions. As the pH in the cathode compartment increases a precipitate of mainly iron hydroxide, $Fe(OH)_3$, forms which can be removed by filtering. There is shown a line **50** which is attached to a pump (not shown). A part of the treated water, and especially that part containing $Fe(OH)_3$ precipitate, is pumped back through the pretreatment reservoir where a large percentage of the iron hydroxide precipitate is removed by the sludge trap. The filtered water as well as newly added acid mine water is then pumped back into the cathode compartment for additional treatment.

By this recycling step, water of acceptable purity is produced and released downstream. Such water exits from output drain **52** and is thereafter passed through a sand filter to remove any remaining iron hydroxide or passed through a second, similar type cell for further treatment. After filtering, the exit water is potable. It can be controlled by the instant treatment process to have a pH of 6 to 8 and an iron content of less than 1 ppm, depending partly on the pH.

Since the precipitate contains a large percentage of iron in the form of iron hydroxide, it may be collected at the various filters and further treated in a reducing furnace in the presence of hydrogen at approximately 400°C. This reduces the $Fe(OH)_3$ to high-content iron filings, giving off steam at the same time. The iron filings may then be extracted magnetically and sold as a by-product. Alternatively, iron and other metals can be recovered from the precipitate as electrolytic iron by electrolysis at about pH 1.6 and ambient temperature.

The hydrogen source for the reducing furnace can be supplied by the evolved hydrogen from the cathode compartment **14**. During electrolysis, hydrogen is evolved at the hollow cathode **16** and may be collected from line **54**. It is possible to feed the hydrogen directly into the reducing furnace, or if desired, it may be collected and sold as a by-product.

In the anode compartment **30**, electrolysis drives the pH more acidic as oxygen gas is evolved leaving behind an excess of hydrogen ions. At the same time, sulfate ions are attracted to the anodes increasing the sulfuric acid content in the anode compartment. A decrease in the pH of the anode compartment from 2.6 to 0.6 or lower during electrolysis has been observed.

The sulfuric acid concentrated in the anode compartment **30** is collected through drains **56** and **58**. This acid can be further concentrated by electrolysis or fractional distillation to obtain a marketable by-product from the process. In particular, the reclaimed acid, even in its nonpure state, appears to be applicable as a pickling acid for the iron and steel industry. Reclaimed sulfuric acid from the anode water has an acid content of 24 pounds per 1,000 gallons (pH 1.2), nearly a tenfold increase in concentration over the entering acid mine water.

The oxygen evolved in the anode compartment **30** can be collected through lines **60** and **62**. It is possible to sell the electrolytic oxygen as collected as a by-product. It is also possible to use it, or ozone produced from it, to increase the rate of precipitation of iron hydroxide by feeding it back into the pretreatment reservoir.

REFERENCES

(1) H.G. Glover, Personal Communication, Autotrophic Acidophilic Iron Oxidizing Bacteria, The Pennsylvania State University, April 1969.

(2) H.G. Glover, "The Control of Acid Mine Drainage Pollution by Biochemical Oxidation and Limestone Neutralization Treatment," *Proceedings 22nd Industrial Waste Conference,* Purdue University, pp. 823-827, 1967.

(3) W.W. Leathen, H. Kensel, and S.A. Braley, "*Ferrobacillus ferrooxidans:* A Chemosynthetic Autotrophic Bacterium," *J Bact,* 72:700-4, 1956.

(4) M. Hutchinson, K.I. Johnstone, and D. White, "Taxonomy of the Acidophilic Thiobacilli," *J Gen Microbiology,* 44:373, 1966.

(5) L.B. Whitesell, R.L. Huddleston, and R.C. Alfred, The Microbiological Oxidation Products of Ferrous Iron in Mine Drainage Water, Preprints, Div Fuel Chem, Am Chem Soc, 13:2:79, 1969.

(6) L.B. Whitesell, Jr., R.L. Huddleston, and R.C. Alfred, *The Microbiological Treatment of Acid Mine Drainage Waters,* Continental Oil Company, Report to Environmental Protection Agency, Water Pollution Control Research Series 14010 ENW (1971).

(7) P.C. Singer and W. Stumm, "Kinetics of the Oxidation of Iron," *Second Symposium on Coal Mine Drainage Research*, Preprints, Pittsburgh, pp. 12-34, 1968.

ION EXCHANGE PROCESSES

The material in this chapter is excerpted from EPA-430/9-73-011, PB 227,734 and U.S. Patent 3,725,259 (see Bibliography, p 390).

Ion exchange in water treatment can be defined as the reversible interchange of ions between a solid medium and the aqueous solution. To be effective, the solid ion exchange medium must contain ions of its own, be insoluble in water, and have a porous structure for the free passage of the water molecules. Within the solution and the ion exchange medium, a charge balance or electroneutrality must be maintained; that is, the number of charges, not the number of ions, must stay constant.

Ion exchange materials usually have a preference for multivalent ions, therefore, they tend to exchange their monovalent ions. This reaction can be reversed by increasing the concentration of monovalent ions. Thus, a means exists to regenerate the ion exchange material once its capacity to exchange ions has been depleted.

The earliest ion exchange materials were either natural or synthetic zeolites—mineral produced from mixtures of aluminum salts and silicates. In the 1930s plastic materials called resins were developed which expanded the applications of ion (cation) exchange. In 1949, an anion exchange resin was developed which enabled the process to be used for total demineralization of water.

In the present day technology of ion exchange, the resins available can be classified as strong acid cation, weak acid cation, strong base anion, and weak base anion types. The affinity or selectivity for the various ions each type of resin will remove is given in Table 5.1.

In water treatment applications, whether for softening, demineralization or for specific ion removal, the different ion exchange resins are generally used as follows.

Strong Acid Cation Resins: The sodium form is used for the removal of hardness cations, namely calcium and magnesium. The hydrogen (acid) form is used for the removal of all cations.

TABLE 5.1: TYPICAL ION SELECTIVITY OF ION EXCHANGE RESINS

ION EXCHANGE RESIN	RESIN ION SELECTIVITY DECREASING ORDER OF PREFERENCE
Strong-Acid Cation (Sulfonic Acid Type)	Ba^{++} > Ag^{+} > Pb^{++} > Hg^{++} > Sr^{++} > Ca^{++} > Cu^{+} > Ni^{++} > Cd^{++} > Zn^{++} > Fe^{++} > Mg^{++} > Mn^{++} > K^{+} > Na^{+} > H^{+}
Weak-Acid Cation (Carboxylic Acid Type)	H^{+} > Cu^{++} > Ca^{++} > Mg^{++} > K^{+} > Na^{+}
Strong-Base Anion (Quaternary Ammonium – Type I)	I^{-} > HSO_4^{-} > NO_3^{-} > Br^{-} > CN^{-} > HSO_3^{-} > NO_2^{-} > Cl^{-} > HCO_3^{-} > OH^{-} > F^{-}
Weak-Base Anion	OH^{-} > SO_4^{--} > CrO_4^{--} > NO_3^{-} > PO_4^{---} > I^{-} > Br^{-} > Cl^{-} > F^{-}

Source: EPA-430/9-73-011

Weak Acid Cation Resins: The hydrogen (acid) form is used for the removal of cations associated with alkaline anions. Hardness cations associated with bicarbonate alkalinity are removed, whereas cations associated with chloride or sulfate anions are not.

Strong Base Anion Resins: The strong base form is used for the removal of all anions, although carbonic acid is formed in total demineralization. This can be removed physically by decarbonation.

Weak Base Cation Resins: The weak base form efficiently removes entire strong acid molecules, for example, hydrochloric acid (HCl) or sulfuric acid (H_2SO_4).

ION EXCHANGE METHODS

Combinations of the available resins have been used in systems for treatment of different waters for specific purposes. The applications of these systems to the treatment of acid mine drainage has been studied mainly to produce potable water where a reduction in the total dissolved solids is required. These systems include: (1) the Desal Process; (2) the Sul-biSul Process; (3) the Modified Desal Process; and (4) the Two Resin system.

Sul-biSul Process: The Sul-biSul Process was developed by the Nalco Chemical Company but is now assigned to the Dow Chemical Company. The process employs a two or three bed system. Cations are removed by a strong acid resin in the hydrogen form, or by a combination of weak acid and strong acid resins. A strong base anion resin operates in the sulfate to bisulfate cycle and removes both sulfate and hydrogen (acid) ions during the exchange reaction. Following this, the effluent water is decarbonated to remove carbon dioxide formed in the process. A flow diagram for the Sul-biSul Process when used in a potable water process is shown in Figure 5.1.

FIGURE 5.1: SUL-BISUL CONTINUOUS ION EXCHANGE FLOWSHEET

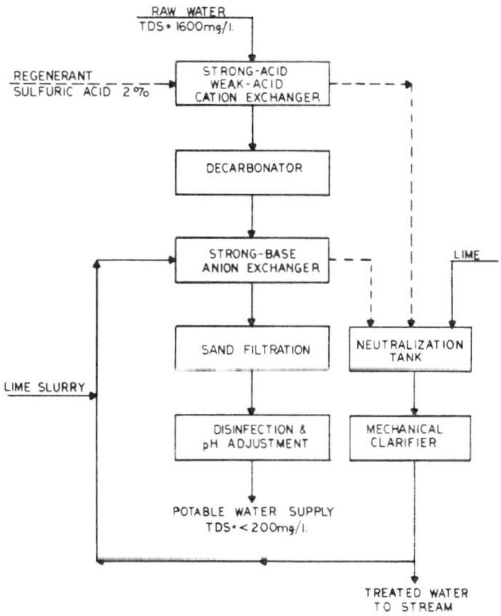

Source: EPA-430/9-73-011

Regeneration of the cation exchange bed is accomplished with either hydrochloric or sulfuric acid. The anion bed regeneration process is unusual; the bisulfate anions are converted back to the sulfate form by the feedwater. The addition of lime slurry to the regenerant will speed this part of the process.

The Sul-biSul Process can be used to demineralize brackish water containing predominantly sulfate anions. This process will be used to treat mine drainage to produce potable water. The process can be applied to waters with a dissolved solids content of up to 3,000 mg/l. The raw water should have an alkalinity content of about 10% of the total anion content with a sulfate to chloride ion ratio of at least 10 to 1. The process is especially suited to alkaline waters containing calcium sulfate such as those contaminated by mine drainage.

Limitations to the process center around the anion exchange resin's low exchange capacity and its inefficient method of regeneration. The exhausted anion resin can be regenerated by the raw water itself; however, this requires a considerable volume of water and takes a significant length of time if the sulfate content is low.

The addition of a cheap alkali such as lime is reported to improve the regeneration; however, a recent study showed poor results. One problem is the requirement for disposal of this large volume of regenerants.

The Sul-biSul Process has been demonstrated as a process that will demineralize brackish waters containing high sulfate concentrations. The process is to be used at Smith Township, Pennsylvania, to produce a potable water from a stream contaminated by mine drainage. The typical raw and finished water quality projected for this plant is given in Table 5.2.

TABLE 5.2: TYPICAL RAW AND FINISHED WATER QUALITY–SUL-BISUL PROCESS

	AVERAGE QUALITY	
Parameter	Raw Water	Finished Water
pH	6.5 - 8.4	8.0
Alkalinity, mg/l	76	10 - 30
Dissolved Solids, mg/l	1500 - 2000	300
Sulfates, mg/l	400 - 1300	50 - 100
Hardness, mg/l	1600	< 150
Chlorides, mg/l	16	2

Source: EPA-430/9-73-011

The Smith Township, Pennsylvania plant was constructed with a capacity to treat 1,893 cubic meters a day (0.5 mgd) at a capital cost of $828,000. Projected operating costs were estimated to be in the range of 10 to 13 cents per cubic meter (40 to 50 cents per 1,000 gallons).

Desal and Modified Desal Processes: The Desal Process employs a three bed system consisting of a weak base cation resin in the bicarbonate form, a weak acid cation resin in the hydrogen form, and another weak base anion resin, but in the free base form. In the first bed, anions are removed and replaced with bicarbonate ions.

Cations are removed in the second unit and replaced by hydrogen ions. In the third bed, carbonic acid is removed by hydroxide ions which converts the resin to the bicarbonate form. In practice, when the system is regenerated, the flow is reversed and the third bed becomes the first.

The Desal Process is ideally suited to saline waters which are alkaline and contain little iron. Most metallic salts are converted into soluble bicarbonates and do not precipitate in the resin beds; however, ferric iron cannot be tolerated. These salts must then be removed by coagulation and sedimentation techniques following this ion exchange process. Regeneration of the beds is very efficient.

The Modified Desal Process is an adaption of the Desal Process for use in treating acid mine drainage. In the Modified Desal Process only the first step of the

Desal Process is employed, that is, the use of a weak base anion resin in the bicarbonate form. This resin effectively removes the sulfate anion as well as any free mineral acidity. The effluent water is then aerated to remove carbon dioxide gas, treated with lime to remove the metallic salts, and filtered as is normally required for producing potable water. Of interest is that the calcium and magnesium are in the bicarbonate form which enables them to be removed as insoluble compounds through the lime softening process. A flow sheet for the Modified Desal Process appears in Figure 5.2.

FIGURE 5.2: MODIFIED DESAL PROCESS FLOW DIAGRAM

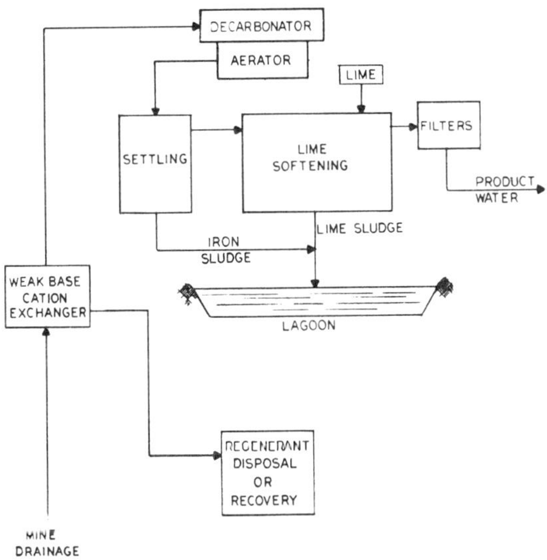

Source: EPA-430/9-73-011

The Desal Process has inherent problems if considered for treating acid mine drainage. The economics of the process lie in the recovery of carbon dioxide for use as a regenerant. The process also requires alkaline waters with little iron content. The process has been successfully used to produce potable water where these conditions were met. In general, the process is not applicable to the treatment of acid mine drainage.

The Modified Desal Process uses the first principle of the Desal Process, the removal of the sulfate ion by a weak base anion resin in the bicarbonate form. This resin also removes free mineral acidity. The presence of iron in the ferric form may present fouling problems through the formation of insoluble precipitates in the anion bed. Product water quality from laboratory tests on the Modified Desal Process have been summarized and are presented in Table 5.3.

Some cost information is available for the Modified Desal Process. These are incomplete estimated costs, although the Commonwealth of Pennsylvania has undertaken the construction of a potable water production facility utilizing this process at Hawk Run near Philipsburg. The Hawk Run plant was designed to treat 1,893 cubic meters per day (0.5 mgd) at a cost of $2,643,000. Operating costs are not available.

TABLE 5.3: TYPICAL WATER QUALITY MODIFIED DESAL PROCESS

Parameter*	Raw Mine Drainage	Weak Base Effluent	Final Effluent
pH	2 - 4	6 - 7	8 - 9
Alkalinity	0	600	25 - 50
Acidity, free	600	0	0
Iron, ferrous	180	100	0
Iron, total	200	130	<0.1
Calcium	180	180	15 - 25
Magnesium	30	30	10 - 20
Manganese	8	8	<0.05
Aluminum	15	5	0
Sulfate	1500	100	75

*All results expressed in mg/l, except pH.

Source: EPA-430/9-73-011

A study by the Culligan International Company provided capital cost estimates for unerected plants as follows.

Capacity, m³/day	Equipment Costs
378.5 (0.1 mgd)	$156,000
1892.5 (0.5 mgd)	$323,000
3785.0 (1.0 mgd)	$465,000

The chemical costs for producing water by this process were estimated at 13 cents per cubic meter (48 cents per 1,000 gallons).

Two Resin System: In a study by the Culligan International Company, a standard two resin system was investigated. In the first step of this system, acid mine drainage is passed through a strong acid cation resin in the hydrogen form for removal of metallic cations.

The water is then passed through a weak base anion resin in the free base (hydroxide) form for removal of the sulfate anions and the free mineral acidity. The demineralized water is then processed through a standard coagulation-filtration process for the production of potable water.

The study cited was a laboratory investigation conducted on synthetic acid mine drainage. The system showed reasonable success although there appeared to be a potential problem with ferrous iron fouling in the cation bed. Hydrochloric acid was found to be a better regenerant than sulfuric acid, but its higher cost (45 to 60%) and problems with the chlorides in the spent solution discounted its use.

The process significantly reduced cations and anions in the two beds, but chemical coagulation and filtration are required to reduce the iron and manganese to levels acceptable for potable use. Based on the chemical costs presented, the Two Resin Process appears to have higher operating costs than either the Modified Desal or Sul-biSul Processes.

The study conducted by Culligan in which this process was proposed presented chemical costs of 17 cents per cubic meter (63 cents per 1,000 gallons). This did not include the chemical costs required for treatment of the spent regenerants. If hydrochloric acid is used to regenerate the cation unit, the chemical costs would be about 21 cents per cubic meter (78 cents per 1,000 gallons).

The Burns and Roe, Inc. Study: The goal of the research described below which was conducted by Burns and Roe, Inc. was the study of economical methods for conversion of acid mine drainage (AMD) wastes to potable water by ion exchange processes. The investigation consisted of laboratory and pilot plant research on the Rohm & Haas Modified Desal, Nalco (Dow) Sul-biSul, and conventional ion exchange processes. The results of the investigation were to be applied to operation of the Hawk Run AMD treatment demonstration plant at Philipsburg, Pennsylvania utilizing the Modified Desal Process.

Two major pilot plant studies of the use of carbonated resins and carbonated feeds in the Modified Desal Process and a bench scale study of the sulfate-bisulfate exchange cycle (the Sul-biSul Process) were carried out. Most of the project effort was directed at obtaining reliable ion exchange run data on different ion exchange processes with different classes of AMD water.

THE MODIFIED DESAL PROCESS–BURNS AND ROE STUDY

The Modified Desal Process is one of the promising ion exchange processes which have been investigated for treatment of AMD wastes to recover potable water. As previously described, the process uses a weak base resin in the free base form which is converted to the bicarbonate form to treat the AMD wastes. The weak base resin exchanges sulfate (or other anions) for bicarbonate, allowing the cations to pass through the bed according to the reaction:

$$MSO_4 + 2R''HCO_3 \longrightarrow R''_2SO_4 + M(HCO_3)_2$$

where R" is the weak base exchange group on the resin matrix and M represents a divalent alkaline earth or transition metal ion. The solution of metal bicarbonates is aerated to oxidize ferrous ion to ferric ion and to sweep out carbon dioxide gas. The effluent is then treated with lime to precipitate metal hydroxides. The solution of metal bicarbonates is aerated to oxidize ferrous ion to ferric ion and to sweep out carbon dioxide gas. The effluent is then treated with lime to precipitate metal hydroxides.

Ammonia is used as the alkaline regenerant to displace sulfate from the exhausted resin. Lime is used to precipitate calcium sulfate from the regeneration wastes and to release the ammonia regenerant for reuse. In this way, ammonia is recycled in the process. It is possible to recover the carbon dioxide and lime used in this process by roasting lime sludge wastes in a kiln.

If this is done, all of the principal chemicals used in the process would be recycled. The net result would approach a zero discharge process with fuel, power, and makeup quantities of chemicals fed into the plant and only potable water, iron hydroxide, and calcium sulfate sludge discharged from the plant.

Experimental

The research on application of the Modified Desal Process to treatment of AMD waters comprises three study areas: the exhaustion process, the regeneration process, and long-term accelerated exhaustion tests.

Exhaustion Studies: Weak anion exchange resins are preconditioned and converted to the bicarbonate form before use with raw feed, or they are allowed to remain in the free base form for use with carbonated feeds. Natural AMD or synthetic AMD mixtures are then pumped through the resin bed at predetermined flow rate and pressure. With the resin in the free base form carbon dioxide is metered directly into the natural or synthetic AMD.

The system is maintained under adequate pressure to assure complete solubility of carbon dioxide in the feed. The effluent stream is periodically sampled to spot test for sulfate breakthrough and to collect samples for effluent analysis. The effluent is collected in one of the large storage tanks and, at the end of the run, a composite sample is taken to characterize the processed water. A schematic flow sheet of a pilot plant for the Modified Desal Process is given in Figure 5.3.

Chemical analysis of the samples is made within the next few days but conductivity, pH, and temperature of samples are determined immediately after collection. The analytical results of each run are plotted, showing change in effluent composition for each parameter as a function of running time. These plots are referred to as breakthrough curves.

The initial or toe breakthrough is the point at which sulfate begins accelerated leaking through the resin bed, and it is identified by extrapolation of the steep portion of the breakthrough curve to zero sulfate in the eluate. The operating time for this point is noted and is used to calculate the volume of water passed through the column at that point for constant flow rate.

The 250 ppm sulfate breakthrough point is identified on the breakthrough curve as the operating time at which the 250 ppm sulfate level is reached in the eluate. These data are reported to assess the useful operating time of the exhaustion run; the initial break signifies onset of sulfate leakage, and the 250 point is the maximum sulfate concentration allowed in drinking water by U.S. Public Health Service recommendation.

The composite sample collected up to the 250 point always contains a much lower sulfate concentration than 250 ppm. However, this point is significant in that it marks the point in the run where good product water is blended with water exceeding the Public Health Service sulfate limit.

FIGURE 5.3: MODIFIED DESAL PROCESS

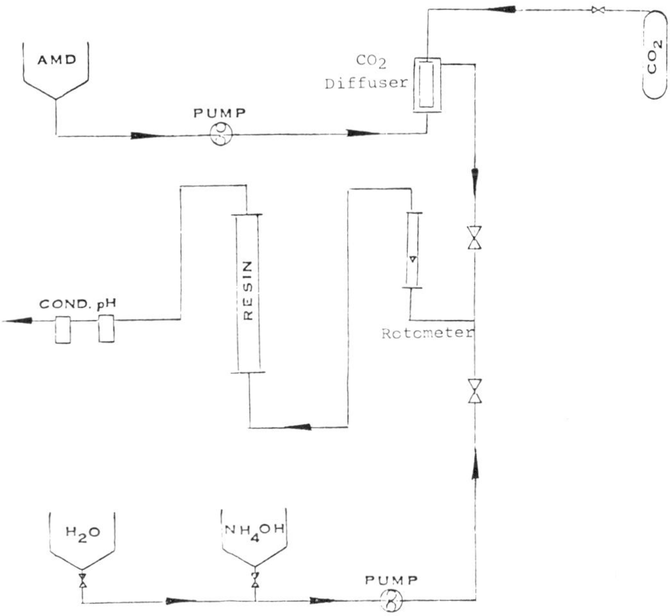

Source: PB 227 734

Amberlite IRA-93 Resin – Flow Rate and Flow Direction: Among the 150 ion exchange runs made with Amberlite IRA-93 resin, most were carried out in the upflow mode. Several were run in the downflow mode to determine if operation in this mode was practical and, if so, how performance would be affected.

These runs were further classified as to whether they were made with bicarbonate or free base resin. Of the upflow runs, thirty-seven were made with bicarbonate form resin, while thirty-one were made with free base resin. In the downflow mode five runs were made with bicarbonate form resin and two were made with free base form resin.

The above runs were made over a broad range of flow rates (from about 0.07 to 0.7 bed volume per minute), and these data were used to determine the effect of feed flow rate on column performance. A number of runs were made comparing upflow and downflow operation in the automatic pilot plant system. These runs supplied additional data on the effect of flow direction on column performance.

Bicarbonate and Free Base Forms of Resin: A total of 103 runs are available to compare the performance of bicarbonate form resin (53 runs) with free base form resin (50 runs). Of the latter, 31 runs, made with sufficient carbon dioxide and at adequate pressure, were selected to give a fair comparison with the bicarbonate form resin runs.

Water Production: A complete water balance was made on three pilot plant runs (7-13, -14 and -15), which considered gross production, as well as the water required for regeneration of the ion exchange columns. The results of these runs were averaged and used as a basis for estimating service water requirements and net production.

Carbon Dioxide Optimization: Thirty-four AMD exhaustion runs were made using Amberlite IRA-93 resin in the free base form. These runs were selected among the 50 carbonated AMD exhaustion runs on the basis of a flow range from 0.20 to 0.30 bed volume per minute. Each run was made under sufficient pressure to dissolve carbon dioxide in AMD. This pressure varied from 20 to 75 pounds per square inch gauge.

The amount of carbon dioxide fed to the influent AMD was closely monitored throughout the run and the amount fed ranged from 0.264 to 10.994 times the equivalent amount of sulfate in the AMD water. The addition of carbon dioxide continued throughout the run or it was terminated at some intermediate point in the exhaustion run.

Other Resins – Nine alternate weakly basic ion exchange resins were evaluated as possible replacements for Amberlite IRA-93 resin. Of these, seven (Rohm and Haas IRA-68, Diamond Shamrock ES-340, Lewatit MP-62, Dow WGR, and Ionac A-305, A-260, and MG-1 resins) were evaluated in the manual system, and three (Rohm and Haas XE-295, Ionac MG-1, and MWA-1) were evaluated in the automatic system.

These resins were used in the free base form, with carbonated AMD feed, under rather fixed operating conditions. A few experiments were also made at different carbon dioxide flow rate levels and pressures. The performances of these alternate resins were compared to Amberlite IRA-93 resin to determine their relative utility in the Modified Desal Process.

Modified Desal Process with Various AMD Feeds – Four classes of AMD wastewaters have been defined. The chemical characteristics of these waters are given in Table 5.4. Representative samples were obtained of Classes 1, 2 and 3, and these were treated using the Modified Desal Process.

Thirty-one runs were selected from those made with Hawk Run water (Class 1 AMD) because they had suitable operating parameters to intercompare them with runs made with waters from other locations.

Four runs were made with Jeddo Tunnel water (Class 2 AMD), one run made with Burgettstown water (Class 3 AMD), and four runs made with limed Altoona water (Class 3 AMD) under similar operating conditions. The suitability and limitations of the Modified Desal Process to treat these other AMD classes were determined, using free base form resin and carbonated AMD feed.

TABLE 5.4: MINE DRAINAGE CLASSES

	Class 1 Acid Discharges	Class 2 Partially Oxidized and/or Neutralized	Class 3 Oxidized and Neutralized and/or Alkaline	Class 4 Neutralized and not Oxidized
pH	2-4.5	3.5-6.6	6.5-8.5	6.5-8.5
Acidity, mg/l ($CaCO_3$)	1,000-15,000	0-1,000	0	0
Ferrous Iron, mg/l	500-10,000	0-500	0	50-1,000
Aluminum, mg/l	0- 2,000	0-20	0	0
Sulfate, mg/l	1,000-20,000	500-10,000	500-10,000	500-10,000

Source: PB 227 734

Ion Exchange Regeneration Experiments: Regeneration studies are carried out on resin beds exhausted in a previous run. In these runs, ammonia solution is passed through the bed, and the concentration of sulfate discharged is determined as a function of regeneration time. The results of these analyses are plotted as sulfate concentration versus regeneration time.

The exhausted resin is regenerated by removing the sulfate with ammonia or other alkali. When the column effluent shows no indication of sulfate, the ammonia step is stopped, and the water rinse begins. After several bed volumes of rinse, the column is treated with water saturated with carbon dioxide gas to recarbonate the resin. If the subsequent run is to be made in the free base form, the regeneration phase is stopped after treatment with aqueous ammonia and several bed volumes of rinse, and the recarbonation is omitted.

The carbonated regenerant is normally discarded after passing through the column at 150 to 200 ml per minute rate in pilot plant work but would be added to the plant production in the demonstration plant. The time required for carbonation varies with the ability of the resin to carbonate and the degree to which the ammonia is rinsed from the resin. The carbonation periods normally vary between one and two hours if the resin is well rinsed, but at times went to five hours.

The ammonia regeneration is usually done on a once-through basis. This procedure was adopted because of its simplicity and because only one ammonia storage tank is available. Later studies showed the once-through regeneration as the preferred mode of regeneration.

Ammonia was recycled in several runs, and the buildup in sulfate level in the regenerant did not at first appear to be detrimental to the regeneration. Therefore, further studies to determine the maximum sulfate level were undertaken to improve the economics of the process, but they showed sulfate in regenerant inhibited resin regeneration.

Ammonia Optimization – A series of eight regeneration runs were made with Amberlite IRA-93 resin to determine the optimum utilization of aqueous ammonia regenerant. The regenerations were made in the downflow mode using accurately measured volumes of known regenerant concentrations. After each limited regeneration, a second regeneration with a large excess of regenerant was made to determine the total amount of sulfate on the resin. These runs were made with quantities of regenerant equivalent to from 0.51 to 2.40 times the stoichiometric quantity of sulfate absorbed on the resin.

A number of runs were made varying the amounts of ammonia regenerant in the automatic pilot plant system. These runs supplied additional data on the optimization of the quantity of ammonia regenerant needed to remove sulfate ion from the resin.

Other Regenerants – Seven runs were made with Amberlite IRA-93 resin in which other chemicals were used in place of ammonia. Two runs were made in the downflow mode using a solution of sodium hydroxide as regenerant. Three runs were made using a solution of sodium carbonate in the downflow mode. Two regeneration runs were made with lime slurry, one in the upflow mode, the second downflow. In each regeneration run an excess of regenerant chemical over the stoichiometric equivalent of sulfate on the resin was used.

Recycled Regenerant – Two series of regeneration runs were made with Amberlite IRA-93 resin using recycled aqueous ammonia regenerant. The regenerations were made in an upflow mode using a measured volume of regenerant solution of known concentration. After each regeneration experiment, the resin was regenerated completely with a large excess of fresh ammonia regenerant to determine the quantity of sulfate remaining on the resin.

In series 1, four regeneration runs were made with recycled regenerant initially containing 1.4% ammonia. In series 2, five regeneration runs were made with recycled regenerant initially containing 3.1% ammonia.

Long Term Test: Automatic pilot plant ion exchange columns were used to carry out long term tests on a number of resins. The purposes of these experiments were to monitor: (1) degradation of resin capacity, (2) changes in resin bead size distribution due to fracture of beads, (3) fouling of resin or equipment with scale or iron oxide precipitates, and (4) other deleterious changes in the resin which might occur as a result of extended service in AMD waters. In addition, these runs were used to provide additional data on regeneration, mode of operation, and bed expansion studies.

More than a year's operation of the Modified Desal Process was simulated using Amberlite IRA-93 resin. Sulfate balances were made on selected cycles of the extended runs, approximately once a week, to evaluate the performance of the ion exchange resin. During this extended test, variations were made in the regeneration conditions for the long term test column.

For example, the concentration or volume of ammonia regenerant was reduced over a number of cycles, to observe the effect of these variables on resin performance. When the regeneration conditions were restored to their normal values, the improvement in performance was documented. After the extended run, chemical analyses were made to determine the amount of iron precipitated on the resin. In addition, sieve analyses were made to assess the extent of resin bead fracture.

Results and Discussion

The numbering system used for the ion exchange runs is as follows. Runs are numbered to indicate which charge of resin is used in the run and its previous history. The first number of the two-part code signifies the charge of resin used. The second number indicates the consecutive run number on that resin. Thus, 3-1 signifies the first experiment with resin charge #3, while 3-2 signifies the second experiment with resin charge #3. Data on the runs can be found in PB 227 734.

Exhaustion Studies: Exhaustion studies were carried out with Amberlite IRA-93 and IRA-68, Ionac MG-1, A-260, and A-305, Duolite ES-340, Dowex WGR and Lewatit MP-62 resins. Some of these resins were used to treat AMD waters of different classes as defined by the Federal Water Pollution Control Administration (FWPCA), and the performances achieved are given.

Amberlite IRA-93 – Effect of Flow Direction and Rate on Utilization of Resin: Carbonated Resin – Plots for a typical carbonated resin run 8-10 are shown in Figure 5.4. The rinse volume for run 8-10 is standardized at 9.12 bed volumes. This rinse volume is used in all figures showing exhaustion runs. When overall utilization for both upflow and downflow trials using carbonated IRA-93 resin are plotted against flow rate the data show little significant difference between the two flow directions.

The average utilization for 37 upflow runs was 36.6% while that for downflow was 35.8%. The upflow data for runs with carbonated resin show a moderate increase in utilization with flow rate over the fairly narrow range of 0.074 to 0.246 bed volume per minute (BV/min), and the downflow data indicate no change over the range of 0.145 to 0.609 BV/min.

A least squares fit of the combined data yields a line which superimposes over the other two lines. This shows that there is a small change in utilization from 0.36 to 0.39 with flow rate change from 0.074 to 0.609 BV/min (equivalent to 0.55 to 4.56 gpm per cubic foot of resin).

It was concluded from these data that since flow direction does not affect utilization, the decision of whether to use upflow or downflow through a resin bed depends on the turbidity of the feed. The flatness of the utilization curve indicates that resin utilization cannot be improved by flow increases although greater production per unit time can be obtained; that is, the carbonated resin system should be run at the highest possible rate to obtain greatest production per unit time.

Carbonated Feed – Plots for a typical carbonated AMD run 7-14 are presented in Figure 5.5.

Effect of Flow Rate – Overall resin utilization in carbonated feed trials shows a moderate decrease with increasing flow rate. When thirty-one trials with carbonated feed are plotted as overall utilization vs flow rate in bed volumes per minute, or specific time, min^{-1}, the resulting line has a negative slope which indicates that overall utilization falls as flow rate increases. In other words, utilization falls as exposure time to the resin decreases.

FIGURE 5.4: BREAKTHROUGH, pH, AND CONDUCTIVITY CURVES FOR A TYPICAL CARBONATED RESIN AMD RUN

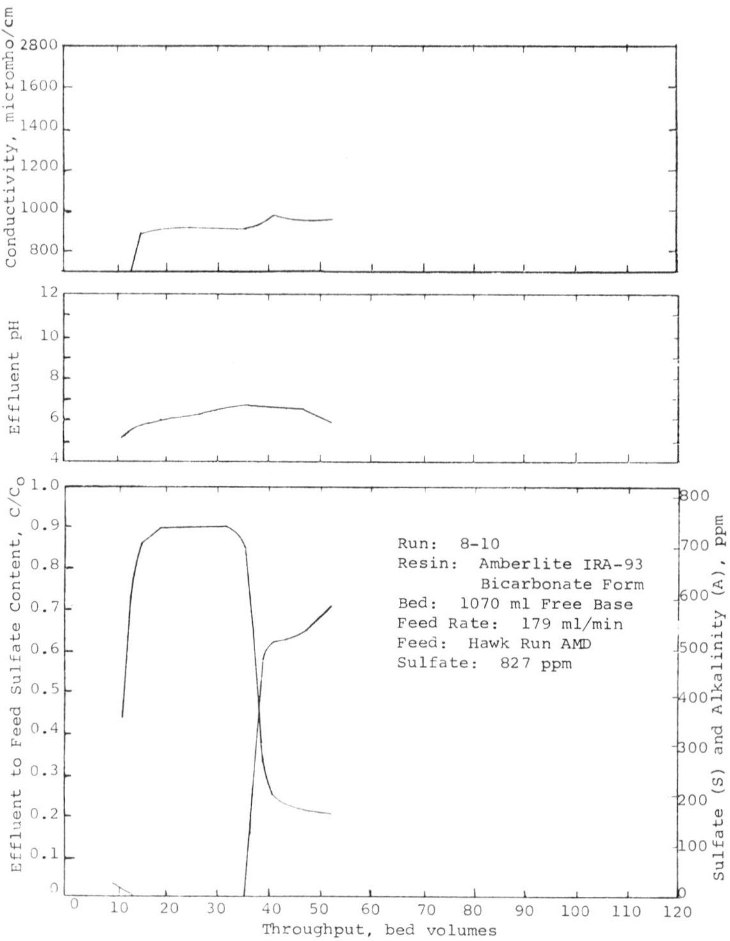

Source: PB 227 734

The normal operating flow for the pilot plant units was in the 0.19 to 0.25 bed volume per minute range which is equivalent to the flux range 2.3 to 3.0 gpm per square foot. A projection of the above line sets maximum utilization at 85%; therefore, the selected flow rate range is a reasonable compromise between utilization and production. The conclusions drawn from the trials with carbonated resin are that flow rate has no significant effect on utilization, and from trials with carbonated feed, that there is a moderate decrease in utilization with flow. The range 0.19 to 0.25 bed volume per minute is optimum.

FIGURE 5.5: BREAKTHROUGH, pH, AND CONDUCTIVITY CURVES FOR A TYPICAL CARBONATED FEED AMD RUN

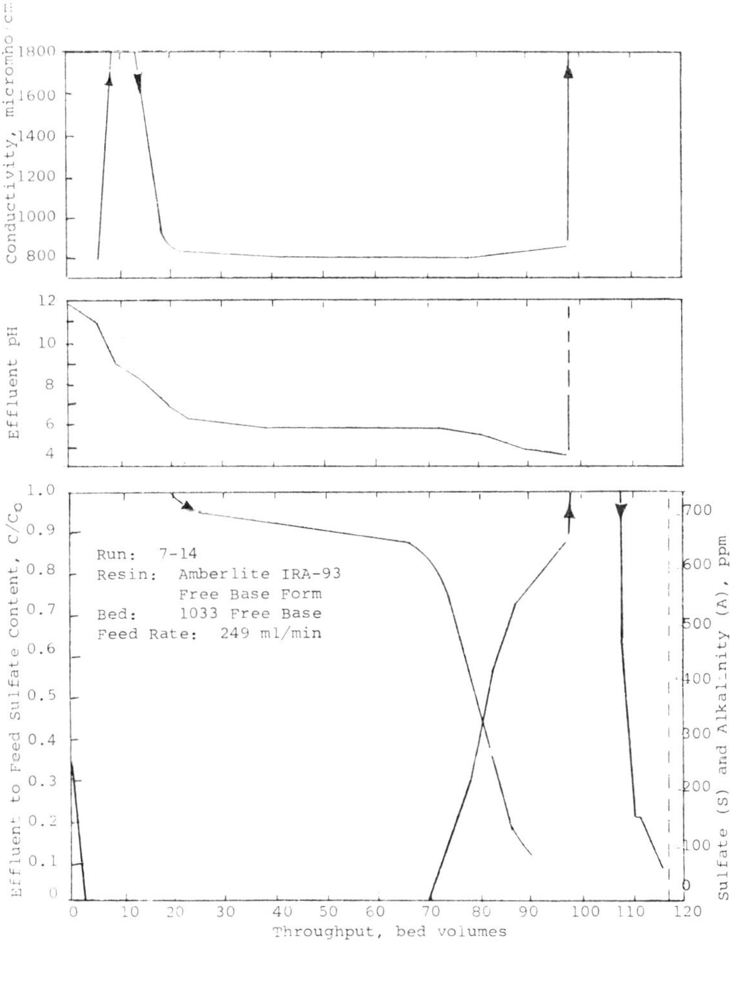

Source: PB 227 734

Effect of Flow Direction – Within the optimum flow rate range, there is no significant difference when exhausting IRA-93 FB with carbonated feed in either the upflow or the downflow modes. This fact is shown by extended trials, made at comparable flow rates on the automatic pilot plant, in which there was no difference in bed capacity, that is, resin utilization, between modes.

Effects of Flow Direction at High Rates – Although there was no positive improvement at the moderate flow rates, two runs were made at very high flows to examine the effects near the flow limits and to determine if there is a significant difference between high flow upward and high flow downward. The results from parallel runs 7-12 and 8-12 show that there was considerable leakage in both runs, little overall difference in resin utilization between these runs, and some operational advantages to downflow.

Pressure losses through the two beds at various times during the runs and their corresponding expansions were determined. The flow rates for these runs were held constant at 0.7 and 0.74 BV/min, respectively, about three times normal.

In upflow, the resin bed expanded by fluidization from its settled height of 23¾ inches and filled the five foot column. The fines and small beads in the bed formed a packed section at the upper retention screen (60 mesh) under the force of the feed flow. This packed section was initially 8 inches high, but during the 120 minute run it grew to 8⅝ inches.

It was concluded that this expansion was due to an increase in bed size, not the collection of additional resin. One unexpected result was that during the run, the pressure drop through this packed section increased (from 0 to 10 psig), while that through the downflow fixed bed remained constant at 1 psig. The pressure drop through the 8 inch bed of fines was greater than that for downflow through a 23 inch fixed bed.

Both runs were made with AMD containing 797 ppm sulfate and with a 3.65 pH; the AMD was heavily carbonated (carbon dioxide rate, 0.6 standard liters per minute) to assure sufficient gas was present to preserve the low pH at the high flows. The carbonation appeared adequate because both runs had high initial alkalinities, which, as usual, decreased as the sulfate leakage increased.

The slope of the breakthrough curve for the upflow trial 7-12 was somewhat flatter than the one for 8-12 and also suggested that the upflow run had sulfate leakage. Analyses of the effluent samples corroborated that these beds had high leakage. The first two samples at 21 and 35 BV showed sulfate contents of 77 and 135 gpm respectively.

For run 7-12 initial breakthrough occurred at 15.6 BV which corresponded to a toe utilization of only 0.186. Since the curve was relatively flat, the effluent took 40 BV to reach the 250 ppm point and gave a reasonable utilization of 0.475. The overall utilization of 0.605 for this run is in line with the toe and 250 ppm values and corroborates them.

For 8-12, the downflow run, the initial breakthrough occurred at 17.7 BV and corresponds to a 0.21 toe utilization. Because the break in this curve was steeper than in 7-12, the effluent reached the 250 ppm point sooner, at 32 BV, which resulted in a lower utilization, 0.381. The overall utilization, however, was the same (0.602) as the other run so, in effect, these two runs were equivalent.

The conclusion drawn from these runs is that there is little difference in ion exchange capacity between the upflow and downflow modes, but that at high rates the downflow mode has mechanical advantages such as low pressure drop and flow stability. Because of the high flow rate (3½ times normal) the utilizations in either mode were only fair but they were still better than the best runs with carbonated resin. High flow rates are more desirable because of their greater productivity from each unit volume of resin.

Comparison of Modes of Exhausting Resin: The Modified Desal Process has two possible modes of exhaustion operation. The usual method (1) is to pass the feed over weakly basic resin in the bicarbonate form. The resin is prepared for exhaustion by regeneration with ammonia or an alkali to convert it to the free base form, followed by a rinse to remove excess base and then a wash with carbon dioxide-saturated water to convert the resin to the bicarbonate form. After carbonation the bed is ready for exhaustion.

In the alternate method, (2), carbon dioxide gas is mixed with the feed under moderate pressure and is passed through a resin bed in the free base form. The bed is prepared by treatment with base to convert it to the free base form and a water rinse to wash out excess base. During this project both methods were tested extensively, and the second, which utilizes the carbonated feed, is about twice as productive and efficient as the first.

One hundred and three exhaustion runs made with IRA-93 were considered for the comparison of the two modes of operation. Of the fifty trials done with carbonated feeds, 31 that were made under comparable conditions were grouped together and yielded a 74.4% average overall utilization; the 53 trials with carbonated resins gave an average utilization of only 36%. In terms of water production, these figures mean that over twice (2.06 times) as much water can be treated by carbonating the feed than the resin.

Water Production: Modified Desal via Carbonated Feed – The Modified Desal Process, when utilizing free base form IRA-93 to treat carbonated AMD, produced a gross product of 637 gal/ft^3, of which 75% (or 480 gallons) is net product. The feed contained 730 ppm sulfate, and the composite product had 110 to 150 ppm in runs designed to maximize production. A higher quality product could have been attained by reducing slightly the throughput; that is, productivity.

Three closely monitored runs, 7-13, 7-14, and 7-15, were made to determine the productivity of a cubic foot of IRA-93 resin (free base form) when treating carbonated AMD. The data show that a cubic foot of resin will treat 627 to 650 gallons of AMD containing 110 to 150 ppm sulfate content.

Run 7-14 was a typical trial, except it underwent two water rinses following the regeneration instead of the usual one. One liter of IRA-93 (free base form) was used to treat Hawk Run AMD which contained 753 ppm sulfate. The feed rate was 0.249 BV/min, and carbon dioxide was added to the feed at 0.3 standard liter per minute (SLPM). Initial breakthrough (toe) occurred at 64.4 BV, which is equivalent to a utilization of 0.72.

The effluent reached the 250 ppm sulfate concentration of 69.8 BV, and this volume corresponds to a 0.783 utilization. The overall utilization calculated from the weight of sulfate retained on the resin for the total run was 0.83, one of the highest utilizations achieved.

The sulfate content of the resulting product composite was 112 ppm. This run was one of the most efficient trials made and is a good example of the capability of the Desal Process on Hawk Run AMD. Increased net water production is possible by optimizing the regeneration and rinse volumes. Reduction from the volumes used in the pilot plant are certainly expected because the 2-inch-diameter Plexiglas columns were designed extra long for experimental purposes and contain a two-bed volume free space above the settled bed. In comparison, the free space in the Hawk Run Demonstration Plant is only one-bed volume, and provision is made not to fill this portion with regenerant or rinse.

The two rinses were used in run 7-14 to remove all traces of regenerant and to allow effective exchange of sulfate at the outset of the run; that is, reduce startup leakage. As a result, the sulfate content of the composite was 112 ppm, whereas for the other two runs which used half the volume of rinse, the sulfate content averaged 143 ppm. Spot samples taken during rinsing of run 7-14 showed virtually no sulfate in the last 2.5 BV of rinse, indicating too much rinse was used.

Comparison of Desal Systems – Comparison of water production data in carbonated AMD run 7-15 and in carbonated resin run 9-3 shows that the former was superior in productivity and in low service water requirements. Comparison of the two systems shows that the carbonated feed system produced almost twice the gross amount of desulfated water (84.9 vs 38.1 BV) and, when the runs are normalized for the difference in feed sulfate concentration, production is twice as great.

In addition, the precarbonation step took 180 minutes of potential exhaustion time to achieve reasonable state of carbonation of the resin. The carbonated resin system, however, is more conservative of carbon dioxide since it uses only one-third of the amount of carbon dioxide and produces 0.58 liter per gram carbon dioxide against 0.44 liter per gram carbon dioxide by carbonated feed. The product quality from the carbonated feed run was better. The composite product from run 7-15 contains only 134 ppm sulfate, while the composite product from 9-3 has 259 ppm.

Carbon Dioxide Optimization: The optimum carbon dioxide usage for treatment of carbonated Hawk Run AMD with the free base form of Amberlite IRA-93 was determined. The optimum amount of carbon dioxide is 1.02 times the sulfate equivalent in the AMD feed, or 2.04 mols carbon dioxide per mol feed sulfate, or 0.935 pound carbon dioxide per pound feed sulfate.

The corresponding overall utilization for this optimum point was 0.765, but the practical operating range is broad. Doubling the carbon dioxide usage from 0.86 to 1.75 only increases utilization from 0.70 to 0.77. Variation of operating pressure showed a small positive effect on utilization. Over the pressure range investigated, from 20 to 75 psig, the utilization showed only a 13% increase.

The carbon dioxide usage was optimized against overall resin utilization in a series of 34 runs in which utilization ranged from 0.43 to 0.84. Three parameters were varied over narrow ranges of operating conditions. The parameters studied are the flow rate (range 0.20 to 0.30 BV/min), the ratio of carbon dioxide to sulfate (range 0.26 to 10.99) and the operating pressure (range 20 to 75 psig).

Utilization was plotted against these three variables, and only the carbon dioxide to sulfate ratio γ shows a significant effect, a sharp break in the curve at 0.75 utilization. The overall utilization vs γ ratio (carbon dioxide to sulfate) curve shows a rapid increase in overall utilization with increasing carbon dioxide up to γ ratio = 1, the point at which the carbon dioxide and the sulfate in solution are stoichiometrically equal.

Beyond this point order of magnitude increases in carbon dioxide yield small increases in utilization. This fact is demonstrated by the flat portion of the curve which has only a 2% increase over a 10-fold change in carbon dioxide concentration. The γ ratio indicates the optimum usage is 2.04 mols carbon dioxide per mol sulfate or 0.935 pound carbon dioxide per pound sulfate. The plot shows some experimental scatter at the knee of the curve, but the range of reasonably attainable utilizations is 0.70 to 0.77 over a γ ratio spread of 0.86 to 1.75.

Effect of Amount of Carbon Dioxide on Sulfate Breakthrough – Insufficient carbon dioxide added to the feed causes sulfate leakage and premature breakthrough of sulfate. The effect of carbon dioxide to sulfate ratio on sulfate breakthrough curves, where effluent to feed sulfate content is plotted against throughput, illustrates the shapes of the sulfate breakthrough that are obtained at various ratios of γ from 0.577 to 10.99.

Run 7-20, with γ = 0.577 shows steadily increasing leakage from the beginning to the end of the run, with no discernable break. The curves of run 7-16, with γ = 1.13, and run 7-19 with γ = 1.66 show progressively lower leakages (38 and 20 ppm, respectively) and more pronounced breakthroughs as the carbon dioxide concentration increases.

The curve of run 7-13 which had a 10-fold excess of carbon dioxide (γ = 10.99), shows only 18 ppm leakage and a definite, rapid and complete breakthrough. All were made under the same conditions using Hawk Run AMD containing 711 to 737 ppm sulfate fed at 250 ml/min (0.24 BV/min).

Limited Carbon Dioxide Input – A series of seven runs was made in which carbon dioxide flow was stopped prematurely. The idea was to determine if the same productivity could be achieved with less carbon dioxide usage. Each of these runs was continued until 85 bed volumes of AMD were processed. In these runs carbon dioxide gas was injected at a rate corresponding to almost two (1.96) times in the stoichiometric equivalent of sulfate in the feed. In each of the runs in this series the carbon dioxide gas flow was stopped at a chosen fraction of the total feed flow to be processed. The fractions of throughput at which carbon dioxide flow was stopped for each run are as follows.

Run Number	Stopped at Fraction of Throughput
8-13	0.37
8-14	0.23
8-15	0.74
8-16	0.74
8-18	0.61
8-19	0.63

The results with these runs were compared to similar runs made with uninterrupted gas flow and (1) excess gas, (2) insufficient gas, and (3) excess gas with high sulfate feed mixtures. A study of these data shows that the earlier the gas flow is interrupted, the poorer the resin utilization achieved. Also, when less than the stoichiometric amount of gas is injected, sulfate breakthrough is premature. Therefore, it is necessary to inject at least the stoichiometric quantity of gas during the run up to the initial sulfate breakthrough point. The data bear out the intuitive feeling that halting the gas injection at this point is economical and that further gas injection beyond this point is wasteful.

Effect of Operating Pressure on Resin Utilization – The effect of operating pressure on resin utilization is small so long as the amount of carbon dioxide is adequate. Eleven runs were made at pressures varying from 20 to 75 psig and the overall utilization changed only 13%, from 0.70 to 0.79. To make the study meaningful, the runs chosen for this comparison had a γ ratio of about 2.

Evaluations of Other Resins for the Desal Process – The performances of weakly basic resins from five manufacturers were compared to IRA-93 and all proved equal or superior in water productivity and quality. The other resins evaluated in short term tests or manual system were: (1) Rohm and Haas IRA-68, (2) Diamond Shamrock ES-340, (3) Lewatit MP-62, (4) Dowex WGR, (5) Ionac A-305, A-260, and MG-1. All runs were made under conditions comparable to a typical IRA-93 run such as 7-14.

Ionic MG-1 showed the best performance, producing 158.5 BV (1,186 gal/ft^3 resin) effluent with an average sulfate content of 74 ppm. It did, however, require more carbon dioxide per equivalent of sulfate feed. Two other resins, Dowex MWA-1 and Rohm and Haas XE-295, were tested on the automatic system. Their overall productivity and product quality indicate that the performance of XE-295 is better than IRA-93, and that of MWA-1 is about equal to IRA-93.

The basis used for comparing these resins was an average of 31 standard runs in which IRA-93 resin was used to treat carbonated Hawk Run AMD. The operating conditions for the averaged runs fell within the flow rate range of 0.2 to 0.3 BV/min with sufficient operating pressure to assure carbon dioxide solubility. The average overall throughput volume of the IRA-93 resin in the standard runs is 82.5 BV (617 gal/ft^3 resin), and the average product composite contained 145 ppm sulfate.

The seven other resins were tested under experimental conditions comparable to IRA-93. They were regenerated with an excess of ammonia to assure complete conversion to the free base form. Productivity of this group when measured at the completion of the run, ranged between 83 BV (621 gal/ft^3 resin) and 158.5 bed volume (1,186 gal/ft^3 resin). The sulfate content for the composite samples of these runs was 74 to 148 ppm.

Ionac MG-1 performed the best, showing an overall productivity of 158.5 BV with average effluent sulfate content of 74 ppm. The better results were obtained at higher carbon dioxide inputs and operating pressures. In contrast to the IRA-93 runs, the productivity of MG-1 increased with carbon dioxide feed rate. For example, with 0.2 SLPM carbon dioxide, 156 bed volumes of effluent were produced, and with 0.4 SLPM carbon dioxide, 165 bed volumes were produced, and with 10.8 SLPM of carbon dioxide, 188.4 bed volumes were produced.

The resins evaluated on the auto system were also run at comparable conditions to the IRA-93 trials, but only overall capacity and effluent volume data were taken (because of the overly long cycles). These resins yielded high throughputs, 135 BV (1,010 gal/ft^3 resin), but they showed varied product quality. Dowex MWA-1 had an effluent with high sulfate, and the effluent from Rohm and Haas XE-295 resin contained low sulfate. Ionac MG-1 did not perform well in this test since insufficient carbon dioxide was used for this high capacity resin. Ionic MG-1 showed a 31% fall-off in retained bed capacity and a 22% drop in sulfate removal over an 800 hour test.

These rapid decreases were caused by ammonia leakage which neutralized the carbon dioxide and inhibited the exhaustion cycle. Therefore, the test was not a valid evaluation of the rate of capacity loss, but it did confirm the high level of capacity, that is, 1.3 to 2.0 meq/ml resin.

It is concluded that some weakly basic ion exchange resins perform better than IRA-93 in the Desal Process on the basis of short-term tests. Of the seven alternate resins evaluated Ionac MG-1 showed twice the productivity with half the sulfate leakage when compared against an average of 31 IRA-93 runs. There are indications that Rohm and Haas XE-295 resin may have performance capabilities as high as MG-1, but this hypothesis is based on only one automatic system test.

Resins with lower productivities such as IRA-68, Diamond Shamrock ES-340 and Lewatit MP-62 have economic merit, since they require less carbon dioxide than MG-1 and still produce 30 to 40% more effluent than IRA-93. The Ionac MG-1 appears to require more carbon dioxide to obtain the higher yields noted in the trials. It appears that at least four times the stoichiometric amount of carbon dioxide was consumed to yield a third more effluent and eleven times as much to give 50% more.

With IRA-93 it was found that the optimum ratio of carbon dioxide to sulfate in the feed is 1.02 and that, when more gas is added, the increase in yield is negligible. MG-1 was evaluated in the upflow mode, and no trials were run downflow. Other considerations of choice of resin are long term stability and resin capacity, resistance to fouling (with iron oxides), and cost.

Modified Desal Process with Various AMD Feeds – The Modified Desal Process with Amberlite IRA-93 resin in the free base form effectively removes sulfate from carbonated acid mine drainage (AMD) in Class 1, Class 2, and Class 3 categories with an average overall utilization of 0.703. The utilization varies from class to class in the range of 0.647 to 0.757 with lower utilizations for Class 3 waters.

Forty AMD runs were completed using IRA-93 resin in the free base form; 31 runs were made with Hawk Run AMD (Class 1) giving an overall utilization of 0.739; 4 runs were made with Jeddo Tunnel AMD (Class 2) giving an average overall utilization of 0.757; 1 run was made with Burgettstown AMD (Class 3) giving an overall utilization of 0.647; and 4 runs were made with limed Altoona AMD (Class 3) giving an average overall utilization of 0.671.

Sixteen runs were made treating aged AMD and synthetic components using IRA-93 resin in the bicarbonate and free base forms. The average overall utilization for the bicarbonate form was 0.332, and for the free base form 0.844. The conclusions drawn from these runs were:

> The Modified Desal Process with IRA-93 resin in the free base form and carbonated AMD feed can effectively treat Class 1, Class 2, or Class 3 AMD to an average utilization of 0.703.
>
> IRA-93 in the bicarbonate form is not as effective as the free base form with carbonated feed.
>
> Class 3 AMD requires a stoichiometric ratio of carbon dioxide to sulfate of about three times greater than that required for comparable Class 1 and 2 waters to obtain 0.70 utilization of resin.

Class 4 AMD was not tested because no water of this type was available. Since Class 4 water is essentially the same as Class 3 except for absence of iron, its behavior during treatment with the Desal Process should be very similar to Class 3 water.

Regeneration Studies: The regeneration studies are divided into three parts: (1) determining the optimum use of ammonia to regenerate exhausted resin; (2) evaluating alternate regenerants as substitutes for ammonia; and (3) recycling spent regenerant to determine its reuse value.

Optimization of Ammonia Regenerant Once-Through – IRA-93 in the sulfate form is effectively regenerated by use of between 105 and 130% of the theoretical ammonia requirement, where 96.7 to 98% of the sulfate on the resin is removed. The concentration of the regenerant in the range 2 to 8% has no effect on completeness of regeneration providing an excess is used. However, at 0.6% ammonia a reduced efficiency was noted. The regenerant efficiency is 92.4% at the 105% level.

The consumption of ammonia regenerant was optimized in a series of eight regeneration trials which showed the optimum point to be at 105% of equivalent amount and 96.7% sulfate removal. Five regenerations (runs 7-17 through 7-21) were made in IRA-93 (sulfate form) with 4.4% ammonia using amounts ranging from 51 to 157% of theoretical and they showed corresponding increases in sulfate removal with increases in ammonia amounts. Removal ranged from 49.9 to 97.8%. The optimum point is identified as the point at which the curve of sulfate removal vs regenerant has an inflection point.

Three trials (runs 8-18, 8-19 and 8-20) were made with 2 and 8% ammonia to determine whether regenerant concentration has an appreciable effect on completeness of regeneration. No significant difference in removal was seen. Runs 8-18 and 8-20 which used excessive amounts (240% equivalent) of 2.1 to 2.3% ammonia, gave 97.2 to 98.9% removal and the one trial (run 8-19) with an excess (190%) of 7.8% ammonia, gave 98.7%.

When the percent sulfate removal is plotted against the percent stoichiometric requirement of ammonia used the plot shows the extrapolated knee of the curve is at 105 and 96.7% sulfate removal and also illustrates a small improvement in sulfate removal after the inflection point is passed.

The regenerant efficiency at the inflection point is 92.4%. At 100% of equivalent requirement, 92.6% of the sulfate is removed; at 130% of theoretical, 97.0% sulfate is removed. Additional regenerant removes little sulfate; for example, increasing the regenerant amount from 130 to 240% only increases removal 1.4%.

Alternate Regenerants – Sodium Carbonate: Sodium carbonate in aqueous solution is effective in changing Amberlite IRA-93 resin from the sulfate (exhausted) form to the free base (regenerated) form and could be used to replace aqueous ammonia as the regenerant. Three regenerations were made using an aqueous solution of sodium carbonate as the regenerant: run 8-25R, run 8-26R, and run 8-27R. In run 8-25R, 96.8% sulfate removal was obtained using sodium carbonate at 2.66 times the equivalent amount of sulfates on the resin. In run 8-26R, 94.7% sulfate removal was obtained using 1.78 times the equivalent amount, and in run 8-27R, 96.1% sulfate removal using 1.77 times the equivalent amount.

Carbon dioxide may be recovered from the regeneration step. During the addition of aqueous sodium carbonate to the bed of Amberlite IRA-93 resin, gas release was noted. Addition of aqueous sodium carbonate at a level of 2.66 times the equivalent amount did not result in improved sulfate removal over lower levels of sodium carbonate. Since the amount of sulfate removed is in agreement with the amount of sulfate retained, essentially complete regeneration is obtained using sodium carbonate at 1.8 times the equivalent amount of sulfate on the resin.

Sodium Hydroxide: Sodium hydroxide in aqueous solution is effective in changing Amberlite IRA-93 resin from the sulfate (exhausted) form and it also can be used to replace aqueous ammonia as the regenerant. Two regenerations were made using an aqueous solution of sodium hydroxide as the regenerant; run 8-21R and 8-24R. In run 8-21R, 95.4% sulfate removal was obtained using sodium hydroxide at 1.50 times the equivalent amount of sulfate on the resin. In run 8-24R, 92.7% sulfate removal was obtained using 1.37 times the equivalent amount.

Sodium hydroxide may be useful as an iron removing agent. During the addition of aqueous sodium hydroxide of approximately 1.6 normality to the bed or resin, it was noted that the iron coating was removed from both the column and resin. This is apparently due to hardening of the iron coating, causing it to flake from the column and resin surface. Similar iron separation was noted with iron coated Amberlite IRA-93 resin stored in aqueous ammonia for several months. Since the amount of sulfate removed is in agreement with the amount of sulfate retained, essentially complete regeneration is obtained using sodium hydroxide at 1.5 times the equivalent amount of sulfate on the resin.

Lime: Lime in aqueous solutions is not efficient in changing Amberlite IRA-93 resin from the sulfate (exhausted) form to the free base (regenerated form. Excessive volumes of regenerant are necessary to complete the regeneration. Two regenerations were made using an aqueous solution of lime as the regenerant: run 8-28R and run 3-14R, downflow resulted in immediate blockage and consequently, no sulfate removal. Run 3-14R, upflow, resulted in 36.2% sulfate removal using lime at 1.72 times the equivalent amount of sulfate on the resin.

The difficulty in using an aqueous solution of lime as a regenerant is that lime is soluble to 1,650 ppm in water. Thus the volume of liquid necessary to allow for efficient exchange is excessive. The formation of calcium sulfate does not aid in further dissolution of lime since calcium sulfate has a solubility of 2,000 parts per million in water at 25°C.

Recycled Ammonia Regenerant – Although several regenerations can be made with one batch of regenerant, recycling regenerant is inefficient. Large excesses of ammonia (two to seven times stoichiometric) must be recycled to achieve 90 to 94% regeneration levels and the regenerations appear to be adversely affected by the sulfate content of the regenerant.

In the tests run, regeneration levels dropped below 90% when the sulfate concentration increased beyond 10,000 ppm. In recycle operations regenerant consumption was about twice the stoichiometric requirement with losses to piping and equipment being considerable. Nine recycled regenerations were made using two different concentrations of ammonia regenerant. There appears to be an effect of regenerant concentration at the levels (1.4 and 3.1% ammonia) investigated. The initial runs in each series, which are comparable since they are equivalent to once-through runs, show only 85% sulfate removal with 1.4% ammonia while 91% was obtained with 3.1%. This data contrasts with the 97 to 99% removal found with the once-through regenerations by 2.2% ammonia. The poorer regenerations in recycle operation are probably due to the law of mass action which says that sulfate removal from the resin is inhibited as its concentration in the solution builds up.

The comparison of the first runs of each series with once-through also suggests that in order to achieve with recycled regenerant the same sulfate removal levels as those in once-through, the initial ammonia concentration must be increased two to four times.

Long Term Tests: The ion exchange pilot plant contains four ion exchange columns which can be operated in a completely automatic mode. A cam timer allows variation of time cycles for exhaustion and regeneration. This timer also operates pumps for AMD feed and regeneration chemicals. Upflow or downflow operation is determined by programming pneumatically operated valves.

All runs were made using free base form resin and carbonated feed. Most of the runs were made in the upflow mode with a single charge of Amberlite IRA-93 resin as part of a long term accelerated evaluation of this resin. A few runs were made with another charge of this resin in the downflow mode.

During the upflow and downflow runs, pressure drop across the ion exchange columns was monitored to determine if resin fouling would interfere with normal operation of the beds. Iron precipitation in the beds was observed visually during the long term tests. At the end of the resin evaluation the extent of iron precipitation was estimated by analysis of acid washes of the beds.

During the first 160 or so cycles of the long term resin evaluation runs, resin fines and iron oxide floc were impinged on the upper containment screen during upflow operation. At the start of the run plugs up to 6 inches thick of floc and fines formed. After about 26 cycles this plug was reduced in size to 4 inches. By about the 160th cycle the plug was gone.

Approximately once a week, sulfate and water balances were made for a complete exhaustion and regeneration cycle. The results of these balances were compared during the course of the long test, and the effects of varying regeneration conditions on resin performance were noted. Regeneration conditions were varied to determine the optimum range of concentration and total ammonia necessary for economical operation.

At the end of the long test, studies were made to determine loss of anion exchange capacity, and changes in the bead size distribution caused by fracture of resin beads. A few runs were made with other resins in shorter tests to evaluate their performance over a number of cycles.

Amberlite IRA-93 Resin – Exhaustion Studies: Upflow vs Downflow Operation – Based on the comparison of about 380 upflow runs and 55 downflow runs in the automatic ion exchange system, it is concluded that there are no significant differences in ion exchange performance between the alternate modes. There were, however, some mechanical difficulties in upflow operations which were not encountered in downflow operation.

Resin fines, along with iron oxide floc, packed against the upper column screens in early runs, impeding flow. In most cases, only an increased pressure drop was noted, but on some occasions complete stoppage of flow was caused. In these instances, a brief reversal of flow direction was made manually which usually corrected the problem.

It was determined that the pressure drops were generated at the upper screen by the resin fines and iron floc, rather than by the expanded resin bed. After about 1,300 hours of operation, the problems associated with blockage of upper screens by resin fines disappeared. At this point, the pressure drop was about one psig, which is equivalent to that observed during downflow operation, and is negligible.

The accumulation of iron floc in the bed used for the long term resin evaluation was documented. In spite of heavy coatings of iron oxide floc on the resin beads (5.8% of the total resin mass), there was no serious loss of resin capacity noted from its presence.

At the end of the long run, the resin was soaked, first in a mixture of one part concentrated hydrochloric acid to four parts water, then in a mixture of one part sulfuric acid to eight parts water, and finally in a fresh mixture of both acids. The rinse solutions were then analyzed for iron content. A total of 148 grams of iron was found in the acid extractions.

Regeneration Studies: After each exhaustion cycle the resin was regenerated automatically with an ammonia solution, and rinsed. In most regenerations excess ammonia was used. After about 25 cycles of normal operation, the 8% regenerant solution strength was reduced to 4%, with no noticeable loss of efficiency in the subsequent exhaustion cycles.

It should be noted that the quantity of ammonia used in regeneration in these cycles varied from about 8 to 90 times the stoichiometric requirement. At about the 100th cycle the ammonia strength was reduced to 2% and this strength regenerant was used up to the 200th cycle. In this period, depending on the volume of regenerant used, the ammonia used was from 4 to 22 times the stoichiometric requirement based on the sulfate retained on the resin. No loss in exhaustion efficiency was noted after these regenerations.

After the 200th cycle, the concentration of ammonia regenerant was reduced gradually to 0.6%. At this point the ammonia in the regenerant solution was about 140% of the stoichiometric quantity needed to completely regenerate the resin. A loss in subsequent exhaustion efficiency was noted, indicating that this quantity of ammonia at this concentration was not regenerating the resin completely.

It should be noted that in pilot plant studies in the manually operated unit using 4% ammonia, stoichiometric regeneration was observed. Therefore it appears that 0.6% ammonia is too dilute to achieve completely effective regeneration, but that 4% ammonia can be used to achieve effective regeneration with near stoichiometric amounts (105%). After the 280th cycle the regenerant concentration was raised to 1.8% and then to 2% until the 370th cycle. In this period the ion exchange system recovered most of its original capacity. In the next 10 cycles, the ammonia concentration was increased to 4% with no concurrent improvement in exhaustion efficiency. A review of the entire long range test shows that a net loss of about 10% of the original capacity occurred. A change in bead size distribution was also observed.

It is a usual phenomenon that ion exchange resins lose a small fraction of their initial capacity early in the resin life, and thereafter stabilize. Therefore, it is anticipated that the loss of capacity in a second year's use of this resin would probably not be as high as 10%. Nevertheless, in the economic study of the Modified Desal Process a 10% per year loss of resin is assumed.

Other Resins Used in Long Term Tests – Ionac MG-1 resin was used in an extended test lasting about 800 hours. Data for part of this run were obscured by failure of a valve, but the resin was cycled between regeneration and exhaustion modes during the equipment failure. The resin utilization data collected during this extended run show considerable scatter. The run confirmed the very high exchange capacity of this resin for sulfate ion.

There is a trend in the utilization data collected as a function of operating time which suggests that perhaps 30% of the exchange capacity was lost during the run period. Because of the appreciable scatter of data, this trend is not well defined, nor are the causes identified.

Rohm and Haas XE-295 resin was run for about 170 hours in the automatic unit, and during this period, one mass balance was made. During the run, considerable difficulty was experienced with iron floc and fines which tended to block flow in the bed. Dowex MWA-1 was run for 500 hours in the automatic unit, showing good exchange properties and no serious loss of capacity during the run.

Other Uses of the Automatic Pilot Plant System – The automatic pilot plant ion exchanges system was also used successfully to condition resin, and to carry out other ion exchange runs using either automatic or manual modes of operation.

Waste Disposal: The Modified Desal Process as installed at the Hawk Run Demonstration Plant results in three separate solid waste products which require ultimate disposal: iron oxide, calcium carbonate, and calcium sulfate. Ammonium sulfate, an intermediate possible waste product, is converted to calcium sulfate and ammonia which is reused. The choice between ammonium sulfate and calcium sulfate as the ultimate disposal waste depends on economic and environmental pollution considerations.

Iron Oxide Wastes – The iron content of Hawk Run AMD corresponds to the production of about 800 pounds of iron oxide as Fe_2O_3 per day. The commercial value of iron oxide compounds varies widely depending on its physical properties, especially those related to pigment uses. The value of iron oxide ranges from about \$0.06 to \$0.20 per pound for common grades to over \$0.80 per pound for high quality transparent pigment grades.

The plan for disposal of iron oxide plant waste consists of discharging it to a settling basin as a sludge, and eventually dredging the solids and using them as a nontoxic land fill. However, sale of this waste for commercial use at, for example, \$0.20 per pound could probably realize at least a \$0.10 per pound profit. At this level, sale of this waste could reduce the cost of water by about \$0.12 per thousand gallons.

Calcium Carbonate – The calcium content of Hawk Run AMD is converted to calcium carbonate in the water softening step. It is estimated that about 1,800 pounds of calcium carbonate will form daily. The plan is to dump this solid into the waste lagoon and eventually dredge it for nontoxic landfill use. It is possible to calcine this calcium carbonate to make lime and carbon dioxide for use in the plant.

An alternative use for the calcium carbonate sludge would be to feed it to nearby polluted streams to reduce the acidity of these streams. This use eliminates the problem of disposal of this chemical, but it also increases the hardness of the receiving stream.

Ammonium Sulfate – After regeneration of the Demonstration Plant ion exchange columns, a dilute solution of ammonium sulfate is recovered. Ammonium sulfate could not be discharged to the local streams because sulfate is a pollutant and ammonia is a nutrient. Accordingly, provision was made at the Hawk Run Plant to recover the ammonia and precipitate calcium sulfate as an alternate waste product.

Calcium Sulfate – Calcium sulfate (gypsum) is obtained from the ammonia recovery step. It is estimated that about 7,000 pounds of calcium sulfate cake will be collected daily. The plan for disposal of calcium sulfate plant waste consists of collecting it in cake form and using it as a nontoxic landfill. However, sale of this waste for commercial use at, for example, $0.03 per pound could probably realize a $0.01 per pound profit. This could reduce the cost of water by about $0.14 per thousand gallons.

Conformance of Hawk Run Waste Streams with State and Federal Laws – Two liquid waste streams, described below, are discharged from the Hawk Run Plant. The overflow from the lagoon is of potable quality and is not a problem. The still bottom wastes would not meet Federal Effluent Standards for the 1985 compliance date.

However, these wastes can be recycled into the plant with feed AMD and thus would be reprocessed instead of being discharged. If this modification were made, the Hawk Run Plant would meet the 1985 requirements by achieving essentially zero discharge.

Liquid wastes from the Hawk Run Demonstration Plant are lagoon overflow and ammonia stripper still bottoms. The solid wastes are iron oxide sludge, calcium and magnesium carbonate sludge and gypsum cake. The volume of water lost in a 0.5 mgd plant amounts to about 167,000 gallons per day. This volume is divided between lagoon overflow and still bottoms. The lagoon overflow is a saturated calcium carbonate solution with traces of iron and an alkaline pH.

Since the saturation level of calcium carbonate is less than 50 ppm, the lagoon overflow is virtually a potable water. The alkaline pH is acceptable since this water is being discharged to an acid stream. The still bottom is a saturated calcium sulfate solution with traces of ammonia and an alkaline pH. Since the saturation level of calcium sulfate is about 2,000 ppm this water is not of potable quality. However, it is suitable for discharge because it is an alkaline water and contains no iron. The solid waste sludges of iron oxide and calcium and magnesium carbonate are to be impounded in the lagoon.

The lagoon is not expected to be dredged more often than once every few years. When this occurs, the sludges could be used as a nontoxic landfill. Proper operation of the landfill site should assure that no solids will be discharged into surface waters.

To avoid potential pollution problems, still bottom wastes could be treated to remove sulfate in the Modified Desal Process. By recycling the still bottom, the only liquid waste would be the lagoon overflow which is water of potable quality. All other wastes from the Hawk Run Demonstration Plant would be solid.

Economics of the Modified Desal Process: An economic model has been developed which combines fixed with variable costs to estimate the cost of potable water production from AMD by the Modified Desal Process. Fixed costs include capital, operational, maintenance, electrical power and building heat. The variable costs are for chemicals such as lime, carbon dioxide, ammonia, as well as resin replacement, power for carbon dioxide evaporation and steam for the ammonia stripper.

Lime required for softening was fixed at 120% of stoichiometric sulfate level, carbon dioxide required for ion exchange was set at 125% of stoichiometric and resin replacement was estimated at 10% per year for a 750 ppm sulfate feed. All other variable costs were determined on the basis of the sulfate level of the raw water, and service water was taken as a fixed volume per regeneration. Costs per 1,000 gallons of product water were determined on the basis of variable feed sulfate level, plant productivity, lime required for ammonia recovery, percent ammonia loss, and plant size.

Effect of Sulfate Level and Plant Productivity on Cost of Product Water – On the basis of a 0.5 mgd plant with 1% ammonia loss and a three times stoichiometric lime requirement, the cost of 1,000 gallons of potable water was found to increase sharply as sulfate level is increased.

Using the fixed costs for the 0.5 mgd plant, but assuming the productivity could be increased to 1.0 mgd, the cost of water is significantly reduced. For a 0.5 mgd plant 1,000 gallons of product from a 750 ppm sulfate feed would cost $2.53. For 1.0 mgd production from the same plant with the same feed the cost would be $1.85 per 1,000 gallons.

The doubled productivity of the 0.5 mgd plant was investigated because pilot plant work indicates that higher plant productivity is possible. The economic results presented here based on production of 0.5 mgd lead to calculated costs higher than what might actually be experienced in the Hawk Run Demonstration Plant.

Effect of Lime Usage in Regeneration – By increasing the lime requirement by as much as four times, the price of 1,000 gallons of product is only increased by 6% for a 750 ppm sulfate feed. This indicates that lime dosage is a relatively unimportant economic factor.

Effect of Ammonia Loss – The efficiency of the ammonia recovery system is found to have little effect on the cost of water. For a 0.5 mgd plant using two times stoichiometric lime with a 750 ppm sulfate feed, a 1% ammonia loss yields a product water cost of $2.53 per 1,000 gallons.

Increasing the ammonia loss to 100%, increases the cost 4.7% to $2.65.

Size of Plant – Based on a 0.5 mgd plant with 1% ammonia loss and two times stoichiometric requirement of lime, the plant size was scaled up to 1.0, 5.0 and 10.0 mgd. Price reductions were found to be large. For example, with a twenty fold scale up from 0.5 to 10 mgd, the cost of water was decreased from $2.57 to $1.05 per thousand gallons for a 750 ppm sulfate feed.

Other Variables – The individual parameters contributing to the cost of producing potable water can be divided into three categories. First there are costs that are varied in the model to assess their effects on the price of product water. The next group of parameters are those which may be varied but are kept constant in the model. The quantity of lime used in softening, the amount of carbon dioxide needed for ion exchange, the ammonia required for regeneration and the rate of resin replacement are all variables that change the cost of producing potable water.

The expected values for these parameters, based on pilot plant experience and accepted methods, that is, the quantity of lime needed for softening, were used and considered constant. In practice these variables can affect the cost of water. However, the relative impact of changing these parameters over the range likely to be encountered will be small.

The fixed costs represent the third group of parameters. In the calculations using the 0.5 mgd plant basis, these parameters are kept constant. Where the model is scaled up to 1.0, 5.0 and 10.0 mgd, labor costs are kept constant, electrical power is increased proportionately and capital, maintenance and operating costs are increased.

Increasing these parameters for larger scale plants yields costs that are conservative. This results from the fact that capital costs for larger plants are relatively less.

Specific Limitations of the Modified Desal Process: The results of nine representative runs in which Amberlite IRA-93 resin was exhausted with water of varying sulfate, total dissolved solid (TDS) and pH levels show that at about 2,200 ppm sulfate the volume of product water is less than the volume of water necessary for regeneration purposes. This indicates that the Modified Desal Process is limited in application to waters containing less than 2,200 ppm sulfate.

The Modified Desal Process is not limited by total dissolved solids or pH levels; however, large quantities of carbon dioxide are required to achieve good resin utilization for high total dissolved solids or alkaline feed waters.

Nine exhaustion runs were completed in which feed waters with sulfate concentrations ranging from 350 to 2,169 ppm sulfate were treated. The 250 point resin utilizations ranged from 0.605 to 0.825 with water production ranging from 1,058 gal/ft^3 for the 350 ppm sulfate feed to 144 gal/ft^3 for the 1,983 sulfate water. A plot of the feed sulfate concentration versus the reciprocal of the water production for each run, or the cubic feet of resin per gallon of product water, yields a straight line. A least squares analysis of this plot gives the equation: $Y = 2.91 \times 10^{-6} X - 0.186 \times 10^{-3}$. On the basis of a constant requirement of 161 gal/ft^3 of water for regeneration as determined in closely monitored water balance studies (run 7-13, 14, 15), it was found that at a feed concentration of 2,198 ppm sulfate there would be no net production of water.

For feedwaters containing sulfate in excess of 1,000 ppm it is possible to pretreat these waters with lime, thereby reducing the sulfate level to the limit of calcium sulfate solubility, which is about 1,000 ppm sulfate. Therefore, it is conceivable that very high sulfate waters could be pretreated with lime and then treated by the Desal Process to make potable water. The limitation of the combined pretreatment and ion exchange process is the final cost of the product water.

THE SUL-BISUL PROCESS–BURNS AND ROE STUDY

The Sul-biSul Process employs two resins, a strong cation exchange resin and a strong anion exchange resin. The AMD feed is first passed through the cation exchanger (hydrogen form resin) which removes the metal ions and exchanges these for hydrogen ions. The reaction is:

$$MSO_4 + 2HR \longrightarrow MR_2 + H_2SO_4$$

where R represents strong acid ion exchange groups on the resin. The product water from this exchange reaction contains sulfuric acid. This eluate is then passed through a strong base exchange resin R' in the sulfate form. Because of the low pH (high acidity) of the feed, sulfate ions in solution and on the resin are converted to the bisulfate form. This conversion of bivalent sulfate to monovalent bisulfate allows twice the amount of sulfate to be stored on the resin, and sulfate is thus removed from solution. The reaction occurs as follows:

$$R'SO_4 + H_2SO_4 \longrightarrow R'(HSO_4)_2$$

Removal of the sulfate results in a good quality product water. When the anion resin is exhausted it is regenerated by rinsing with water or an alkaline water regenerant. As the higher pH water enters the resin, bisulfate is converted to sulfate in the reversal of the above equation, sulfuric acid is discharged with the regenerant solution, and the anion resin is converted back to the sulfate form. The unusual feature of this process is the removal of sulfate from feedwater by anion exchange using only water or water with a little added alkali as the regenerant.

Experimental

The cation exchange process received only limited attention. Most of the experimental effort was directed at investigating the exhaustion of the sulfate resin to the bisulfate form, and the regeneration of the bisulfate form resin back to the sulfate form.

Thirteen pilot plant runs were made with natural AMD and synthetic mine waters using this process. Because the results obtained were unexpectedly poor, a detailed bench scale study was initiated. The bench scale tests were carried out on two small charges of Dowex SBR-P resin, designated as BS1 and BS2. Synthetic AMD feed solutions were subjected to a pretreatment with Amberlite IR-120 resin in the hydrogen form prior to treatment in the sulfate form, anion bed.

Exhaustion Studies: There were 13 pilot plant exhaustion runs, of which 10 were made with Dowex SBR-P and 3 were made with Dowex 11. Thirty-three bench scale runs were made with Dowex SBR-P. All but four of these runs used the sulfate form of the resin. The four other runs were made with the hydroxyl form of the resin. All of the bench scale runs were carried out in the downflow mode at less than 0.1 bed volume per minute.

Regeneration Studies: A careful bench scale study of the regeneration of the exhausted Sul-biSul anion beds was made. Regenerations were carried out with distilled water, tap water, ammonia, and 2% sodium hydroxide. The concentration of ammonia regenerant was varied from about 5 to 9% in these runs. Regenerations were carried out in the downflow mode at a flow rate of less than 0.1 bed volume per minute. At the end of each regeneration, the bed was rinsed with distilled water, and the rinse analyzed to complete the mass balance.

Results and Discussion

Exhaustion Tests: Initial bench tests were used to determine the capacity of Dowex SBR-P, the anion resin used in the Sul-biSul Process. The results show that when the pH of the feed is relatively high (1.65) the resin removes approximately the expected amount of sulfate (1.39 meq/ml versus the manufacturers capacity of 1.2 meq/ml).

When the pH is low (0.26 to 0.44) about twice the capacity is obtained. This indicates that the sulfate-bisulfate cycle is in operation. When the feed pH is high mostly sulfate is present, and one divalent sulfate displaces two hydroxyl ions. At low pH, monovalent bisulfate dominates and twice the apparent capacity is found since one bisulfate replaces one hydroxide.

Results for the bench scale runs in which sulfate form SBR-P resin was exhausted with various concentrations of sulfuric acid show that increasing the feed sulfate level, and consequently reducing the pH, produced increasingly better resin utilizations. Tests were also conducted to determine if partial utilizations were a result of insufficient time to reach equilibrium. The results show that equilibrium is reached quickly, and that feed composition (cation concentration, sulfate concentration and pH) regulate the resin performance.

The bench scale exhaustion tests provide data for Sul-biSul performance over a broad range of feed sulfate and pH levels. In the context of these data the 13 pilot plant tests are more easily understood. On the basis of the feed sulfate level and influent pH, the observed utilizations fit into the trend found in the bench tests. The pilot plant data show that low utilizations are expected on the basis of low sulfate concentration which results in a high pH in the anion resin feed.

The bench tests also show that as the sulfate level in the feed approaches 15,000 ppm, or as the anion bed influent pH reaches 0.60, maximum utilization of the sulfate form resin is achieved. However, higher resin utilization does not signal better water production for the Sul-biSul Process. For example, a run with 9,910 ppm sulfate feed gave a utilization of 0.806 and a run with 1,641 ppm sulfate feed gave a utilization of 0.317. For a cubic foot of resin the high sulfate run processed 85 gallons of water. It is evident that more water can be processed from low sulfate feeds in spite of low resin utilizations.

Regeneration Tests: Bisulfate form SBR-P resin can be restored to the sulfate form by raising the resin pH. This may be accomplished by using various alkalies, or by rinsing with an alkaline water. When tap water regeneration was made, less water was needed to regenerate partially exhausted resins than fully exhausted resins. Since low sulfate feeds produce more gross water and require less water for regeneration than high sulfate feeds, they are more economical sources to win potable water.

If the Sul-biSul Process is used to produce potable water from AMD, it is counterproductive to use tap water as the regenerant. Data show that even in the case of low sulfate feed, tap water regenerant uses more water than is produced. Regeneration required 343 gal/ft^3 more water than was produced. Using a more concentrated alkali, such as ammonia does not improve the situation. With 8% ammonia large deficits of water were encountered. Using a caustic regenerant reverts Sul-biSul to a conventional two bed demineralization process.

The Sul-biSul system is applicable where a surplus of alkaline low sulfate feedwater is available. If raw water can be used as the anion bed regenerant, and if regeneration wastes can be disposed of in the stream, a net production of water can be attained. This indicates that Sul-biSul is not applicable to treat most AMD wastes.

If the AMD water is acidic, no net production of water is possible. Those few coal mine drainage waters that are alkaline usually contain calcium in the form of gypsum. Regeneration with this type of water can cause serious fouling because of calcium sulfate precipitation in the bed.

Waste Disposal: The Sul-biSul system generates two wastes. The cation resin produces a spent acid regenerant waste. Anion resin regeneration can be accomplished by using an alkaline raw water. In this case the sulfate on the resin is released to the raw water.

Most AMD sources are not sufficiently alkaline to permit raw water regeneration. In these cases a weak alkali, such as ammonia, is used producing a spent alkaline waste. The acidic and alkaline wastes produced in this process could be neutralized by blending the two.

Because of the large volume of 8% ammonia necessary to regenerate the anion bed as compared to cation regeneration acid requirements, the acid wastes in this process could be neutralized, leaving excess spent alkaline regenerant. This waste would require additional neutralization, and would represent a liquid waste disposal problem.

Economics: The economics of most ion exchange processes are dependent on the net production of water and the chemical requirements for regeneration. For Sul-biSul, alkaline raw water anion bed regeneration would be ideal. This is not always possible, and other alkalies are usually needed. With ammonia regeneration, between 21 and 36 bed volumes of the reagent are needed to convert the anion bed from the bisulfate to the sulfate form. Using five times stoichiometric quantities of regenerant, only 10 bed volumes of one normal caustic soda would be necessary to put the resin completely in the hydroxyl form. Since maximum water production is the objective, conventional two bed demineralization, that is, regeneration with caustic, would be more advantageous than the Sul-biSul Process. A hydroxyl form resin would uniformly yield high resin utilization and would not be limited to a particular pH feed water. Except where excessive alkaline raw water is available, Sul-biSul cannot economically produce potable water from AMD. For ammonia and tap water regenerations net losses of water ranged from 192 to 343 gal/ft^3.

Specific Limitations of the Sul-biSul Process: The applicability of the Sul-biSul Process for the treatment of AMD is dependent on the availability of large quantities of alkaline water. This water must be sufficiently alkaline and abundant so that it may be used as a regenerant and then discharged to the stream. If the raw water cannot be used as the anion bed regenerant, other alkalies must be employed. When this happens all tests have indicated that there is a negative net production of water. The low sulfate, low TDS feedwaters give the greatest gross water production, although the low concentration feeds give reduced anion bed utilizations.

CONVENTIONAL ION EXCHANGE–BURNS AND ROE STUDY

Experimental

Two bench-scale ion exchange experiments were made using conventional cation exchange processes, the first using Amberlite IR-120 in the hydrogen form, the second using the same resin in the sodium form. An ion exchange column was prepared containing 57.9 ml of preconditioned IR-120 resins which was converted to the hydrogen form with hydrochloric acid. After rinsing, it was exhausted with dilute AMD containing 193 ppm sulfate at a flow rate of about 0.25 bed volume per minute. This run was numbered 21-1.

After this run, the same resin was converted to the sodium form by elution with excess hydrochloric acid, followed by a water rinse, and then treated with 10% sodium hydroxide solution. After rinsing, this column was exhausted at the same flow rate with dilute AMD containing 266 ppm sulfate. This run was numbered 21-2. The eluates from these runs were collected in fractions and analyzed for iron, hardness, pH and conductivity. Composite samples from each run were analyzed for methyl orange acidity as well as the parameters noted above.

Aliquots of the two composites, and a dilute AMD feed (containing 185 ppm sulfate) were treated with lime slurry and then analyzed for the above parameters. These neutralized aliquots provided data on lime required for neutralization, as well as quality of product water attainable by conventional ion exchange or lime treatment.

Results and Discussion

Conventional ion exchange processes are limited to the treatment of waters containing less than about 500 ppm total dissolved solids because of high cost. For lower total dissolved solids feeds, different treatment processes vary in merit depending on the nature of the feedwater and the intended use of the product. When acid mine drainage (AMD) discharges into receiving streams, the diluted mixture is generally not potable. The detrimental constituents in the diluted water may be heavy metal ions (especially iron and manganese) and free acidity, rather than sulfate ion.

In this case, possible cost savings could be made if lime treatment or conventional ion exchange were used. Several conventional ion exchange methods are available to reduce the concentrations of these ionic constituents to acceptable levels. These processes were reviewed and compared to alternate chemical treatment processes for purifying water.

Hydrogen Form Cation Exchange Resins: The passage of AMD through a strong cation exchanger resin in the hydrogen form would produce an acidic effluent. If the raw water contains less than 250 ppm sulfate and a total dissolved solids not exceeding 500 ppm, it could be made potable by raising the pH.

In run 21-1 a 57.9 ml bed of IR-120 in the hydrogen form was exhausted at 0.25 BV/min with a dilute AMD containing 193 ppm sulfate. Since the major objectionable constituent in the raw water is iron, the run was stopped when the effluent iron content reached 0.3 ppm. Run 21-1 produced 365 bed volumes of effluent (2,730 gal/ft^3) which had a pH of 2.40.

Bringing the pH of this water to 8.3 required 4.63 meq/l of lime. The product water, although potable and containing no iron, was hard and had a high TDS level. The amount of lime required was high and, because the resin required an acid regeneration, an acidic waste was produced. Equipment costs for the hydrogen form cation exchange are high since, in addition to the ion exchange vessels, neutralization equipment is needed for both the product and waste streams. The neutralized regenerant wastes present a liquid waste disposal problem.

Sodium Form Cation Exchange Resins: Treatment for AMD to produce potable water with a sodium form cation resin also requires a feed with less than 250 ppm sulfate and less than 500 ppm TDS. Run 21-2 used the same bed of IR-120 resin as above, converted to the sodium form, and it was run at the same flow rate. In this run 242 bed volumes (1,810 gal/ft^3) of effluent were collected before the 0.3 ppm iron limit was reached.

The composite pH was 3.64, but the water had a phenolphthalein acidity of only 15 ppm as calcium carbonate. Consequently, very little lime was necessary to raise the effluent pH to 8.3. The product water was soft, containing only sodium sulfate, but the process did not significantly reduce the TDS level. Regeneration required a concentrated brine, and the spent regenerant was acid.

Equipment requirements for this process would also include ion exchange vessels and neutralization apparatus for the waste stream. Chemical costs are smaller since product neutralization requires less lime, and the spent regenerant would be only moderately acid. The neutralized regenerant from this process also represents a liquid waste disposal problem.

Anion Exchange Resins: Weak or strong anion exchange resins used separately in the free base or hydroxyl form are not applicable for the treatment of acidic AMD. Iron precipitation in the resin bed causes serious fouling problems, limiting the use of anion resins in these forms.

Two Bed Ion Exchange: Run 11-2/9-7 represents a typical two-bed ion exchange process. The cations are removed in the first bed (Amberlite IR-120 in the hydrogen form), releasing an equivalent amount of hydrogen ions. The effluent from this column passes through the anion exchange resin (Amberlite IRA-93, free base form), where the sulfate should be absorbed on the resin. Since the effluent from this ion exchange column is essentially sulfuric acid, and the pH of this effluent is very low, one would expect very efficient utilization of the resin. This run gave a utilization of 0.238, or 19.8 bed volumes of product (148.1 gal/ft^3) for the 806 ppm sulfate feed.

Run 11-3/9-9 was similar to the above run in every way except it was run with the sodium form of IR-120 resin in place of the hydrogen form. Resin utilization was extremely poor, only 0.078, yielding 6.39 bed volumes of product (47.8 gal/ft^3) for the 821 ppm sulfate feed. The sodium form of the IR-120 resin converts the sulfuric acid and iron sulfate in AMD to sodium sulfate. The acidity in the cation bed effluent is low. Since weak base resins in the free base form require acidic feeds, resin utilization was low.

Run 11-4/12-1 used IR-120 in the hydrogen form followed by Dowex 11 (strong base exchanger) in the hydroxyl form. Resin utilization was good, 0.896, with 61.85 bed volumes (460.6 gal/ft^3) of water produced.

Product water quality from these three processes differs materially. Hydrogen form cation exchange followed by free base weak anion exchange produces deionized water from AMD. Sodium form cation exchange in conjunction with free base anion exchange is not a reasonable process, but to the extent that it does function, the effluent contains sodium hydroxide. The hydrogen form cation resin followed by the hydroxyl form anion exchange resin produced deionized water.

These three conventional ion exchange systems have similar chemical, equipment and waste disposal costs. Each requires acid and alkali (or brine) regeneration producing waste streams with either high or low pH. In addition to dual ion exchange vessels, neutralization apparatus for spent regenerants is necessary.

Neutralization vs Ion Exchange Treatment of AMD: In some cases neutralization of acidic, low sulfate AMD with alkalies can produce water of similar quality as produced by ion exchange but at reduced cost. An experiment was carried out to compare conventional cation exchange treatment with lime neutralization of low sulfate AMD (185 ppm sulfate). In this experiment 2.25 meq/l of lime was used to change the pH from 4.01 to 9.92 and produced a hard, iron-free, potable water.

This process is more economical than ion exchange for this water because it requires less chemicals, less process equipment and produces only solid iron oxide wastes. The main limitation of lime neutralization is that it cannot produce potable water from AMD containing more than 250 ppm sulfate. Lime neutralization can be combined with soda ash treatment. This produces a softer product and additional solid waste (calcium carbonate sludge).

The extra cost of soda ash treatment may be offset by the preference of soft water. A combination of lime and limestone treatment may be more economical than lime treatment alone. Chemical neutralization of AMD polluted water can produce potable water from waters containing less than 250 ppm sulfate. For higher sulfate waters one must look to other treatment methods which reduce the sulfate and TDS in the water.

AEROJET-GENERAL PROPRIETARY PROCESS

D.O. De Pree; U.S. Patent 3,725,259; April 3, 1973; assigned to Aerojet-General Corporation describes a method for the removal of Group I-b, II-b, III, IV, V, VI, VII and VIII metal salts from acidic waste streams such as acid mine drainage by contacting the stream in countercurrent fashion with a cation exchange bed containing a cation exchange resin either in the H^+ or alkali or alkaline earth substituted form to deposit metal ions from the salts thereon, neutralizing the acidic effluent from the cation exchange resin, and regenerating the cation ion exchange resin with a solution of a chelating agent and precipitating the metal from the solution by treatment with a base or with a base and carbon dioxide.

The chelating agent is subsequently recovered for recycle by precipitation of the transition metal with a base or with a base and carbon dioxide. The neutralization of the acidic effluent from the cation exchange resin is preferably carried out with lime or calcium carbonate.

The process is particularly applicable to the polymethacrylic cation exchange resin such as Amberlite CG-50 and others such as polyacrylic and polysalicylic acid-formaldehyde polymers. The preferred chelating agents are the β-diketones such as acetylacetone, 1-phenylbutane-1,3-dione, dibenzoyl methane, 2-acetylcyclohexanone and alkyl, aryl or halogen derivatives thereof, the important consideration being that the metal ion to be chelated form an organic soluble chelate with any of the above ligands.

The preferred solvent for the chelating agents is methanol. Other solvents which may be employed include alcohols other than methanol, aliphatic and aromatic ethers and polyethers, aromatic hydrocarbons, aliphatic hydrocarbons and mixtures of any of the above. The chelating agents are usually present in an amount sufficient to form a 0.1 to 5 N solution. The upper limit of the chelating agent is determined by the solubility of the metal chelate in the particular solvent employed. Ferric acetylacetone, for example, is extremely soluble in methanol forming clear solution concentrations of 3 and 4 normality.

In the example the parts and percentages are by weight unless otherwise indicated.

Example: One liter of a synthetic acid mine water was prepared by dissolving the following salts in 0.01 N sulfuric acid and diluting with 0.01 N sulfuric acid to one liter.

	Weight, grams
$CaSO_4 \cdot 2H_2O$	0.344
$MnSO_4 \cdot H_2O$	0.024
$Al_2(SO_4)_3 \cdot 18H_2O$	0.186
$FeSO_4 \cdot 7H_2O$	0.997
$MgSO_4 \cdot 7H_2O$	0.246

The pH of the solution was found to be 2.4. The metal ion composition, as determined by emission spectroscopy, was found to be as follows in ppm: Fe, 205; Al, 20; Mn, 10; Mg, 27; and Ca, 95. The pH of 100 ml of the synthetic acid mine water was adjusted to 4.5 with finely divided calcium carbonate. The addition was made very slowly because CO_2 evolution causes a gradual raising of the pH.

This solution was then passed through a ¼" x 9" chromatographic column containing 2.5 grams (dry weight) of water swollen Amberlite CG-50 resin (H^+ form). Three milliliter samples of eluate were collected and found to have a pH of 3.1 and were analyzed for the various cations by emission spectroscopy. The data contained are shown in Table 5.5.

TABLE 5.5

Fraction Number	Cation Conc., ppm				
	Fe	Al	Mn	Mg	Ca
1	<15	<15	--	----	------
5 and 6	157	<1	9	27	450
9 and 10	155	<1	8	22	300
15	187	<1	8	16	285
20	187	<1	7	18	370
25	182	<1	8	21	327
30, 31 and 32	168	<1	8	10	215

Source: U.S. Patent 3,725,259

These data indicate that all of the aluminum, none of the manganese and approximately 25% of the iron were removed by a single pass. The increase in acidity resulting from the exchange of H^+ ion from the column reduces the efficiency of the transition metal removal. The eluate fractions were combined, the pH readjusted to 4.5 and the treatment repeated using the same resin. An equivalent reduction in transition metal ions was achieved by the recycle as indicated by the analysis shown in Table 5.6. Twenty-five milliliter fractions of eluate were collected.

TABLE 5.6

Fraction Number	Cation Conc., ppm				
	Fe	Mn	Al	Mg	Ca
1	145	7.5	<2	19	265
2	124	8.0	<2	25	400
3	111	7.5	<2	24	360
4	140	7.5	<2	20	360

Source: U.S. Patent 3,725,259

The spent cation exchange resin from the treatment (two cycles) described above was treated in the chromatographic column with 100 ml of a 2 N solution of acetylacetonate in methanol. The solution was prepared by diluting 20 grams of acetylacetonate with methanol to 100 milliliters.

Three milliliter fractions of eluate were collected and analyzed by emission spectroscopy for iron, aluminum, manganese, magnesium, and calcium. The results shown in Table 5.7, indicate a very effective regeneration of the column by removal of the iron and aluminum. As manganese and little calcium or magnesium were deposited on the column, these ions are not present in the eluate.

TABLE 5.7

Fraction Number	Cation Conc., ppm				
	Fe	Mn	Al	Mg	Ca
1	12	<1	<1	<1	<10
2	600	70	<1	<1	<10
3	59	45	<1	<1	<10
4	13	50	<1	<1	<10
5	9	24	<1	<1	<10
6	8	14	<1	<1	<10
7	17	14	<1	<1	<10
8	4	11	<1	<1	<10
15	8	8	<1	<1	<10
20	4	7	<1	<1	<10

Source: U.S. Patent 3,725,259

The chelate solution (dark red in color) from fraction 2 (approximately 2 ml remained after removing analytical sample) was treated with concentrated NaOH in methanol containing a trace of water (2%). A dense precipitate of $Fe(OH)_3$ settled leaving a straw colored solution. The precipitate was readily removed by filtration. The water eluates from the cation exchange resin treatment of the synthetic acidic mine effluent were treated with calcium carbonate and adjusted to a pH of 6.0.

REVERSE OSMOSIS

The material in this chapter was excerpted from EPA-430/9-73-011; EPA-670/2-73-100; PB 230 756 and U.S. Patent 3,795,609. (see Bibliography, p 390).

Natural osmosis occurs when two solutions of different concentration but in a common solvent are separated by a permeable membrane. If the membrane is permeable by the solvent and not the solute, then the solvent will flow from the dilute solution into the more concentrated solution until an equilibrium of equal concentration is established. In the reverse osmosis process, the direction of solvent flow is reversed by the application of pressure to the more concentrated solution. As a result, the concentrated solution's strength increases and this is termed the solute or brine. The solvent or permeate is the product from the process.

Reverse Osmosis Membranes: The development of reverse osmosis membranes has been significant during the last decade. There are three types of reverse osmosis systems commercially available; they are the hollow fiber, spiral wound and tubular membranes (Figure 6.1). Membrane construction centers around the use of cellulose acetate.

FIGURE 6.1: REVERSE OSMOSIS SYSTEMS

a.

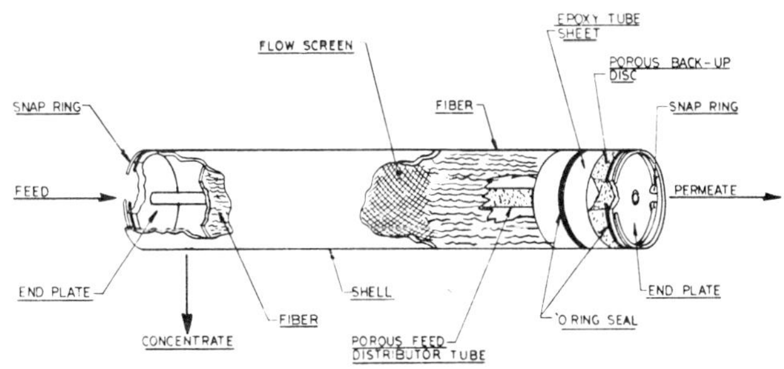

Hollow Fiber Module

(continued)

FIGURE 6.1: (continued)

b.

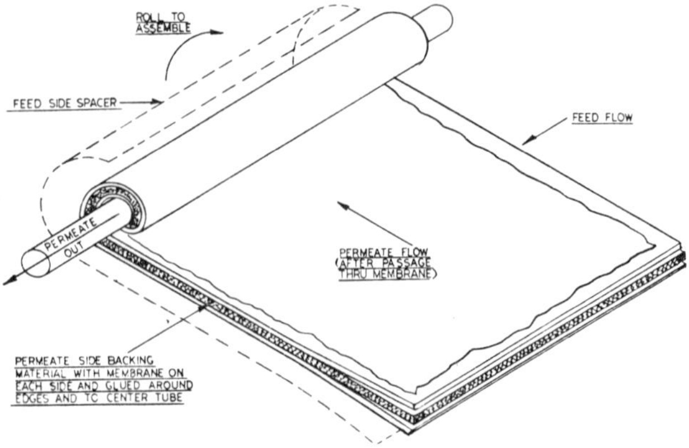

Spiral Wound Membrane

c.

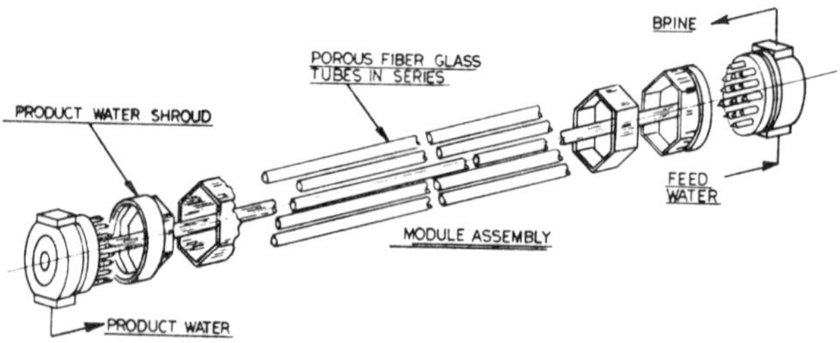

Tubular RO Module Configuration

Source: EPA-430/9-73-011

Reverse Osmosis Process: The reverse osmosis process has been studied for use in treating acid mine drainage by the Federal Government. Studies were conducted using the spiral wound, tubular, and hollow fiber membranes in 37.85 cubic meters per day (10,000 gpd) test units.

Mine drainage was first filtered to remove suspended solids and then processed through the test units at 600 psi operating pressures. Brine from the units was partially recycled to increase the feed water dissolved solids concentration and increase the product water output. The flowsheet for this system is shown in Figure 6.2.

These studies demonstrated that the reverse osmosis process is highly effective in removing nearly all of the dissolved solids in acid mine drainage. Recoveries of 80 to 90% of the feed water volumes were obtained. Problems with membrane module fouling occurred from calcium sulfate formation but this could be flushed from the units by operating at lower pressures for a short period.

FIGURE 6.2: FLOW DIAGRAM OF A REVERSE OSMOSIS SYSTEM

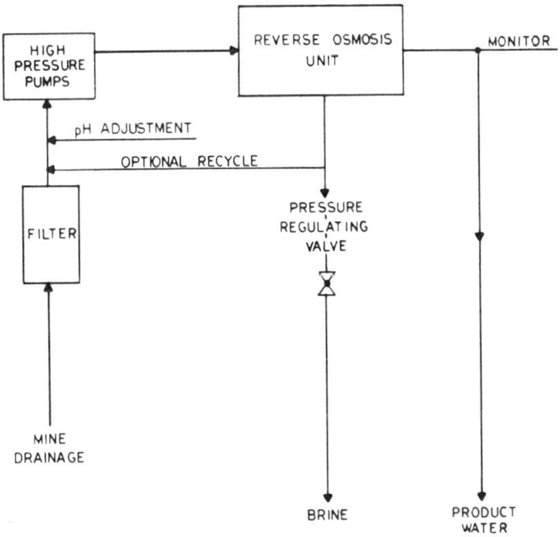

Source: EPA-430/9-73-011

Iron fouling occurred in one test but was prevented in later studies by lowering the pH of the feed water to less than 3.0. Acid mine drainages containing both ferric and ferrous iron forms were successfully processed. Mine drainage containing high concentrations of dissolved solids including sulfates, caused operating problems with excess calcium sulfate formation and could not be processed.

Biological oxidation of ferrous iron was prevented by ultraviolet disinfection. There was no advantage to first treating the raw water to remove the iron, acidity, and other parameters. Product water from the unit usually contained less than 70 mg/l of dissolved solids. Typical quality of the product water is given in Table 6.1.

Water of this quality however, is not acceptable for potable uses due to its pH, acidity, iron and manganese content. Further treatment by chemical neutralization, coagulation, filtration and disinfection would produce a potable quality water.

The brine or reject water produced when processing acid mine drainage presents a disposal problem. In the tests conducted, brine volume was usually about 10 to 15% of the raw water feed. This results in an increase of 8 to 12 times in the concentrations of the various ions that were present in the raw water. Methods for disposal of the brine would include the various neutralization processes, evaporation ponds, deep well injection, and the neutrolosis process.

Numerous studies and demonstration plants have been completed by the Office of Saline Water on the desalting of brackish and saline waters. Cost information available from these sources can be applied to estimating capital and operating costs for treating acid mine drainage.

TABLE 6.1: TYPICAL PRODUCT WATER QUALITY BY REVERSE OSMOSIS SYSTEMS*

Parameter**	Raw Water Quality	Product Water Quality
pH	3.1 - 3.4	4.2 - 4.4
Specific Conductance	1000 - 2000	17 - 46
Acidity	210 - 460	32 - 46
Calcium	125 - 260	0.4 - 2.2
Magnesium	92 - 170	0.3 - 1.4
Iron, total	77 - 110	0.4 - 1.2
Iron, ferrous	61 - 71	0.3 - 1.0
Aluminum	12 - 15	0.2 - 1.0
Manganese	14 - 43	0.1 - 0.5
Sulfate	660 - 1340	0.9 - 4.6

* Synopsis of tests conducted on spiral wound, tubular, and hollow fiber membrane systems.

** All units are in mg/l, except specific conductance (μ mhos), and pH.

Source: EPA-430/9-73-011

In the tests conducted on mine drainage, product water output was about one-third less than the output in processing seawater. Consequently, the capital cost estimating figures presented here have been increased 50% to compensate for this.

Capital costs will vary from 18 to 28 cents per liter per day ($0.68 to $1.05/gpd) for a plant capacity of 3,785 cubic meters per day (1.0 mgd). Operating costs for plants of this size will range from 13.2 to 18.5 cents per cubic meter treated (50 to 70 cents/1,000 gal). When processing mine drainage, the cost for brine disposal and final treatment of the product water must also be included.

Neutrolosis Process: During evaluation of the reverse osmosis process in treating acid mine drainage, methods were investigated for the economical treatment of the reject water or brine. This would amount to 10 to 15% of the volume of raw water processed, and would contain an 8 to 12 times increase in the concentration of the raw water's dissolved solids. The treatment methods considered centered around the neutralization processes. As a result the Neutrolosis Process was developed.

The Neutrolosis Process consists simply of treating the brine from the reverse osmosis unit by a conventional neutralization process, and returning the clarified treated water to the reverse osmosis unit's feed stream. As a result the Neutrolosis Process produces product water and sludge.

The mine drainage treated by this neutrolosis pilot plant contained iron mostly in the ferric form. In neutralizing the resulting brine, the pH was raised to 4.5, at which point most of the iron and aluminum was removed. To control solids formation in the reverse osmosis unit, the pH is adjusted to less than 4.0.

Therefore, it is an advantage to have the treated brine water at a pH as low as possible. When ferrous iron is dominant in the raw mine water and the brine, it will be necessary to raise the pH much higher for treatment. Consequently, additional acid will have to be added to this treated water to maintain the pH below 4.0 in the raw water feed stream.

Continued recycle of the treated brine may cause the dissolved solids to increase to a point where a periodic blowdown may be required. Dissolved solids at these levels are not expected to have an effect on the operation of the reverse osmosis unit. The sludge produced by the Neutrolosis Process consists mostly of ferric hydroxide and calcium sulfate. The sludge volume was about 1% of the raw water feed volume.

The Environmental Protection Agency and The Office of Saline Water have conducted studies of reverse osmosis treatment of acid mine drainage. Two of these studies are excerpted below.

PROCEDURES—EPA STUDY

Test Equipment

Two spiral wound type reverse osmosis units were used in the course of these studies. The initial unit was a 37,850 liter per day (10 K—10,000 gallon per day nominal product flow) unit consisting of five 4 inch x 10 foot Schedule 40 steel pipe pressure vessels which had been lined with an epoxy to prevent corrosion. A Moyno high pressure progressive cavity stainless steel pump was used to supply water to the unit at pressures up to 5,517 kilonewtons per square meter—kN/m^2 (800 psig).

Each of the five pressure vessels contained three spiral wound modules (Figure 6.3 and 6.1b). Each module was 0.92 meter long, 9.53 centimeters in diameter, and contained 4.65 m^2 of modified cellulose acetate membrane. The three modules were placed in series in each pressure vessel (tube). There were two effluents from each module (or tube of modules), i.e., brine and product water.

Typically, the brine from one module (or tube of modules) served as the feed to the next module (or tube of modules) and was progressively dewatered in this manner. Each vessel therefore contained 13.96 m^2 of membrane for a five vessel unit total of 69.8 m^2.

FIGURE 6.3: SPIRAL WOUND MODULE CONFIGURATION

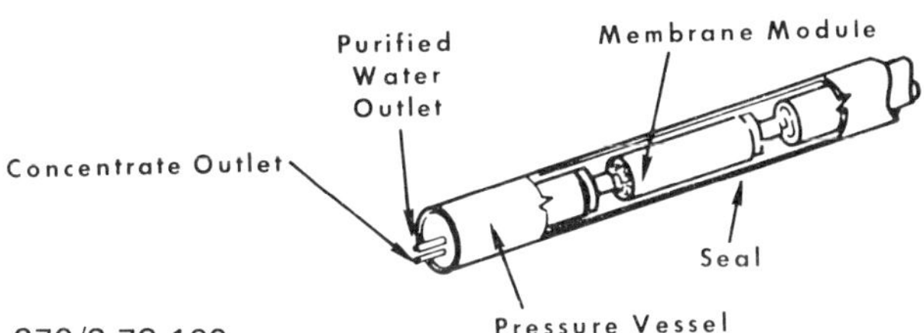

Source: EPA-670/2-73-100

The flow diagram for the 10 K unit is presented in Figure 6.4. Pressurized sand filters and 10 micron cartridge filters were used to remove suspended solids from the feed water before the water entered the RO unit. Filtration was necessary as suspended solids can clog the narrow brine channels and foul the RO membranes.

For these studies, the pressure vessels were arranged in a 2-2-1 array in order to maintain, as nearly as possible, a uniform brine flow through each vessel. The first two vessels were in parallel. Brines from these vessels were combined to serve as the feed to vessels 3 and 4, which were in parallel. Brines from vessels 3 and 4 were then combined to serve as feed to vessel 5.

In order to increase recovery beyond 70% and yet maintain sufficient brine flow velocity and turbulence in each module to prevent boundary layer precipitation, it was necessary to recycle a portion of the brine into the feed to the RO unit.

In this case, the spiral unit membranes were actually treating a blended feed (mixture of raw water and recycled brine) rather than the raw water.

Inlet pressure and pressure drops across each tube were measured on pressure and differential pressure gauges. Internal flow measurements were made using venturis and differential pressure flow gauges. External flows were measured volumetrically.

As product flow was strongly temperature dependent, careful temperature measurements were made using a laboratory thermometer. The spiral 10 K unit was not mobile and therefore was used solely at Norton.

The second spiral wound unit produced 15,140 liters per day (4,000 gpd) of product water (4 K unit) and consisted of three 4 in x 10 ft pressure vessels arranged in a 2-1 array. Operation and construction of this unit was the same as the 10 K unit previously described.

This 4 K unit was especially suited to portable field testing. Several feed pumps were tried on the 4 K system with the ultimate selection of a Goulds MB 13600 stainless steel-ceramic, multi-stage, centrifugal pump which had been especially designed for reverse osmosis applications. For field application, the 4 K unit was mounted in a portable metal building. In addition to tests at Norton, the studies at Morgantown, Ebensburg, and Mocanaqua were made on this 4 K unit.

For the Mocanaqua study, a hollow fiber unit was tested alongside the spiral wound 4 K system. Initial Mocanaqua hollow fiber tests were made on a single permeator. EPA later purchased two additional permeators to enlarge the hollow fiber system to 6 K (22,710 lpd or 6,000 gpd of product flow) in order to more effectively compare hollow fiber results with spiral wound.

Approximately 139.6 m^2 of B-9 modified nylon hollow fiber membrane were packed in each 15.2 cm x 1.22 m stainless steel pressure vessel as illustrated in Figure 6.1a. The three permeators were arranged in a 2-1 array. All flows were measured volumetrically.

FIGURE 6.4: FLOW DIAGRAM FOR SPIRAL WOUND REVERSE OSMOSIS UNIT

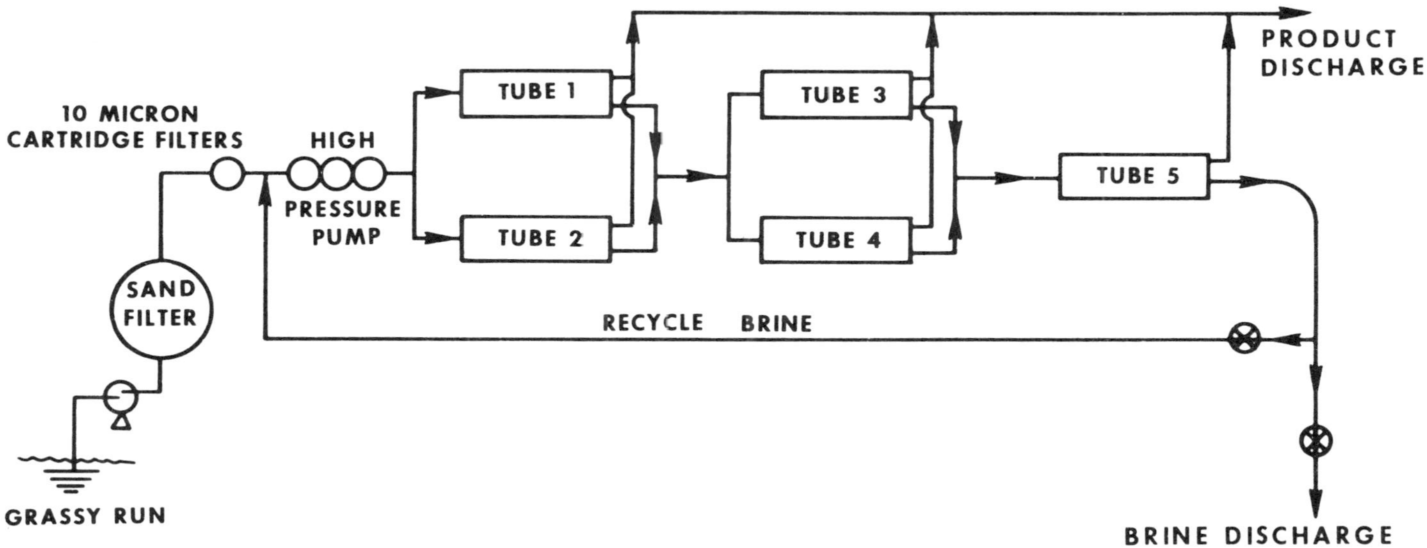

Source: EPA-670/2-73-100

Calculations

Recovery was defined as the percentage of water resulting as product as compared to the initial volume entering the unit. Therefore, recovery equalled product flow divided by feed flow expressed as a percentage. Since accurate measurements could be made of the brine and product flow rates, feed flow was calculated as the sum of the brine and product flow.

Since very little of the original pollutants appear in the product stream, virtually all remain in the brine stream. As recovery increases, progressively less water is available for dilution of these pollutants. Above 75% recovery, small changes in recovery level produce large changes in brine concentration.

The standard way of expressing membrane performance is flux rate; i.e., gallons of product flow per square foot of membrane per day at a specified temperature and net pressure. To obtain these values, product flow was corrected from the observed temperature to 77°F by a correction chart supplied by each manufacturer.

The 77°F temperature was used since it is the standard of the reverse osmosis industry; however, a more realistic value for mine drainage studies would be 50°F. In general, product flow increased roughly 1.5 to 2% per degree Fahrenheit.

Average pressure was determined by reading gauge pressures in and out of each vessel and averaging them. Osmotic pressure, which in this study was mathematically correlated to conductivity, was subtracted from the average applied pressure to give net driving pressure. Therefore, flux rates were determined according to the following equation:

$$\text{Flux (gal/ft}^2\text{/day at 77°F and desired net pressure)} = \frac{\text{Product flow (gpm) at observed temperature} \times \text{Correction factor to 77°F} \div \text{Average pressure} - \text{osmotic pressure (psi net)}}{\text{Membrane area (ft}^2) \times 1{,}440 \text{ min/day} \times \text{desired net pressure}}$$

Salt rejection measures the ability of a membrane to reject specific ions; therefore, rejection is a measure of the efficiency of a membrane in separating pollutants from the product water by being relatively impermeable to the passage of salts.

This measurement is generally made in reference to ion concentrations in the water entering the reverse osmosis unit as compared to the treated water discharged. For these studies, percent salt rejection = (concentration of influent - concentration of product) x 100 ÷ concentration of influent. In the case of the hollow fiber unit at Mocanaqua, the influent was raw acid mine drainage. For all spiral wound studies, blended feed was the influent to the unit.

Loss in flux has been observed on RO membranes which were treating high purity water. This flux loss phenomena in pure water systems is due to compaction of the membrane and/or its porous support structure. In systems free of fouling, this flux loss follows a linear log-log plot in respect to time.

Under actual treatment conditions, some fouling occurs in every application and the flux trend may approximate a straight log-log line for varying lengths of time until a significant buildup of fouling occurs. At that time, an obvious deviation occurs as the flux line begins to curve downward. Also, the slopes of flux loss trends are steeper than for compaction losses alone.

The log-log slope is very important in analysis of reverse osmosis operation as it is an indicator of the severity of fouling and a rough predictor of membrane life. Flux decline slopes in excess of -0.05 are felt to be intolerable.

Therefore, membrane performance, deterioration, and fouling were evaluated by plotting observed flux values on log-log paper and calculating log-log flux decline slopes from the line of best fit. Taking values from the graph, the slope was calculated by the following equation:

$$\text{Slope} = \frac{\text{Log Flux}_2 - \text{Log Flux}_1}{\text{Log Time}_2 - \text{Log Time}_1}$$

RESULTS—EPA STUDY

Maximum Recovery Studies, Reverse Osmosis and Neutrolosis

The acid water source for these studies was Grassy Run, a surface stream of which 90% of the flow emanated from abandoned coal mines. Due to surface flow and biological influence, the ferrous iron was rapidly oxidized to ferric. By the time the water reached the Norton Facility, more than 90% of the iron was in the ferric state. As a surface flow, wide variations in temperature were observed in relatively short periods of time. In addition to the AMD, Grassy Run also contained sewage from approximately 30 residences upstream of the EPA Facility.

The 10 K spiral wound reverse osmosis unit was used for these tests. As shown in Figure 6.4, sand filtration and 10 micron cartridge filtration served as pretreatment for the RO unit. Osmotic pressure was assumed to equal 70 kN/m^2 (10 psi) per 1,000 micromhos conductance. Five tests were made during the high recovery stage of RO testing.

Test Number One (Reverse Osmosis): This 100 hour reverse osmosis test was made at 91% water recovery and was monitored 24 hours per day. The operational and chemical data are given in Tables 6.2 and 6.3. Salt rejections were uniformly above 99% on all multivalent ions. Figure 6.5 presents the history of membrane performance during the test run.

Both the unit flux and the tube five flux values decreased with time as the run progressed due to membrane fouling. An explanation for the up and down flux variation is not available. The tube five fouling rate was somewhat more severe than that for the total unit as tube five was the last one and was subject to the highest ion concentrations.

A greater risk of precipitation would be expected. Calcium sulfate precipitation appeared to be the predominant fouling mechanism as $CaSO_4$ crystals formed in brine samples within one hour after sampling.

TABLE 6.2: OPERATING PARAMETERS FOR TEST NUMBER ONE

Parameter	Unit	Value
Raw Water Feed Flow	gpm	5.75
Product Water Flow	"	5.24
Brine Water Discharged	"	0.51
Brine Water Recycled	"	3.69
Minimum Brine/Product Flow Ratio	ratio/module	9:1
Maximum Brine/Product Flow Ratio	"	17:1
Water Recovery	percent	91.2
Recovery of Blended Feed	"	55.5
Feed Pressure	psi	600.6
Feed Temperature	°F	44
Unit Flux, gal/ft^2/day @ 600 psi & 77° F		20.50
*Tube Five Flux, gal/ft^2/day @ 600 psi & 77° F		19.55
Length of Run	hours	100.2
Date of Run		March 9-13, 1970

All Values are Means from 26 Data Sets
* Last tube in unit

TABLE 6.3: CHEMISTRY ANALYSES FOR TEST NUMBER ONE

	Conductance	pH	Acidity	Calcium	Magnesium	Aluminum	Total Iron	Sulfates
Raw Feed	1200	2.7	630	110	33	35	110	810
Blended Feed	4200	2.2	2600	480	180	170	530	4000
Brine	9500	2.0	5900	1100	410	400	1200	9500
Product	250	3.4	115	3.0	0.7	1.1	2.8	17
Rejections[(a)]	94.1%	-	95.5%	99.4%	99.6%	99.4%	99.5%	99.6%

All units are mg/l except pH and conductance (Mmhos/cm)

[(a)] $\text{Rejection} = \frac{\text{Blended Feed Concentration-Product Concentration}}{\text{Blended Feed Concentration}} \times 100$

Source: EPA-670/2-73-100

The rate of fouling was severe enough that once per week flushing would be required to maintain membrane performance. Long-term operation of this system with the incorporation of flushes appeared possible.

FIGURE 6.5: MEMBRANE PERFORMANCE DURING TEST NUMBER ONE

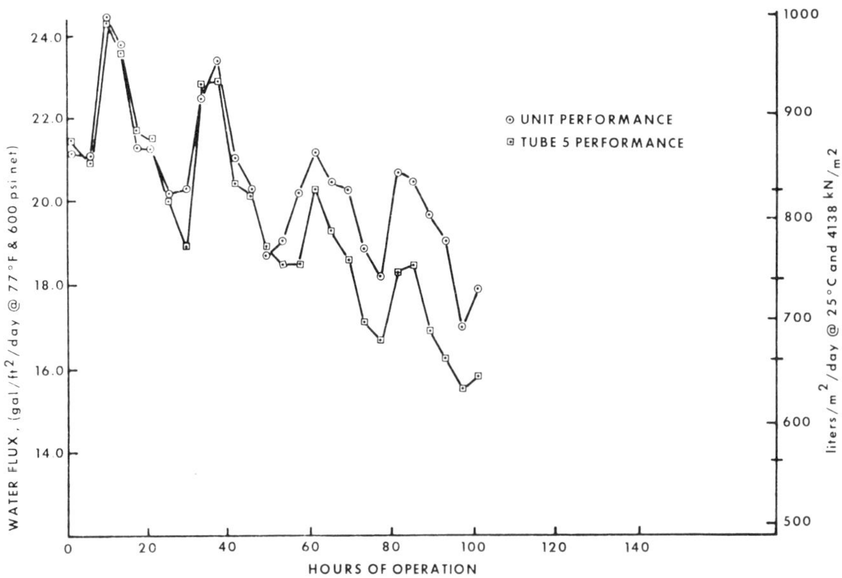

Source: EPA-670/2-73-100

Test Number Two (Reverse Osmosis): During this 96.5 hour reverse osmosis test run, the unit operated at 91% recovery during the day (7 hr/day), was lowered to 85% recovery at night (17 hr/day) and was monitored during the 8 hour work day. Table 6.4 presents a comparison of typical operational data at the two recovery rates. Flux values were higher at 85% than at 91% recovery due to less fouling at the lower recovery.

Chemical analyses show that the salt rejections were virtually identical at both recovery rates and were comparable to the rejections of Test Number One. The concentration of pollutants in the brine at 91% recovery was roughly 1.2 times the respective concentrations at 85% recovery.

As lowering the recovery tended to reduce the fouling, a test was made to determine if the fouling mechanisms could be detected by chemical analyses and subsequent material balances. During the run, samples were taken immediately before and 30 minutes after lowering the recovery from 91 to 85%. At 91% recovery, the quantity of all constituents entering the unit agreed within 10% with the quantity leaving the unit.

When the recovery was lowered to 85%, the mass balance indicated approximately 30% more of each ion (calcium, sulfate, iron, aluminum and magnesium) was coming out of the unit than was going in. These results clearly showed that the in-unit precipitate consisted of more than calcium sulfate alone and that the precipitate could be at least partially flushed from the unit by lowering recovery and increasing brine flow. Other studies indicate that iron deposition may prove to be the long-term threat to flux. However, calcium sulfate precipitation is the immediate threat as massive $CaSO_4$ precipitation was observed in high recovery brine samples within one hour of sampling.

TABLE 6.4: OPERATING PARAMETERS FOR TEST NUMBER TWO

Parameter	91 Percent Recovery	85 Percent Recovery
Hours per Day	7	17
Raw Feed Water Flow, gpm	5.42	5.63
Product Flow, gpm	4.94	4.94
Brine Flow Tube Five, gpm	4.09	4.11
Brine Flow Discharged, gpm	0.48	0.69
Brine Flow Recycled, gpm	3.61	3.42
Water Recovery Percent	91.1	84.4
Recovery of Blended Feed, Percent	54.7	54.6
Feed Pressure, psi	600.0	602.5
Feed Temperature, °F	45°	44°
Unit Flux, gal/ft^2/day @ 600 psi Net & 77° F	18.32	18.79
Tube Five Flux, gal/ft^2/day @ 600 psi Net & 77° F	16.80	17.86
Length of Run, Hours	96.5	
Date of Run	March 24-27, 1970	

All Values are Means from Six Data Sets

Source: EPA-670/2-73-100

Test Number Three (Neutrolosis): Brine disposal from an RO system is a severe problem. Even at 91% recovery, 9% of the water treated resulted as a highly polluted liquid brine. Because of $CaSO_4$ fouling, 91% was near the maximum attainable recovery under the test conditions at Norton. Previously, the Norton Mine Drainage Field Site had conducted extensive research on neutralizing acid water and RO brine. Originally, the plan for controlling brine was to lime neutralize the brine at 91% recovery and then blend the neutralized supernatant with the product water from the reverse osmosis unit, fully realizing that the neutralized supernatant would degrade the high quality product water.

The idea of "neutrolosis" was then conceived. Instead of blending the neutralized brine supernatant with the product, blend the neutralized supernatant back into the feed to the RO unit. In this manner, the only effluents from the process would be RO product of a high quality and a neutralized sludge. The major concern was that the higher pH of the neutralized supernatant would increase the pH of the blended feed to a point where ferric iron would hydrolyze and precipitate (around pH 4).

Test Number Three was an initial 30 hour feasibility test of the neutrolosis concept. The reverse osmosis unit recovery was 91%, the brine was neutralized, and the supernatant did not raise the blended feed pH enough to cause iron precipitation and the overall neutrolosis water recovery was 98.3%.

Test Number Four (Neutrolosis): A complete flow diagram of neutrolosis reverse osmosis-neutralization) is presented in Figure 6.6. For the neutrolosis test, brine from the RO unit passed directly into a 189 liter stainless steel reaction tank where lime was added. The neutralized brine then was pumped from the reactor to a 4,542 liter upflow settling tank.

In the settling tank, the majority of the solids was removed as sludge and the supernatant was filtered and returned to the feed of the RO unit. Thus, the only effluents from the process were RO product water and neutralized brine sludge. For this 99.6 hour test, the RO unit operated at 91% recovery. Total recovery was 99.0%. Table 6.5 presents the operational data for the test run.

As in Tests One and Two, the salt rejections in Test Number Four were better than 99% on all multivalent ions. Lime neutralization removed virtually all the iron from the brine. However, pH 4.7 was not high enough to accomplish total aluminum removal and 460 mg/l of acidity remained in the supernatant. Even though calcium (lime) was added, there was a net loss in calcium and in sulfate indicating that a significant amount of calcium sulfate was removed during the neutralization process.

FIGURE 6.6: FLOW DIAGRAM FOR NEUTROLOSIS TESTS

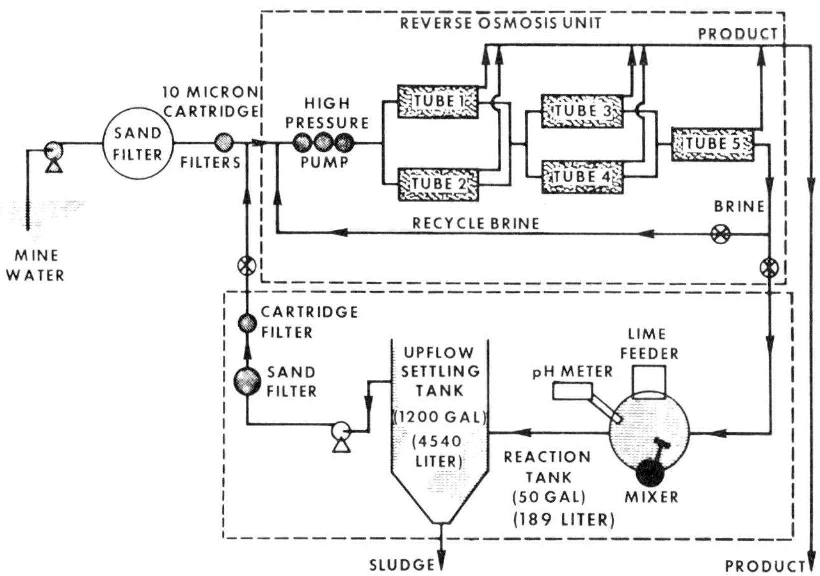

Source: EPA-670/2-73-100

TABLE 6.5: OPERATING PARAMETERS* FOR TEST NUMBER FOUR

Parameter	Unit	Value
Raw water feed flow	gpm	5.02
Product water flow	"	4.97
Brine water discharged (flow to neutralizer)	"	0.48
Brine water recycled	"	3.49
Minimum brine/product flow ratio	ratio/module	6.3:1
Maximum brine/product flow ratio	"	24:1
Neutralized brine supernatant recycled	gpm	0.43
Neutralized brine sludge discharged	"	0.05
Reverse osmosis water recovery	percent	91.2
Recovery of blended feed	"	55.6
Total system recovery - neutrolosis	"	99.0
Percent sludge (by volume of discharged brine)	"	10.4
Feed pressure	psig	602.2
Feed temperature	°F	58
Unit flux, gal/ft^2/day @ 600 psi net and 77°F		15.43
Tube five flux, gal/ft^2/day @ 600 psi net and 77°F		9.83
Length of run	hours	99.6
Date of run		May 11-15, 1970

*All values are means from 82 data sets.

Source: EPA-670/2-73-100

Comparing the product quality of this neutrolosis test with that of Test Number One, the product qualities were virtually identical. The only effects of recycling the nuetralized supernatant were seen in the blended feed concentrations. The neutrolosis blended feed had increased concentrations of magnesium and aluminum due to insufficient removal during neutralization.

Previous studies indicated both these constituents could be removed around pH 7. If the magnesium and aluminum are not removed, a long-term buildup would occur and neutralization should be taken close to pH 7 to alleviate this problem. Any ions which would not precipitate might be removed by occasionally increasing pH and "blowing down" the system.

A sharp decrease in tube five flux in respect to the unit flux indicated the severity of the fouling. Calcium sulfate precipitation was substantially worse in tube five than in the rest of the unit as tube five was subject to the heaviest concentration of pollutants.

Four low-recovery (50%) flushes were required during the first 90 hours of the test to remove the severe fouling. Flush number five was a 3 hour flush at low pressure (690 kN/m^2 or 100 psi) and low recovery (40%) using BIZ enzyme detergent. The BIZ flush was more successful in increasing flux than were low recovery acid flushes. The severe fouling observed in Test Number Four would prohibit feasible operation under these conditions.

Test Number Five (Neutrolosis): By operating the RO unit near 80% recovery and employing the neutrolosis concept, it was hoped that the severe fouling observed in Test Number Four could be reduced to tolerable levels. As lowering the RO recovery increased the volume of brine discharged, the flow of neutralized supernatant also increased and again the possibility of the supernatant increasing the blended feed pH to the point where ferric iron would precipitate was feared.

Table 6.6 presents operational data for the run, The recycled neutralized brine flow was 4.731 l/m at 80% RO recovery as compared to 1.628 l/m in Test Number Four. However, the pH of the blended feed was still only pH 2.5, far from the danger of ferric iron precipitation.

TABLE 6.6: OPERATING PARAMETERS* FOR TEST NUMBER FIVE

Parameter	Unit	Value
Raw Water Feed Flow	gpm	4.83
Product Water Flow	"	4.77
Brine Water Discharged (Flow to Neutralizer)	"	1.31
Brine Water Recycled	"	2.65
Minimum Brine/Product Flow Ratio	ratio/module	6:1
Maximum Brine/Product Flow Ratio	"	23:1
Neutralized Brine Supernatant Recycled	gpm	1.25
Neutralized Brine Sludge Discharged	"	0.06
Reverse Osmosis Water Recovery	percent	78.5
Recovery of Blended Feed	"	54.6
Total System Recovery-Neutrolosis	"	98.8
Percent Sludge (by Volume of Discharged Brine)	"	4.6
Feed Pressure	psig	600.0
Feed Temperature	°F	61.5
Unit Flux, gal/Ft^2/day at 600 psi Net & 77" F		12.28
Tube 5 Flux " " "		9.99
Length of Run	hours	130.6
Date of Run		June 3-8, 1970

*All values are means from 34 data sets.

Source: EPA-670/2-73-100

Total recovery during this 130.6 hour test was 98.8% at 78% RO recovery as compared to 99.0% recovery in Test Number Four. The rate of flux decline was significantly less at 80% RO recovery (Test Five) as compared to Test Four. Both the unit flux values and the tube five flux values declined at approximately equal rates during this 80% test, indicating the fouling was uniformly distributed throughout the unit. In the earlier tests, the tube five fouling rate was noticeably more severe than the unit fouling rate. Comparing this neutrolosis test at 80% RO recovery with Test Number One, the fouling rates were almost identical.

As in previous tests, salt rejections were near 99% on all multivalent ions. A 2.5 hour low-recovery (50%) flush improved the flux rate 7% for the unit and 38% for tube five, indicating that the majority of earlier fouling had occurred in tube five.

Operation of this neutrolosis system at 80% RO recovery and 98.8% total recovery appeared feasible for long-term use if periodic flushes were incorporated to control fouling and sustain membrane performance.

Discussion of Five Maximum Recovery Tests: The chemical analyses of the blended feed (water actually entering RO unit) and brine were compared in an attempt to determine the cause of Test Four's severe fouling as compared with Tests One and Five. These analyses, as shown in Table 6.7, indicated that the water characteristics were identical between Tests One and Four, except for magnesium and sulfate concentrations which were slightly higher in Test Four, yet only Test Four fouled severely.

TABLE 6.7: BLENDED FEED, BRINE AND PRODUCT CHEMISTRY ANALYSES FOR 10 K STUDIES

Test	RO recovery	Total recovery	Cond.	pH	Acidity	Ca	Mg	Al	Total iron	Sulfate
			BLENDED FEED							
#1 Reverse osmosis	91.2	91.2	4200	2.2	2580	480	180	170	530	4000
#4 Neutrolosis	91.2	99.0	5200	2.2	2700	470	270	190	520	4700
#5 Neutrolosis	78.5	98.8	3000	2.5	1150	460	120	91	210	3000
			BRINE							
#1 Reverse osmosis	91.2	91.2	9540	2.0	5900	1100	410	400	1200	9500
#4 Neutrolosis	91.2	99.0	10,000	2.0	6000	1100	550	410	1200	11,000
#5 Neutrolosis	78.5	98.8	5600	2.3	2400	940	250	190	460	6400
			PRODUCT							
#1 Reverse osmosis	91.2	91.2	250	3.4	115	3.0	0.7	1.1	2.8	17
#4 Neutrolosis	91.2	99.0	340	3.4	130	3.0	1.5	1.1	1.6	21
#5 Neutrolosis	78.5	98.8	210	3.7	36	2.1	0.6	1.0	1.2	18

All units are mg/l except for recovery (percent), conductance (Mmhos/cm), and pH.

Source: EPA-670/2-73-100

All constituents were much lower in concentration in Test Five than in either Test Four or Test One as the reverse osmosis unit recovery was only 78% as compared to 91% in the other two tests. Still, Test Five fouled at the same rate as Test One.

The conclusions drawn from these data were that the fouling in all tests was due to precipitation, but this precipitation apparently was not directly related to chemical concentration (as the concentrations in Test Five were significantly less than Test One, yet the fouling rates were similar; and the concentrations were approximately equal in Tests Four and One, yet Test Four fouled severely).

Instead, the physical properties of the recycled neutralized brine supernatant appeared to promote precipitation. Since the neutralized recycle was already above saturation for calcium sulfate, it was assumed that minute $CaSO_4$ seed crystals existed in the recycled supernatant and these promoted precipitation in the unit.

Product water quality, as shown in Table 6.7, was much superior in Test Five as compared to Tests One or Four as the lower RO unit recovery (78%) did not concentrate the pollutants as much as the higher recovery (91%) runs. Since salt rejections were approximately the same in all these tests, lowering the brine concentrations improved product quality. None of the product waters were of potable quality. Further treatment with lime to increase the pH and remove the iron would be required to increase the quality to potable standards.

In conclusion, the feasibility of 91% recovery reverse osmosis operation under the conditions at Norton was borderline. If periodic flushing would continue to control fouling, operation would be feasible. Disposal of the 9% highly polluted liquid brine would still be a major problem.

The use of neutrolosis at 91% RO unit recovery produced a total recovery around 99% and 1% of the initial volume resulted as a dense sludge which could be landfilled. Unfortunately, the fouling rate of the RO membranes under these conditions was too severe for feasible long-term operation.

By employing the neutrolosis concept at 80% RO unit recovery, fouling was minimized to levels where continuous long-term operation appeared feasible. Periodic flushes should be able to control membrane fouling. Total recovery by this system was 98.7%. The remaining 1.3% of the original water resulted as sludge.

The neutrolosis concept is a significant breakthrough in RO technology. Converting 99% of the acid water treated into a high quality product water and the remaining 1% into a dense sludge is a dramatic result. The reverse osmosis problem of brine disposal was eliminated in the Norton application by this neutrolosis process.

This process or modifications of it could be used to treat most types of acid mine drainage. Ferrous iron sites may have to include aeration and higher pH neutralization to remove the iron prior to recycling. At such sites, it may be possible to inject acid to lower the pH of the neutralized brine supernatant before recycling it into the feed.

Sludge dewatering devices would increase the recovery even further. Development of calcium sulfate precipitation inhibitors would reduce membrane fouling problems. In brief, neutrolosis is a very promising method for the treatment of acid mine drainage.

Two-Stage Reverse Osmosis System

In order to utilize sodium hexametaphosphate as a $CaSO_4$ precipitation inhibitor, the water must first be neutralized. Neutralization of the entire volume of water to be treated by the reverse osmosis unit was impractical. A more feasible scheme was to operate the 10 K RO unit on a raw AMD at 91% recovery and to neutralize the brine (9% of the original volume—therefore neutralization costs would be less).

After settling the solids, the neutralized supernatant water would be fed into a smaller, second, RO unit in which sodium hexametaphosphate could be used as a chelating agent since the pH would be close to neutral.

To test this arrangement, a 4 K unit was used as the second-stage RO unit. The 4 K unit consisted of three pressure vessels arranged in a 2-1 array where the brines from tubes one and two combined to serve as the feed to tube three.

Two tests were made of the proposed process. Brine from the 10 K unit operating at 91% RO recovery was collected, neutralized, and stored. From the storage tank, the neutralized water was pumped into the 4 K RO unit which operated at 50% recovery.

Overall system recovery was 94.75%. A continual decrease in product flow was observed during the 2.6 hour 4 K run on brine supernatant water which had been neutralized to pH 4.8. A rapid increase in pressure loss across the unit occurred indicating immediate and severe fouling terminating the test.

Sodium hexametaphosphate was ineffective at pH 4.8 in controlling $CaSO_4$ precipitation. As very little iron was present in the neutralized feed, the inability of the chelating agent to perform successfully was probably the result of the acidic pH (4.8). No further tests were made at higher pHs, however. Salt rejections during this test were unexplainably low.

Morgantown Ferrous Iron Spiral Wound Study

The next series of studies was conducted on ferrous iron. The first ferrous site was the acid mine discharge into Indian Creek located 10 miles south of Morgantown, West Virginia, at the Arkwright Mine of Christopher Coal Company. The flow diagram for the 4 K unit which was used in these tests and support facilities is shown in Figure 6.7. Neither brine recycle nor pH control was used during this study.

Initially, tubes 1 and 2 contained standard-flux, high-selectivity modules and tube 3 contained high-flux, intermediate-selectivity modules. These modules had been in operation approximately 180 hours prior to this study. After the first test run, the modules were badly fouled and were replaced. The loading for the second test was: tube 1, new high-flux modules; tube 2, new standard-flux modules; and tube 3, used high-flux modules.

The first run, lasting 45.7 hours, was begun at low pressure and 40% water recovery. The recovery was gradually increased by raising the pressure until 50% recovery was established. Unfortunately, data were available for only the first 27.8 hours of operation.

FIGURE 6.7: FLOW DIAGRAM FOR 4 K SPIRAL WOUND REVERSE OSMOSIS UNIT

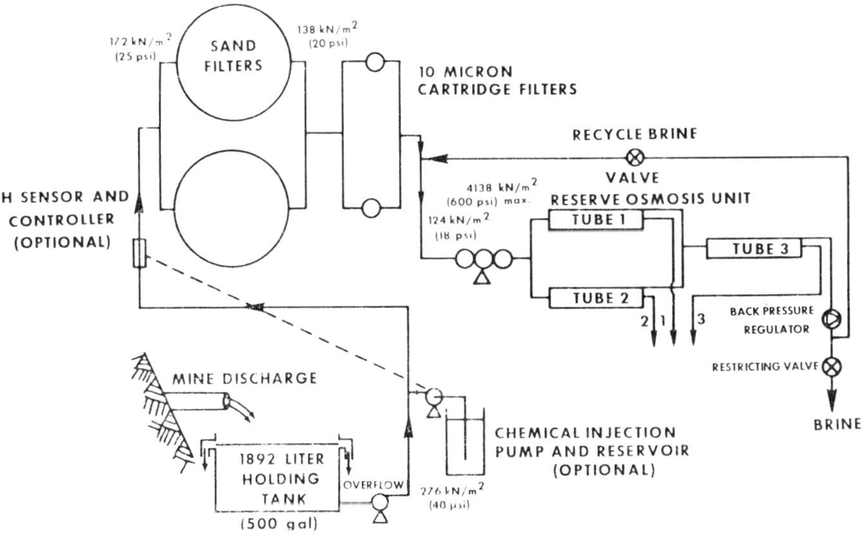

Source: EPA-670/2-73-100

The differential pressure, which indicates rapid fouling, increased initially but levelled off after four hours of operation. However, performance of the membrane, as illustrated by the water flux rate, showed a definite decreasing trend.

The second test was begun at 50% water recovery and lasted 66.1 hours, but data were available for only the first 46.4 hours. At the end of 66.1 hours, the unit was badly fouled by precipitation which occurred when a brass valve corroded and diminished the water supply to the unit, thus the unit was receiving only recirculated brine.

The water flux decreased 21% in 46.4 hours of operation. The flux value for tube 3 (where the precipitation problems are most acute due to the higher concentration of ions) decreased only 19%. A 35% flux loss occurred in tube 1, 5% in tube 2, and 19% in tube 3. The high flux decrease in tube 1 (loaded with new high-flux modules) was difficult to explain, especially since tube 2 was in parallel to tube 1 and was subject to the same conditions.

After termination of the test run, an explanation for the tube 1 flux loss was found. The brine seal on the last module of the tube had failed and allowed water to bypass the module. The module produced progressively less water until it became severely fouled by precipitates.

If results from tube 1 are ignored and flux declines between tubes 2 and 3 are compared, the decrease appears to be a concentration related phenomena, as a much greater decrease was observed in tube 3 where the concentration was the greatest.

There was no significant increase in the differential pressure for the unit or for the individual tubes. This indicated that the rate of fouling was not too severe to prohibit short-term operation on this water at the 50% recovery level. Rejection of the various ions by the membranes were relatively high (98% on all multivalent ions). Rejection of ferrous iron was somewhat lower than the rejection of ferric iron as trivalent ions are better rejected than are divalent ions. Maximum brine/product (b/p) flow ratios during the study were near 9.4:1. Minimum b/p ratios were 5.1:1.

Discussion of Morgantown Study: Chemical composition of the feed water at the Arkwright Site was similar to that of the 91% recovery brine from the RO unit at the Norton Site (Table 6.3). Attempts to further treat the brine from the Norton test by RO had revealed that calcium sulfate fouling occurred at 50% recovery. Therefore, it was not surprising that 50% recovery was the maximum attainable at Arkwright.

The two test runs, though short-lived, were terminated by equipment and facility problems, rather than module or membrane failure. However, membrane fouling was evident. 50% recovery was about the maximum recovery possible before $CaSO_4$ deposition began to rapidly foul the membrane.

Effects of fouling were easily seen during the study by the decrease in flux values for the membranes. Unfortunately, the exact cause of fouling was difficult to determine. All evidence pointed to calcium sulfate deposition, but the role of ferrous iron could not be determined.

During the study, a sample of brine at 50% recovery was collected and a precipitate formed within one hour. Analysis of the precipitates indicated the composition to be virtually $CaSO_4$. Trace amounts of iron were present. A sample of the precipitate that lined the pressure vessel was analyzed. The composition of the in-unit precipitate was similar to the external brine precipitate.

The majority of the precipitation in the unit occurred during the last few hours before the runs were terminated. In both test runs, mechanical problems caused the recovery to soar to undetermined levels (probably greater than 90%) before the unit shut down and thus the brine was very concentrated and precipitation occurred. As calcium sulfate is the first salt to exceed solubility, it was expected that the precipitate would be largely calcium sulfate. The analytical results confirm this conclusion.

However, this predominance of $CaSO_4$ deposition may not be representative of what occurred during successful operation. The possibility of ferrous iron fouling cannot be ruled out nor can it be substantiated by these results.

Contrasting the salt rejections at Arkwright with those at Norton, the rejections at Norton were slightly higher in most cases. This difference in rejection rates was due to the high ionic concentration at Arkwright whose higher osmotic pressure resulted in less product flow while the salt flow remained essentially the same. This conclusion was valid since osmostic pressure is directly related to ionic concentrations and increases in ionic concentrations elevate osmotic pressure.

In conclusion, it does not appear that RO would be a feasible method for treatment of water of this quality based upon the problem of membrane fouling which severely limited recovery (50%) resulting in disposal of an equal volume of brine for each gallon of water reclaimed.

Ebensburg Ferrous Iron Spiral Wound Study

A ferrous iron test site with concentrations more typical of AMD commonly encountered was located near Ebensburg, Pennsylvania, at the 32-33 discharge of Bethlehem Mines Corporation Cambria Division. The 4 K spiral wound test system was installed there. The study included tests to determine if iron fouling could be controlled by pH adjustment, i.e., the lower the pH, the less likelihood of iron precipitation.

An improved version of Gulf Environmental System's standard flux, high-selectivity module was used in this Ebensburg study. In these modules, the membrane was cast directly upon the backing material and the old glue joints were eliminated. The glue joints may have contributed to small "leakage" problems in the past and the manufacturer felt that their elimination should insure high rejection rates.

The system flow arrangement was the same as that at Arkwright (Figure 6.7) except for acid injection which was used during this study. Osmotic pressure was assumed to equal 10 psi per 1,000 micromhos of conductance. Water quality of the Ebensburg site was similar to that at Norton (except for ferrous iron). Four tests were made during this study.

Test Number One: As the pH of the AMD at Ebensburg was 3.6, the possibility of ferric iron precipitation was feared. Sulfuric acid was, therefore, injected automatically into the water before it entered the RO unit in order to lower the pH to 2.5. The test run lasted 191 hours and the amount of acid injected was gradually diminished toward the end of the run. The recovery during the test was maintained near 84% by recycling a portion of the brine into the feed to the unit.

As tube 3 was subject to the most polluted water, it was more prone to fouling than tubes 1 and 2. The pressure drop (Δ P) across tube 3 after 100 hours of operation had increased significantly, thus indicating membrane fouling. The recovery was then reduced for three hours in an effort to remove the fouling. The flush was successful in restoring the ΔP across tube 3 to normal values. Significantly, the ΔP across tubes 1 and 2 decreased during the run while tube 3 increased, indicating that the fouling was a concentration-related phenomenon.

Very little fouling was observed in tubes 1 or 2 as their log-log slopes were -0.043. An excessive flux decline was observed in tube 3. A rapid change in log-log slope for tube 3 occurred after approximately 40 hours of operation on the Ebensburg water.

It is felt that the tube 3 flux loss was due to calcium sulfate fouling. Although acid injection was diminished as the run progressed, very little change occurred in the tube 3 pH. Thus, the decrease in acid injection probably had no significant effect on tube 3 behavior.

Salt rejections were uniformly greater than 99% on all multivalent ions. A module was removed from the downstream end of tube 3 at the end of the test and analyzed. A white flaky precipitate in the module was found to be almost pure gypsum. No tests were made to determine the presence of iron.

Test Number Two: The injection of sulfuric acid to control the influent pH was not used in this test as it was observed that, at 80 to 85% recovery, the recycled brine lowered the pH of the resulting "blended feed" from pH 3.6 to pH 3.1.

This test was plagued with shutdowns. In a total run time of 86.3 hours, the longest continuous run was 22.5 hours and the longest continuous period of data available was for only 17.3 hours. The shutdowns were all caused by pump problems.

Each accidental shutdown acted as a flux rejuvenator. Planned full pressure shutdowns have long been used at Norton to dislodge precipitates and restore membrane performance. In essence, each of these accidental shutdowns served to clean and relax the membrane.

Considering the data between shutdowns as individual runs, the flux values in most cases decreased as the run progressed. As in Test Number One, the flux decline in tube 3 was significantly greater than the decline in tubes 1 or 2.

The ΔP value over the entire test run did not change appreciably for tubes 1 and 2 but did show a continuous increase across tube 3. This was analagous to the results of Test Number One. Flux rates for tubes 1 and 2 were greater at the end of the test than at the beginning. The tube 3 flux decreased 9.2% over the entire run. Overall, there was significantly less fouling in Test Number Two than in Test Number One, due to the shutdowns which acted as precipitation removers. As in Test Number One, the salt rejections of all multivalent ions were greater than 99%.

Test Number Three: This was an 18 hour, 80 to 85% recovery recycle brine run with sodium hydroxide injection. Without brine recycle or acid injection, the RO unit would have to treat pH 3.6 AMD at this site. It was the intent of Test Number Three to simulate pH 3.6 influent conditions to the 4 K unit by injecting sodium hydroxide to increase the pH of the blended feed.

The point of injection was highly critical in this application. Referring to Figure 6.7, if the NaOH were injected into the blended feed, and overtreatment occurred, the ferric iron would precipitate directly in the RO unit. To protect the RO unit, the NaOH was injected into the raw feed prior to the sand filters.

As the recycled brine lowered the blended feed pH from pH 3.5 to pH 3.1, the NaOH increased the raw feed pH to counteract the effect of the recycled brine. This meant the raw feed must be increased to approximately pH 3.9 to obtain the desired result. At pH 3.9, however, the ferric iron rapidly precipitated in the sand filter. The clogged sand filter prevented water flow to the unit so the test was discontinued. The collected data were not sufficient to establish any definite trends.

Test Number Four: By eliminating brine recycle, the pH conditions which Test Number Three was trying to produce were achieved. With no recycled brine, the raw AMD was fed directly into the RO unit at pH 3.6. Unfortunately, without the use of brine recycling, the small 4 K unit was limited to approximately 55% recovery.

The test was run 94 hours with 50 to 55% recovery. The ΔP remained constant throughout the test run. Flux values, though somewhat erratic, were identical for tubes 1, 2 and 3. Flux history did not indicate fouling in any of the tubes. This was quite different from the readily apparent fouling seen in Tests Number One and Two.

Peaks in the flux value graph all occurred during periods of lowest water temperature which indicate some discrepancies may exist between the manufacturer's temperature-flux normalization curve and the behavior of the Ebensburg water. Salt rejections were greater than 99% on all multivalent ions except aluminum.

The downstream module in tube 3, which had been installed at the beginning of Test Number Four, was removed and a post mortem analysis made. Although the module was damaged during shipment, some iron was found to be present on the membrane surface. There was no calcium sulfate present on the membrane.

Discussion of Four Ebensburg Tests: At this site, it was seen that pretreatment of the feed to lower the pH was not necessary for short-term tests as long as the pH was near 3.1 and/or the recovery was near 50%. Product water quality though uniformly good, would require slight addition of lime and filtering to be of potable quality.

As the fouling at 85% recovery could be removed successfully by a weekly 3 hour low-recovery flush, fouling would not prohibit operation at that recovery level. The flux at 50% recovery (Test Number Four), however, was roughly 15% greater than the flux during the 85% recovery runs (Tests Number One and Two). Therefore, obvious improvements in membrane performance can be obtained by operating below the $CaSO_4$ fouling level.

Mocanaqua Ferrous Iron Studies Comparing Spiral Wound, Hollow Fiber, and Tubular Units

To investigate Mocanaqua fouling phenomena, EPA and the Commonwealth of Pennsylvania contracted with Rex Chainbelt in 1971 to evaluate the iron fouling problem in the laboratory, to modify the 1969 tubular system, and to conduct additional studies at Mocanaqua.

In order to make these studies more comprehensive, EPA was to conduct simultaneous studies with the 4 K spiral wound RO unit at the same site. A hollow fiber RO permeator was also obtained from this study, and EPA later enlarged it to a three permeator array for a hollow fiber system of comparable size to the spiral wound system. For the ensuing study, all three systems operated side by side on the same water; the only exception was the acid injection to the feed during the last phase of spiral wound testing. Typical water quality characteristics of the Mocanaqua discharge are presented in Table 6.8.

TABLE 6.8: TYPICAL RAW WATER QUALITY CHARACTERISTICS OF MOCANAQUA DISCHARGE

Parameter	Units	Value
pH		3.4
Conductance	Mmhos/cm	1100
Acidity	mg/l as $CaCO_3$	230
Calcium	mg/l	120
Magnesium	"	90
Total iron	"	80
Ferrous iron	"	68
Aluminum	"	11
Sulfate	"	800
Manganese	"	15
Silica	"	10
TDS	"	1200
Dissolved oxygen	"	1
Temperature	°Fahrenheit	54

Source: EPA-670/2-73-100

It was postulated that the most probable cause of the observed fouling was bacterial oxidation of ferrous iron at the membrane surface and subsequent precipitation of the resulting ferric iron which had been hydrolyzed. For this reason, ultraviolet disinfection lights were installed as part of the pretreatment system for all three field units tested in the belief that the UV light would kill most of the iron oxidizing bacteria and thus inhibit or prevent bacterial oxidation of Fe(II) to Fe(III).

Iron oxidation studies were performed during the actual field study to evaluate the effectiveness of the UV light. Specific conclusions relative to possible oxidation control techniques from the oxidation study were:

- ultraviolet disinfection inhibited oxidation;
- lowering the pH to 2.5 inhibited oxidation;
- some oxidation occurred in blended feed samples which had been exposed to UV and kept in the dark;
- at creek temperature (54°F), virtually no oxidation took place in 24 hours; and
- the oxidation mechanism was bacteria since the above variations in treatment inhibited oxidation.

Since some oxidation occurred in two of the blended feed samples that had been exposed to ultraviolet treatment, it was probable that the UV did not effect a 100% kill. If bacteria were present in the blended feed, it was assumed that significant growth was probable at the membrane surface in the RO units.

Another series of oxidation tests at air temperatures 15.6° to 23.9°C (68° to 75°F) resulted in significantly faster oxidation rates as is typical in most biological growth. Two sets of spiral wound modules were used in the 4 K unit during the study. During Phase I, each module contained 4.65 m^2 of standard perchlorate-modified cellulose acetate membrane for a total unit membrane area of 41.9 m^2. The modules used for Phase II contained 5.77 m^2 of formamide-modified cellulose acetate for a total unit membrane area of 51.9 m^2.

Figure 6.8 presents the system arrangement and flow diagram for the spiral unit.

FIGURE 6.8: SPIRAL WOUND REVERSE OSMOSIS SYSTEM

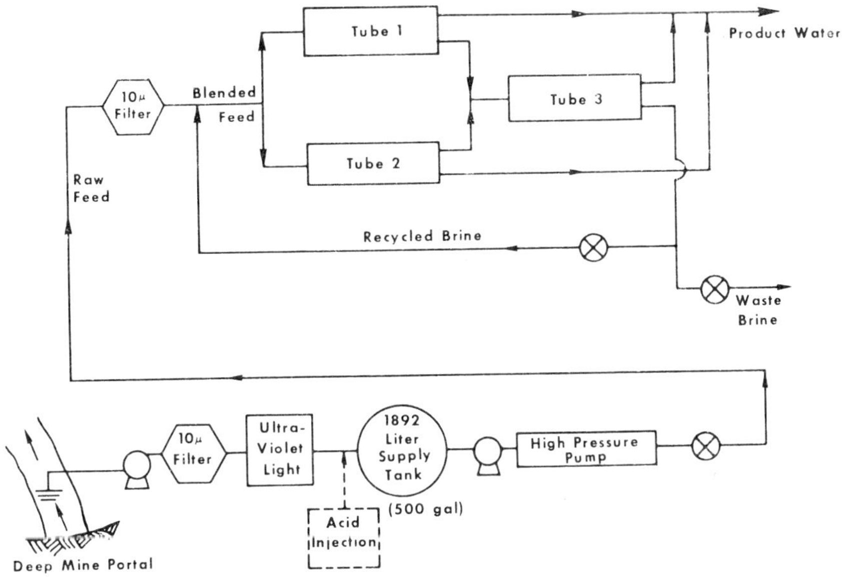

Source: EPA-670/2-73-100

Acid mine water was pumped through ten micron cartridge filters and a UV disinfection light before entering the reverse osmosis unit. During Phase II, sulfuric acid was injected to lower the blended feed pH to 2.9 and thus inhibit both ferrous oxidation and ferric precipitation.

The second unit tested was a hollow fiber system which initially consisted of one permeator. As illustrated in Figure 6.1a, approximately 139.6 m^2 of 8 to 9 hollow fiber nylon membrane were packed in a 15.2 cm x 1.22 m stainless steel pressure vessel.

During Phase I, the single permeator operated at 400 psi from a Moyno high pressure pump which also supplied feed water to the tubular unit. As Phase I tests were successful two additional permeators were added for Phase II and formed a three permeator array, 6 K unit, as shown in Figure 6.9.

The brine from both new permeators in parallel served as the feed to permeator Number 3 (the original Phase I permeator). Flow control orifices were placed in each brine line from permeators 1 and 2 to ensure equal brine flows and prevent system unbalance. Because high brine flows are not required for hollow fiber systems, it was not necessary to recycle brine. Therefore, the unit received raw acid mine water that had been filtered and had passed through the ultra-violet light for disinfection.

Phase I tubular studies utilized a unit consisting of sixty 7.6 cm x 2.39 m plastic tubular #310 modules illustrated in Figure 6.1c and arranged as shown in Figure 6.10. Each module contained eighteen ½" i.d. porous fiberglass tubes that were lined with cellulose acetate membrane and connected in series. Membrane area was 1.57 m^2 per module for a total system membrane area of 94.4 m^2. Pretreatment was 10 micron filtration and UV disinfection.

FIGURE 6.9: HOLLOW FIBER REVERSE OSMOSIS SYSTEM

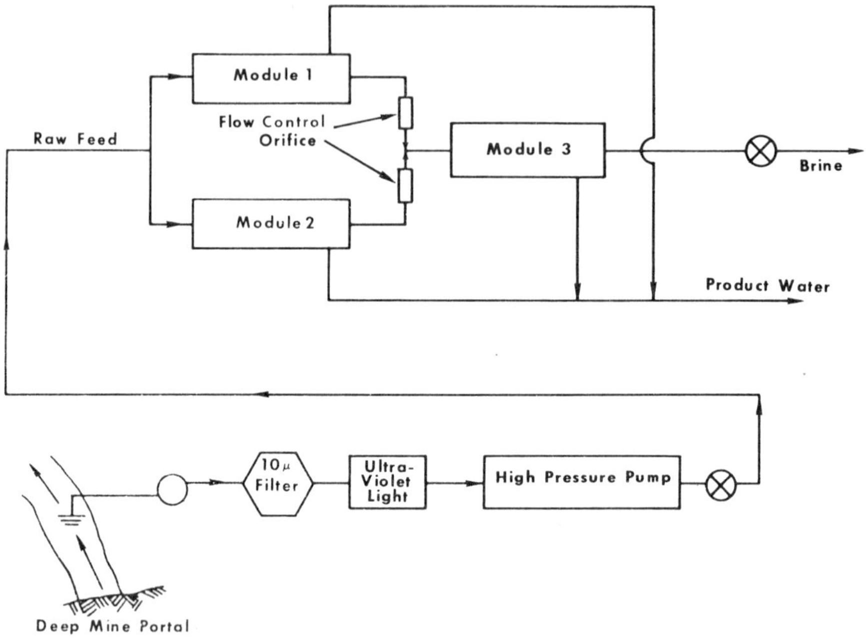

Source: EPA-670/2-73-100

Turbulence promotion rods were placed in modules where brine flow was lowest to inhibit boundary layer precipitation and fouling. Unfortunately, use of turbulence promoters also resulted in high pressure drops across the modules in which they were installed, thus reducing the average applied pressure.

During Phase II, the sixty #310 modules were replaced with five #610 modules for a total membrane area of 7.87 m^2. It was necessary to determine the osmotic pressure of the water in order to compensate for variations in water quality and recovery. The spiral wound unit was modified to enable measurement of osmotic pressure by pressure differences between the brine and product side of the membrane.

By varying recovery, osmotic pressure measurements were made over a complete range of concentrations and a correlation was developed between brine conductivity and osmotic pressure. Since acid injection was utilized in Phase II spiral wound studies and since acid affects conductivity, a different correlation was necessary for Phase II.

FIGURE 6.10: PHASE I MODULE ARRANGEMENT FOR TUBULAR REVERSE OSMOSIS

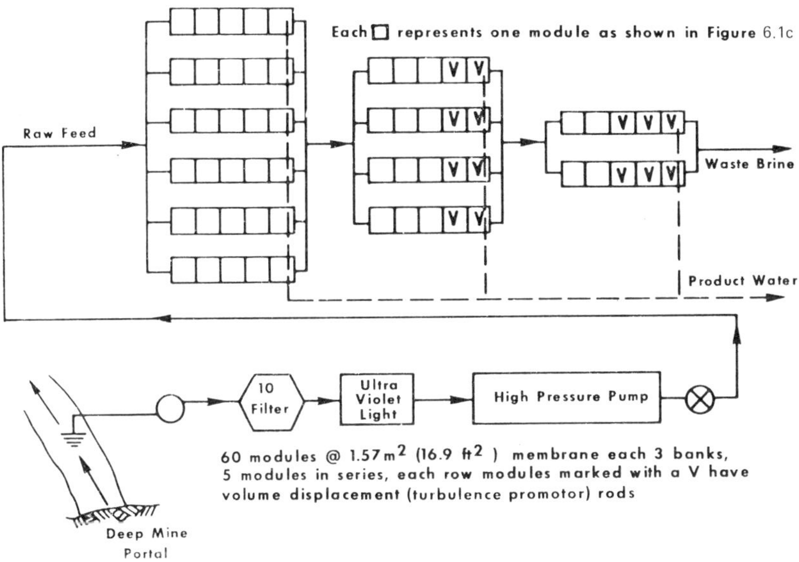

Source: EPA-670/2-73-100

Spiral Wound Phase I Studies: Phase I of the spiral wound studies was made at a unit recovery of 75% and system operating pressure of 600 psi. A summary of operating parameters for this study is given in Table 6.9. Water temperatures of 63°F were caused by the Gould high-pressure pump that imparted energy to the 54°F raw AMD.

The necessity of brine recycle only aggravated the temperature rise since a portion of the water was continually recycled through the pumps. The increased temperature benefitted observable flux rates since product flow was strongly temperature dependent. Since all data were corrected for temperature, the higher operating temperature was not a factor in data evaluation.

In actual operation, however, the higher the temperature, the higher the flux. Overall system recovery during Phase I was maintained at 75% during the first 1,054 hours of operation. The log-log flux decline slope was -0.034 for tube 1 and -0.032 for tube 3 during this period. At 1,054 hours elapsed time, the ultraviolet light was turned off to determine its effectiveness in controlling flux decline.

The increase in flux decline was dramatic with the log-log slope increasing to near -0.2 even though recovery remained at 75%. Following a phosphoric acid flush, the UV was turned on again but recovery was increased to 84%. Flux decline was so severe at 84% recovery (log-log slope greater than -4.0) that the run was terminated after only 70 hours at that recovery level and flushed again with phosphoric acid. Though the acid flush improved flux, it did not stabilize on the original slope, presumably indicating all the fouling had not been removed. Total operating time during Phase I was 1,922 hours.

TABLE 6.9: OPERATING PARAMETERS FOR SPIRAL WOUND PHASE I STUDY AT 75% RECOVERY

Parameter	Value
Raw water feed flow, gpm	6.02
Product water flow, gpm	4.50
Brine water discharged, gpm	1.52
Brine water recycled, gpm	4.26
Minimum brine/product flow ratio (Tubes 1 and 2), ratio/module	5:1
Maximum brine/product flow ratio (Tube 3), ratio/module	12:1
Water recovery, percent	74.8
Recovery of blended feed, percent	43.8
Feed pressure, psig	602
Feed water temperature, °F	62.6
Tube one flux, gal/ft^2/day @ 600 psi net and 77° F	19.56
Tube two flux, gal/ft^2/day @ 600 psi net and 77° F	19.52
Tube three flux, gal/ft^2/day @ 600 psi net and 77° F	18.77
Length of run, hours	1672
Date of run	May 1 - July 12, 1971

All values are means from 73 data sets.

Source: EPA-670/2-73-100

Fouling observed at 84% recovery was attributed to calcium sulfate precipitation. Analysis of the acid flush water confirmed the presence of large amounts of both iron and calcium after the 84% recovery run. A loss of calcium, sulfate and iron was detected in the routine chemical analyses. At 75% recovery and the UV off, it was felt that bacterial-propagated iron precipitation was causing the flux decline.

Operating parameters for the spiral wound unit during the 84% recovery run are given in Table 6.10. Chemical analyses for the 75% recovery operation indicate a 99+% rejection on all multivalent ions. Product waters would still require treatment for iron and manganese and pH adjustment before potable water quality criteria could be met.

Spiral Wound Phase II Studies: The manufacturer suggested that increasing the b/p flow ratio would decrease the log-log slope observed during Phase I. One way of accomplishing a b/p increase was to lower the operating pressure and thereby reduce product flow. Lowering input pressure to 400 psi would also serve to lessen membrane compaction which contributes to flux losses. During Phase I, the maximum b/p ratio per module was 12:1 and the minimum was

5:1. During Phase II, the maximum b/p ratio was 22:1 and the minimum 10:1 which amounted to about 200% increase over Phase I.

TABLE 6.10: OPERATING PARAMETERS FOR SPIRAL WOUND PHASE I STUDY AT 84% RECOVERY

Parameter	Value
Raw water feed flow, gpm	4.92
Product water flow, gpm	4.13
Brine water discharged, gpm	0.79
Brine water recycled, gpm	5.50
Minimum brine/product flow ratio (Tubes 1 & 2), ratio/module	8:1
Maximum brine/product flow ratio (Tube 3), ratio/module	12:1
Water recovery, percent	83.9
Recovery of blended feed, percent	39.6
Feed pressure, psig	600.0
Feed water temperature, °F	66.6
Tube one flux, gal/ft^2/day @ 600 psi net and 77°F	17.03
Tube two flux, gal/ft^2/day @ 600 psi net and 77°F	17.06
Tube three flux, gal/ft^2/day @ 600 psi net and 77°F	16.25
Length of run, hours	70
Date of run	July 13 - July 17, 1971

All values are means from four data sets.

Source: EPA-670/2-73-100

As the earlier oxidation tests had shown pH control to be an effective oxidation inhibitor and as the lower pH would also inhibit precipitation of the ferric iron already present, sulfuric acid was injected into the feed water to lower the blended feed pH to 2.9. The UV light was also used during the majority of this Phase II study. Table 6.11 presents the operating parameters for this 2,454 hour Phase II study at 75% recovery.

Early in the study, leaks were observed in tubes 1 and 2. All leaks were found to be near the ends of the modules where glue connected the membrane to the perforated product tube. The modules were not replaced because considerable operating time would be lost and since modules in tube 3 had no leaks, their product quality would be representative of normal modules. Also, as the leaks in tubes 1 and 2 were very small, they would have no effect on flux studies.

Although unit rejections were lower for Phase II than for Phase I because of the leaks, tube 3 rejections were comparable to Phase I results. Log-log slopes of -0.012 for tube 1 and -0.009 for tube 3 were obtained during the 2,454 hours of operation at 75% recovery.

These values were improvements over the -0.034 slopes from Phase I. The combination of acid injection to inhibit precipitation of ferric hydroxide and the higher b/p ratios and brine flow rates were believed responsible for the flux stabilization. Discontinuing use of the ultraviolet light at 2,002.5 hours elapsed time had no apparent effect on flux stability. A phosphoric acid flush at the end of the 2,454 hour study did not improve flux. Approximately 60 ml of sulfuric acid (98%, technical grade) was required to treat each 1,000 gallons of water entering the RO unit in order to lower blended feed pH to 2.9.

TABLE 6.11: OPERATING PARAMETERS FOR SPIRAL WOUND PHASE II STUDY AT 75% RECOVERY

Parameter	Value
Raw water feed flow, gpm	5.18
Product water flow, gpm	3.86
Brine water discharged, gpm	1.32
Brine water recycled, gpm	6.32
Minimum brine/product flow ratio (Tubes 1 & 2), ratio/module	10:1
Maximum brine/product flow ratio (Tube 3), ratio/module	22:1
Water recovery, percent	74.5
Recovery of blended feed, percent	33.6
Feed pressure, psig	399
Feed water temperature, °F	66.6
Tube one flux, gal/ft^2/day @ 400 psi net and 77°F	12.81
Tube two flux, gal/ft^2/day @ 400 psi net and 77°F	12.55
Tube three flux, gal/ft^2/day @ 400 psi net and 77°F	11.58
Length of run, hours	2,454
Date of run	July 27 - November 9, 1971

All values are means from 100 data sets.

Source: EPA-670/2-73-100

Hollow Fiber Phase I Studies: Phase I hollow fiber testing began with a single permeator operating at 75% recovery at 2,758 kN/m^2 (400 psi) and continued for a total of 838 hours. Operating parameters are given in Table 6.12.

TABLE 6.12: OPERATING PARAMETERS FOR HOLLOW FIBER PHASE I STUDY AT 75% RECOVERY

Parameter	Value
Raw water feed flow, gpm	2.08
Product water flow, gpm	1.55
Brine water discharged, gpm	0.53
Water recovery, percent	74.4
Feed pressure, psig	400.0
Feed water temperature, °F	53.3
Unit flux, gal/ft^2/day @ 77°F and 400 psi net	2.32
Length of run, hours	838

All values are means from 78 data sets.

Source: EPA-670/2-73-100

Rejections of all multivalent ions were in the range of 99% based on raw feed. These rejections, though slightly less than the spiral wound, were still excellent. As with the spiral unit, the product water would require treatment for pH, iron and manganese before potable standards could be met.

The log-log flux decline slope was 0.037 during the 838 hour study. This flux decline slope was comparable to Phase I spiral wound results. A small increase in pressure drop across the permeator was observed which probably indicated minor iron fouling.

Phase II Hollow Fiber Studies: For Phase II studies the permeators were operated in the 2-1 array to allow high recovery experiments and confine the expected calcium sulfate fouling to a single permeator. Generally, hollow fiber units are not used in staged arrays but are arrayed for parallel operation since brine flow requirements for hollow fiber systems are not as critical as for tubular and spiral wound systems.

During Phase II, several variations in operating parameters were made. In order to evaluate the effect of the ultraviolet light on system operation, the light was turned off from 140 hours elapsed time (e.t.) through 595 hours e.t. This time interval represented 978-1433 hour e.t. on permeator number 3 (last stage permeator). No apparent flux trend change occurred as a result of no UV. This was opposite to the dramatic slope change which occurred on spiral wound when the UV was turned off.

At 390 hours through 426 hours e.t., the system recovery was increased to 85%. This increase in system recovery had very little effect on the recoveries of permeators 1 and 2 whose recoveries were still below 65%. Thus, very little change in flux occurred in the first two permeators during this high system recovery operation. Permeator 3, however, which was subject to the most severe pollutant concentrations, suffered a drastic flux loss. Upon lowering recovery again to 75%, permeator 3's flux again stabilized.

At 596 hours elapsed time (1,434 for permeator 3), the UV light was turned on again but system recovery was increased to 85%. These operating conditions were maintained for 111 hours, including operating at 80% recovery for 38 hours, before system recovery was lowered (707 hours e.t., 1,545 e.t. for permeator number 3).

Again, severe flux losses occurred in permeator 3 (log-log slope = –1.77). Although the flux in permeator 3 stabilized, it was now following a significantly steeper flux decline slope (log-log slope = –0.36 at 78 to 80% recovery). A subsequent flush with disinfectant failed to improve flux. Sodium hydrosulfide was used to flush permeator 3 at 2,081 hours (on permeator 3) and successfully restored flux to the original flux decline slope of –0.037 at 78% recovery. Flux immediately dropped again, however, but stabilized at lower values until the end of the run at 2,670 e.t. (permeator 3).

Permeators 1 and 2 appeared to suffer a flux slope change near 600 hours elapsed time after following an initial slope of –0.015 until that time. Following the 85% recovery test which ended at 707 hours elapsed time, first stage permeator flux stabilized following a log-log slope of –0.03 for approximately 540 hours until

1,243 hours e.t. During the last 590 hours of operation, all three permeators lost and then regained flux prior to a final sodium hydrosulfide flush. This flush virtually restored first stage permeator flux to the original -0.015 slope values. Since the tests was terminated at this point, it was not possible to determine if the -0.015 slope could be maintained following the flush. Although the flush improved permeator 3's flux, it did not return the flux to the original slope values. A post mortem analysis of permeator number 3 revealed that the fouling was largely due, as expected, to calcium sulfate precipitation.

In the case of the two first stage permeators whose flux had been restored by the final flush, an additional 31 psi pressure drop had been acquired during the study which was not removed. Since neither of these first stage permeators had ever operated above 66% recovery, it must be assumed that iron fouling caused the flux losses and pressure drop increase as it is unlikely that calcium sulfate would precipitate at that recovery level on the Mocanaqua water.

Operating parameters for Phase II hollow fiber at 75 and 85% recovery are given in Tables 6.13 and 6.14, respectively. Rejections during Phase II were comparable to Phase I results, i.e., the range of 99% based on raw feed.

TABLE 6.13: OPERATING PARAMETERS FOR HOLLOW FIBER PHASE II STUDY AT 75% RECOVERY

Parameter	Value
Raw water feed flow, gpm	6.18
Total product flow, gpm	4.72
Brine water discharged, gpm	1.46
Unit water recovery, percent	76.4
Feed pressure, psig	400
Feed temperature, °F	54.3
Permeator No. 1 product flow, gpm	1.86
Recovery, permeator No. 1, percent	59.9
Flux, permeator No. 1, gal/ft^2/day @ 400 psi net & 77°F	2.78
Permeator No. 2 product flow, gpm	1.83
Recovery, permeator No. 2, percent	59.5
Flux, permeator No. 2, gal/ft^2/day @ 400 psi & 77°F	2.74
Permeator No. 3 product flow, gpm	1.03
Recovery, permeator No. 3, percent	41.4
Flux, permeator No. 3, gal/ft^2/day @ 400 psi net & 77°F	1.92
Length of run, hours	1667
Date of run	June 10 - August 28, 1971

All values are means from 83 data sets.

Source: EPA-670/2-73-100

TABLE 6.14: OPERATING PARAMETERS FOR HOLLOW FIBER PHASE II STUDY AT 85% RECOVERY

Parameter	Value
Raw water feed flow, gpm	5.73
Total product flow, gpm	4.86
Brine water discharged, gpm	0.87
Unit water recovery, percent	84.8
Feed pressure, psig	400
Feed temperature, °F	54.0
Permeator No. 1 product flow, gpm	1.89
Recovery, permeator No. 1, percent	65.5
Flux, permeator No. 1, gal/ft^2/day @ 400 psi net & 77°F	2.84
Permeator No. 2 product flow, gpm	1.85
Recovery, permeator No. 2, percent	65.0
Flux, permeator No. 2, gal/ft^2/day @ 400 psi net & 77°F	2.78
Permeator No. 3 product flow, gpm	1.12
Recovery, permeator No. 3, percent	56.3
Flux, permeator No. 3, gal/ft^2/day @ 400 psi net & 77°F	2.10
Total operating time @ 85 percent recovery, hours	147
Longest run, hours	111
Date of run	6/26-6/28/71, 7/5-7/9/71

All values are means for 11 data sets.

Source: EPA-670/2-73-100

Significant variations were noted in the log-log flux curve of the hollow fiber unit. The slope appeared to increase as brine flow decreased and a possible relationship may exist in the form of a b/p flow limit although insufficient data were available to substantiate this relationship. As brine flow rates decreased, recovery and risk of precipitation increased and this may have caused the observed slope changes. Whatever the reason, flux slopes were minimized when the brine flow rate was in excess of 3.785 l/min (1 gal/min).

Tubular Unit Phase I Studies: The tubular RO system was also operated in two separate phases during this study. Phase I utilized 60 Type 310 modules in a 6-4-2 array with five modules in each series in each row. The last two modules in bank 2 and the last 3 modules in bank 3 contained volume displacement rods (VDR) that increased the brine velocity.

The normal inside diameter in the tubular system is one-half inch which corresponds to a linear brine velocity of 1.64 ft/sec per gpm of brine flow. The volume displacement rods effectively increased this velocity to 2.5 ft/sec/gpm. When utilizing VDR's, headloss through the module increased significantly. In a module without VDR's, the headloss at one gpm brine flow was five psi per module, while with VDR's this increased to 22 psi per module.

Initial Phase I operation was at 600 psi and 75% recovery. During Phase I, the product water flux decreased steadily from 13 gal/ft^2/day to 8.5 gal/ft^2/day in only 480 hours. It was originally assumed that the system had been contaminated with iron oxidizing bacteria because of trouble experienced with the UV light.

At 160 hours, the system was disinfected with quaternary ammonium compound and a slight increase in flux was noted. However, membrane relaxation probably occurred during this time since the unit operated at low pressure. Relaxation generally results in a flux increase for a short period of time.

Immediately after the disinfection, the flux continued to decline rapidly. At this point, it was felt that the brine velocities might possible be too low and that concentration polarization effects were causing the rapid fouling. Minimum brine velocities were increased from 1.2 to 1.4 ft/sec to 2.0 to 2.2 ft/sec.

Tubular Unit Phase II Studies: At the end of Phase I, it was noted that the flux declines experienced with the tubular system were not experienced with the hollow fiber system or the spiral wound system also operating alongside. It was also noted that both the spiral wound system and the hollow fiber system had considerably lower salt passage (higher rejection) as compared to the tubular system.

The high salt passage of the tubular 310 membrane may have had some influences on the flux declines experienced. To test this, five high-flux-low-salt-passage modules (type E610) were installed and put into operation for Phase II. During Phase II, recovery was limited to less than 50% at 600 psi due to the use of only five modules. Operating parameters for this 807 hour study are given in Table 6.15.

TABLE 6.15: OPERATING PARAMETERS FOR TUBULAR PHASE II STUDY AT 43% RECOVERY

Parameter	Value
Raw water feed flow, gpm	1.48
Product water flow, gpm	0.63
Brine water discharged, gpm	0.85
Water recovery, percent	42.5
Feed pressure, psig	617.0
Feed water temperature, °F	55.0
Unit flux, gal/ft^2/day @ 77°F & 600 psi net	15.60
Length of run, hours	807

All values are means from 34 data sets.

Source: EPA-670/2-73-100

Rejections were higher for Phase II (type E610) membranes than for Phase I membranes (type 310). An extremely high initial compaction was experienced during the first few hours of operation. The flux then stabilized on a -0.040 log-log slope for the first 300 hours. A gradual decline then occurred through about 440 hours, at which time the flux stabilized at 13.7 gal/ft^2/day (557 l/m^2/day) for the remainder of the study.

The change in the flux decline slope experienced from near 300 hours e.t. was also noticed to a lesser degree on the hollow fiber unit and was believed caused by a higher-than-normal iron(III) content in the AMD. The decline was probably entirely due to iron fouling since operation at 40 to 45% recovery was well below the $CaSO_4$ fouling range.

The modules were flushed with a sodium hydrosulfite solution (4 weight percent) for 1 hour. This resulted in a dramatic increase in flux. Since additional operating time was not available, it is not known how much of this flux increase was due to cleaning and how much was due to membrane relaxation. It is felt that a substantial gain was accomplished since membrane relaxation alone would not account for increase in flux of about 33%.

The salt passage was calculated based on average brine concentration experienced on the membrane, i.e., the average of feed and brine concentrations. This procedure allows comparison of salt passages while operating the RO system at different product water recoveries.

Generally, salt passage for the type 310 modules ranged from 1 to 1.5% for Ca, Mn, Fe, Al and SO_4. Silica passage was extemely high at an average percent of 54. Type 610 modules had significantly lower salt passage; i.e., 0.4 to 0.5% for Ca, Mg, Mn, Fe, Al and SO_4. Silica passage was also considerably lower than the 310 modules at an average percent of 6.2% passage, No apparent changes occurred in the salt passage throughout the operational period. The mechanical operation of the tubular system was excellent. No module failures were experienced over the entire 2,800 hours operation.

In summary, a number of points can be made regarding tubular system operation. The use of volume displacement rods is definitely not recommended. The price paid in headloss far exceeds the benefits obtained. If higher velocities are required, it appears recirculation of brine would be the preferred alternative. With regard to required velocities, it appears that a minimum velocity of about 1.5 feet per second is desirable, since operation at this velocity with the higher-flux-low-salt-passage modules was satisfactory.

It should be noted, however, that the recoveries during Phase II were quite low (45 to 70%), and that operation at higher recoveries may require higher velocities to offset the concentration polarization effects due to higher brine concentrations.

It is not known if the high initial flux loss experienced during Phase I with the type 310 modules was specific for the modules utilized or a result of the higher salt passage. In any event, low-salt-passage modules are definitely recommended for both flux and product water quality considerations.

Comparing the flux history of Phase I 310 modules to the flux history from earlier field testing, significantly lower flux declines were noted in this study.

For example, greater than 80% of the original flux was lost in 400 hours in the previous study compared to about 45% of the original flux in this study. This would indicate that the pretreatment system did have some effect on the tubular system operation.

Calcium Sulfate Fouling at Mocanaqua: Very little or no calcium sulfate fouling occurred at 75% recovery since both the spiral wound and hollow fiber log-log flux plots had shallow slopes and were linear. In contrast, the severe fouling observed at 85% recovery was due to calcium sulfate precipitation. Therefore, at some point between 75 and 85% recovery, the solubility of calcium sulfate was exceeded and precipitation occurred. It was felt that precipitation occurred in the solubility product range of 35 and 50 x 10^{-5}.

Maximum Recovery Prediction

In each of the spiral wound studies made at the various AMD sites, one result was significant—the limiting factor in high recovery reverse osmosis operation was calcium sulfate precipitation. The following formula was derived to enable prediction of maximum recovery.

$$R = 100 - 0.055 \sqrt{(Ca) \times (SO_4)}$$

where R = maximum recovery (percentage)
Ca = AMD feed calcium concentration (mg/l)
SO_4 = AMD feed sulfate concentration (mg/l)

This method of predicting maximum recovery is felt to be accurate to ±5% recovery on acid mine drainage. The derivation of the formula is described in EPA-670/2-73-100.

COMPARISON OF SPIRAL WOUND, HOLLOW FIBER AND TUBULAR SYSTEMS—EPA STUDY

In order to compare RO systems, it was vital that all comparisons be made on the same basis, i.e., under the same operating conditions. For this reason, water flux, when corrected for osmotic pressure (which corrects for water quality and recovery) and temperature, served as an effective unit for comparing of membrane performance. Water flux, however, did not compensate for efficient packaging of the membranes in the RO system, and therefore was a poor basis for comparing system performance.

System performance was evaluated on the basis of product output per cubic foot of vessel volume per unit of time. To arrive at this factor, water flux was divided by pressure vessel volume.

Table 6.16 presents a summary of observed flux and productivity data from the Mocanaqua studies where spiral wound, hollow fiber, and tubular systems operated on a side-by-side basis. Tubular and Phase I spiral wound data—although actual operation was at 600 psi—were also normalized to 400 psi to enable direct comparison with the hollow fiber unit.

In terms of average flux, spiral wound membranes were clearly superior with 400 psi fluxes of 12.3 gal/ft^2/day for spiral versus 10.40 for tubular and only 2.48 for hollow fiber.

TABLE 6.16: COMPARISON OF WATER PRODUCTION CAPABILITIES OBSERVED DURING MOCANAQUA STUDIES

	Pressure vessel volume ft^3	Enclosed membrane area ft^2	Membrane packing density ft^2/ft^3	Avg. flux $gal/ft^2/day$[(a)]	Total vessel flux/day gal/day[(a)]	Output per cubic foot of vessel volume	
						per day[(a)]	per min.[(a)]
Spiral-wound (Phase I)	1.13	150	133	(19.28 @ 600) 12.86	(2892 @ 600) 1929	(2559 @ 600) 1707	(1.78 @ 600) 1.19
Spiral-wound (Phase II)	1.13	186	165	12.31	2290	2026	1.41
Hollow-fiber (Phase II)	0.65	1500	2308	2.48	3720	5723	3.97
Tubular (Phase II)	0.63	16.9	26.8	(15.60 @ 600) 10.40	(264 @ 600) 176	(418 @ 600) 280	(0.29 @ 600) 0.19

[(a)] At 77°F (50°C) and 400 psi net pressure.

Source: EPA-670/2-73-100

Because hollow fiber permeators packed from 8 to 10 times as much membrane in the same volume as the spiral system and 88 times as much as the tubular system (see Packing Density, Table 6.16), the low specific flux of the hollow fiber was made up for by greater membrane area. Although the hollow fiber flux rate ranged from only 12 to 20% that of the spiral system, depending on the spiral's operating pressure, the hollow fiber product output per cubic foot of vessel volume was 2.2 to 3.3 times spiral output and from 14 to 21 times tubular output, again depending upon operating pressures. The tubular system ranked lowest in both flux and output per cubic foot of vessel volume. In Table 6.17 are the relative costs for each system.

TABLE 6.17: RELATIVE COST COMPARISONS FROM MOCANAQUA STUDY

System	Cost for one pressure vessel and membrane	Observed output (gal. per vessel per day @ 77° F & indicated net pressure)	Initial cost per unit output (gal/day)
Spiral-wound Phase I	$ 850.	2892 @ 600	$0.29
Spiral-wound Phase II	$ 850.	2290 @ 400	$0.37
Hollow-fiber Phase II	$1000.	3720 @ 400	$0.27
Tubular Phase II	$ 265.	264 @ 600	$1.00

Source: EPA-670/2-73-100

Initial costs for purchase of one pressure vessel complete with membrane were used. Dividing the initial cost by the observed output (gallons per day of product) yielded the initial cost-per-unit-output figures shown in the table. Although prices and relationships observed were valid for the Mocanaqua study, extrapolation of these figures for large units would result in large errors.

Other factors such as hardware requirements would vary with each application and manufacturer and were not included in pressure-vessel-membrane-package-cost figures.

A third basis for comparison was product quality. Rejections were calculated by comparing the product concentration with concentrations entering the RO unit. A summation of system rejections is presented in Table 6.18. Spiral wound membranes held a slight edge in rejecting ability.

Leaks during spiral wound Phase II studies were apparent in unit rejections (Table 6.18), but tube 3 rejections were of the same level as those from Phase I spiral studies. Under Mocanaqua conditions, the rejection advantage of spiral membranes would be of little significance as product water from all three units would require additional treatment for pH, manganese, and iron before drinking water standards could be met.

The final basis for comparison was the decrease of productivity with time as evaluated by log-log flux decline slopes. In Table 6.19 are examples of log-log slopes chosen to simulate comparable recovery conditions. In all these examples, recoveries in the respective tubes or units were near or below 60%.

Phase II spiral wound system again had a slight advantage in flux decline slope over the hollow fiber systems; however, the hollow fiber slope was superior to Phase I spiral performance. The tubular system's slope, calculated from the line of best fit of the tubular log-log flux plot, was considerably steeper than either the spiral or hollow fiber slope.

Extrapolation of the flux curve to predict flux values at 3 and 5 years was necessarily based on the assumption that the log-log flux decline slope remained constant. Validity of that assumption is certainly open to debate and only sustained operation over that period of time would prove or disprove it. If the assumption were valid, the flux levels at 3 and 5 years of elapsed time are given in Table 6.19.

Significance of Rejections: Most important of the several factors that determine reverse osmosis product quality are raw water quality, recovery level, and membrane rejection capability. If water of potable quality is required without post-treatment of the product, then significant restrictions are placed upon the mode of operation and characteristics of the RO unit.

Iron and manganese concentrations in potable water are restricted by the U.S. Public Health Service to 0.3 and 0.05 mg/l, respectively. Of all commonly appearing AMD constituents, iron and manganese are the most critical for treatment to potable standards.

Since the membrane rejection ratio is constant, product quality is directly dependent on concentrations on the brine (concentrated) side of the membrane. Increasing recovery serves to increase brine side concentrations and in turn degrades product quality. Therefore, for every raw feed concentration, it is possible to calculate the maximum recovery that can be obtained without the product concentration exceeding the U.S. Public Health Service limit.

A family of operating curves has been developed for iron and manganese and may be used to approximate this maximum recovery (Figures 6.11 and 6.12).

FIGURE 6.18: COMPARISON OF MEMBRANE PERFORMANCE AT MOCANAQUA

System	Flux, gal/ft²/day @ 77° F & indicated pressure	Rejections[a], Percent									
		Conductivity	Acidity	Calcium	Magnesium	Total iron	Ferrous iron	Aluminum	Sulfate	Manganese	Silicon
Spiral-wound Phase I	19.28 @ 600 psi	99.2	91.7	99.8	99.8	99.8	99.8	99.2	98.8	-	
Spiral-wound Phase II	12.31 @ 400 psi	97.6	94.6	99.6	98.5	99.0	98.9	97.6	99.1	99.8	
Spiral-wound Tube 3, Phase II	11.58 @ 400 psi	98.7	-	99.8	99.7	99.7	-	97.7	99.7	99.5	
Hollow-fiber Phase I	2.32 @ 400 psi	-	-	99.5	99.3	99.2	99.1	97.8	99.7	99.4	95.1
Hollow-fiber Phase II	2.48 @ 400 psi	96.9	84.8	99.2	98.8	98.9	98.9	94.7	99.5	-	-
Tubular Phase I		95.6	81.6	98.2	98.5	98.8	98.4	91.7	99.3	-	
Tubular Phase II	15.6 @ 600 psi	96.2	-	99.5	99.5	99.4	99.0	98.2	96.9	99.4	90.0

[a] Rejection = (Influent concentration-product concentration) × 100 ÷ Influent concentration.

Source: EPA-670/2-73-100

TABLE 6.19: PROJECTED MEMBRANE PERFORMANCE[a]—MOCANAQUA STUDIES

Unit	Flux @ 100 hrs.[b] gal/ft²/day	Log-log flux decline slope	Predicted flux after 3 years	Predicted flux after 5 years
Spiral-wound Phase I Tube 1	21.5	-0.034	17.8	17.5
Spiral-wound Phase II Tube 1	13.0	-0.012	12.2	12.1
Hollow-fiber Phase II Tube 1	2.9	-0.015	2.7	2.6
Tubular Phase II	17.2	-0.063	12.1	11.7

[a]Assumptions:
1. Log-log flux decline slope remains constant.
2. Salt rejection is assumed to remain constant.

[b]Taken from log-log graph of flux versus time.

FIGURE 6.11: MAXIMUM OPERATING CONDITIONS FOR IRON LIMITS OF 0.30 mg/l

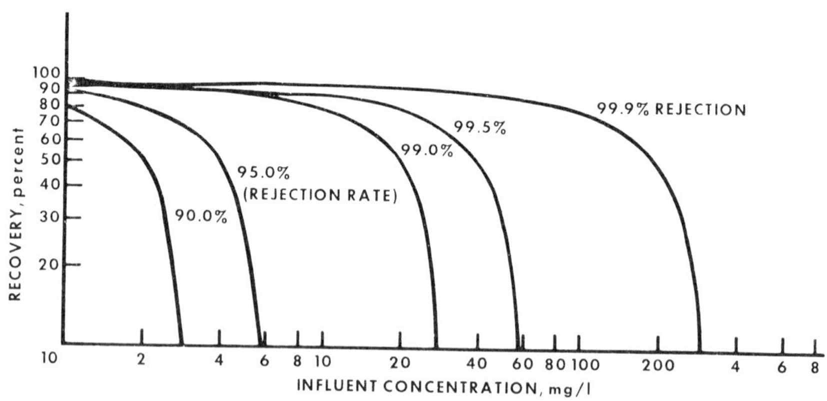

FIGURE 6.12: MAXIMUM OPERATING CONDITIONS FOR MANGANESE LIMITS OF 0.05 mg/l

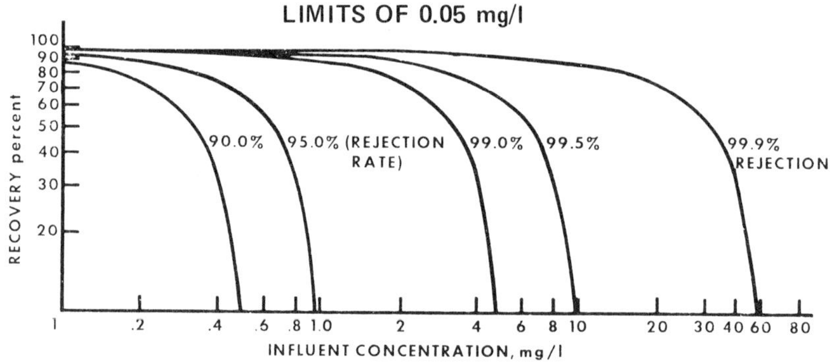

Source: EPA-670/2-73-100

For example, if an influent water contained 50 mg/l of iron and 5 mg/l of manganese, the maximum recovery to meet potable standards would be 30% for the iron criteria when using 99.5% rejecting membranes. The same conditions allow a 66% recovery before manganese limits are exceeded.

It is not necessary for the RO unit to reduce the iron and manganese to this low level since the product water must receive posttreatment in the form of neutralization to increase the pH to an acceptable level. RO-treated water normally has a pH less than 5. Neutralizing to pH 7 and filtering will remove residual iron, and thus, the iron concentrations can be kept in acceptable limits.

If the iron is in the ferrous state, it will rapidly oxidize at pH 7, and then it can be removed by filtering. To remove residual manganese by neutralization, neutralization to pH 9 to 10 is required. This neutralization in itself would exceed normally acceptable pH limits and acidification of the water would be needed following manganese removal.

In conclusion, assuming potable quality product water is required from the RO unit, membrane rejection is generally not significant for iron removal since postneutralization is required anyway. However, for AMD waters containing significant amounts of manganese, the rejection capability is of vast importance in achieving potable quality at minimum cost. Membranes with rejections below 95% would not appear desirable for AMD potable applications.

If the RO system can be operated in a mode where the iron and manganese can be reduced to an acceptable level by the addition of a posttreatment step, then the unit can be operated at a high recovery.

PROPRIETARY PROCESS—NEUTROLOSIS

R.D. Hill, R.C. Wilmoth and R.B. Scott; U.S. Patent 3,795,609; March 5, 1974; assigned to U.S. Administrator of the Environmental Protection Agency describe the neutrolosis process as follows. Referring to Figure 6.13, contaminated acid mine drainage waters entering from source **10** may have to be pretreated to insure suitability for separation by reverse osmosis. A typical pretreatment for acid mine drainage is oxidation of ferrous ions present in the contaminated water in oxidizer **14**, chemically as by ozonation or chlorination, or biologically.

In addition, disinfection to control bacteria and other organisms that cause membrane fouling may be required in disinfecter **15**, chemically, e.g., by chlorination or by ultraviolet light. Suspended solids present in the contaminated mine waters are removed by passage through a filter system **16**.

If the drainage waters are already in an oxidized state and do not contain membrane fouling organisms, the feed stream may be passed by way of line **12** directly to filter system **16**. In any event, the filtered mineral contaminated water is pressurized by pump **18**, then passed through reverse osmosis unit **20** where the water is separated into a product water stream which passes out of the osmosis unit through line **22** and a brine stream which leaves through line **24**.

The purified product water in line **22** is low enough in mineral content for discharge directly to stream **26**.

FIGURE 6.13: NEUTROLOSIS PROCESS

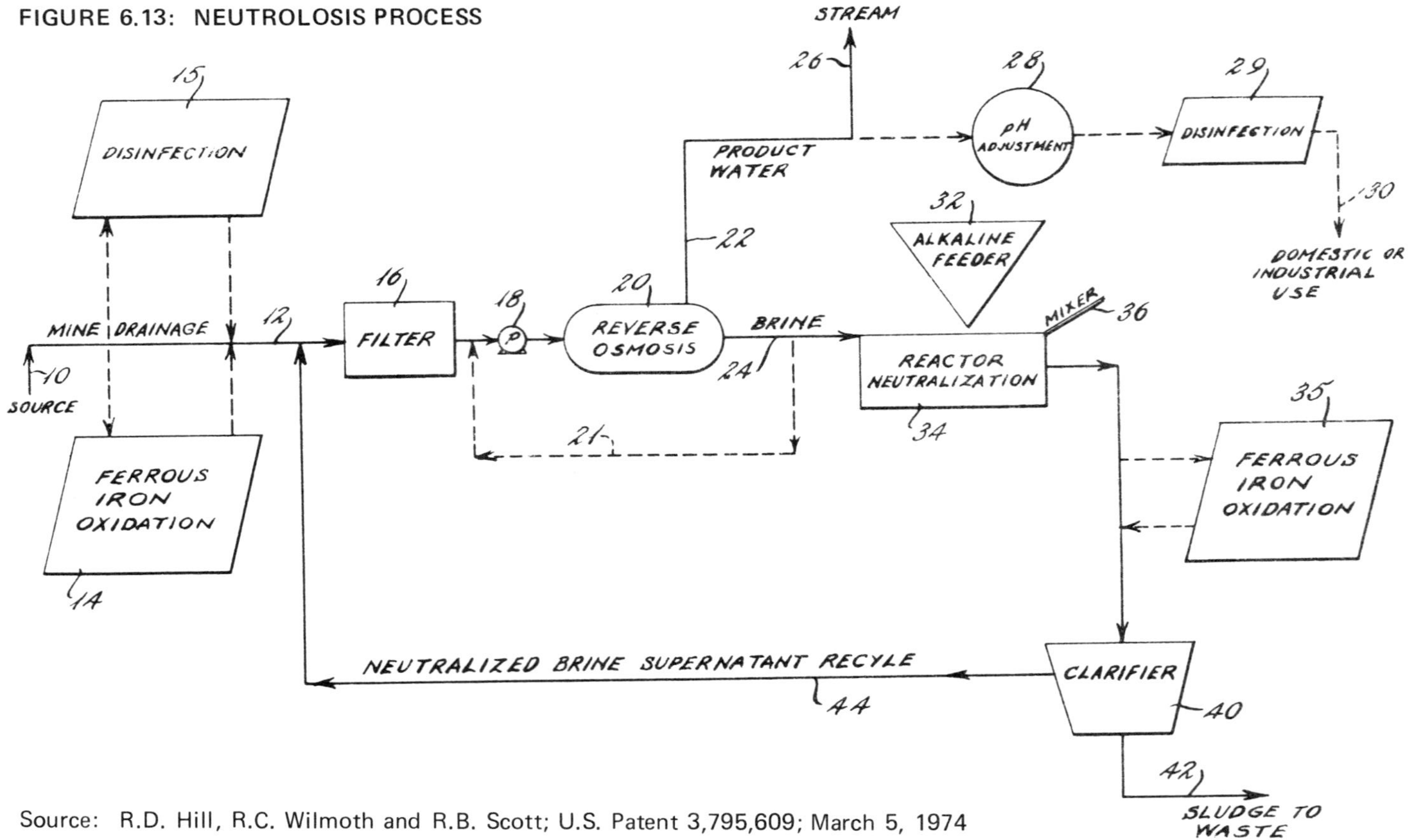

Source: R.D. Hill, R.C. Wilmoth and R.B. Scott; U.S. Patent 3,795,609; March 5, 1974

If the product water is intended for direct domestic or industrial use, it may require further treatment like neutralization in pH adjustment unit **28** and disinfection in unit **29** before discharge via line **30** to its intended domestic or industrial use.

The purified product water stream leaving the reverse osmosis unit **20** by way of line **22** will constitute at least 50% of the water feed entering reverse osmosis unit **20**. However, since an internal brine recycle forms an important part of the system as a whole, the purified product water will constitute at least about 80%, and normally more than 90% of the water input to the system from inlet **10**.

The brine stream which leaves osmosis unit **20** through line **24** will then comprise less than 50%, and usually less than 25%, of the combined feed water and recycled brine entering reverse osmosis unit **20**. The brine enters into (partial) neutralizer **34** where it is reacted with suitable alkaline material, e.g., limestone or lime, fed from hopper **32** under agitation from mixer **36**.

Typically acid mine wastes are relatively high in iron, in aluminum and sulfate and partial or full neutralization with inexpensive neutralizing agents like lime or limestone will cause a considerable proportion of these contaminants to precipitate from the brine. In some instances where the ferrous ion content of the feed stream is high it may be desirable to place a ferrous-to-ferric oxidation step at this stage of the process. In such instances, the slurry and precipitate pass to an oxidation unit **35** where ferrous ion is converted to ferric iron by aeration or some other oxidation process.

The slurry of brine and precipitate passes from the neutralizer **34** or oxidizer **35** into clarifier **40**, where the sludge is separated out from the neutralized brine stream and removed by way of line **42** to disposal. The supernatant brine is recycled from clarifier **40** through line **44** back to repeat passage through reverse osmosis separation. In the flow sequence illustrated by the drawing, recycled brine from line **44** is admixed with the feed water just ahead of filter **16**.

In order to separate out as much of the contaminants as is reasonably possible in the form of an inert solid co-product without upsetting the system, the reaction in neutralizer **34** is adjusted to a pH which is optimum for the system as a whole. In the instance of acid mine drainage, neutralization to a pH of about 4.5 is believed to be optimum.

If the pH were raised much nearer to neutrality some danger exists that the neutralized brine recycled from clarifier **40** by way of line **44** will cause instantaneous iron precipitation when admixed with incoming acid mine drainage water. Any such precipitate might foul filter **16** or the membranes inside reverse osmosis unit **20**.

In the flow sequence illustrated by the figure, the only products of the separation process are the purified product water leaving in line **22** and the waste sludge leaving line **42**. Allusion has already been made that the purified product water preferably constitutes in excess of 90% of the water feed entering from source **10**. The balance of the water, which may be less than 2%, leaves as part of the sludge product from the clarifier **40** by way of line **42**.

The high water recovery rate and the separation out of mineral contaminants as an essentially inert sludge demonstrates how efficiently the procedure of the process can treat contaminated drainage waters.

HIGH PRODUCT WATER RECOVERY SYSTEM—OSW STUDY

This study relates to laboratory feasibility studies conducted on Proctor #2 mine waters at the Hollywood site of Pennsylvania State University, Pennsylvania. A typical analysis of Proctor #2 mine water is as follows: pH 2.98, acidity as $CaCO_3$ 1,800 mg/l, iron (III) 50 mg/l, iron (II) 400 mg/l, aluminum 200 mg/l, calcium 200 mg/l, magnesium 100 mg/l, manganese 8 mg/l, sulfate 2,500 mg/l, conductivity 2,600 μmho/cm, and temperature range, 9° to 13°C.

In order to adequately evaluate various treatment schemes, it was highly desirable to obtain the feasibility data on actual AMD brines. The brines were to be produced at the AMD site via a portable, lightweight, suitcase size, 1,500 gpd hollow fiber RO unit.

Design of the Portable RO Unit: A process schematic of the portable reverse osmosis unit utilized for the investigations during Phase I is shown in Figure 6.14. The pilot system was divided on two pallets. Flexibility has been provided in the design so that brines can be recycled to the suction side of the feed pump for maintaining desired flow velocities through the RO membranes under low flow conditions.l The permeator pallet houses the RO membrane module. In the field, the RO module can be connected to the high pressure pump via quick disconnects. The back pressure is regulated by orifice type Nupro metering valves on the reject and brine recycle lines.

Treatment Investigations of RO Brines

The following test plan for field evaluation was developed for investigations during Phase I of this study:

(a) RO treatment of Proctor #2 water at 50 to 75% recovery. Collect 50 gallon batches of brine samples.

(b) Analyze for raw, product and brine water quality.

(c) Oxidize Fe^{++} in brine samples with hydrogen peroxide prior to neutralization.

(d) Obtain neutralization curve (pH vs limestone and lime dosage). Use approximately 2 times brine acidity for limestone dosage in conjunction with aeration to raise the pH to approximately 5.0. Then add lime to further raise the pH to desired levels (7.5, 10.8 and 12.0).

(e) Limestone and lime neutralization to the following pH levels: 7.5, 10.8 and 12.0. Obtain bench scale data on: settling characteristics, settled effluent quality and sludge volumes.

(f) Save: sludge samples for sludge recycle tests; effluent samples for $CaSO_4$ seeding and softening tests as follows: CO_2 and soda ash softening tests on neutralized effluent (pH 10.8 only); sodium aluminate softening tests on neutralized effluent (pH 12.0 only).

(g) Run optimally softened supernate through Rohm & Haas IR-120 (strong acid type cation exchanger) and determine the extent of calcium removed.

FIGURE 6.14: PORTABLE REVERSE OSMOSIS PILOT UNIT

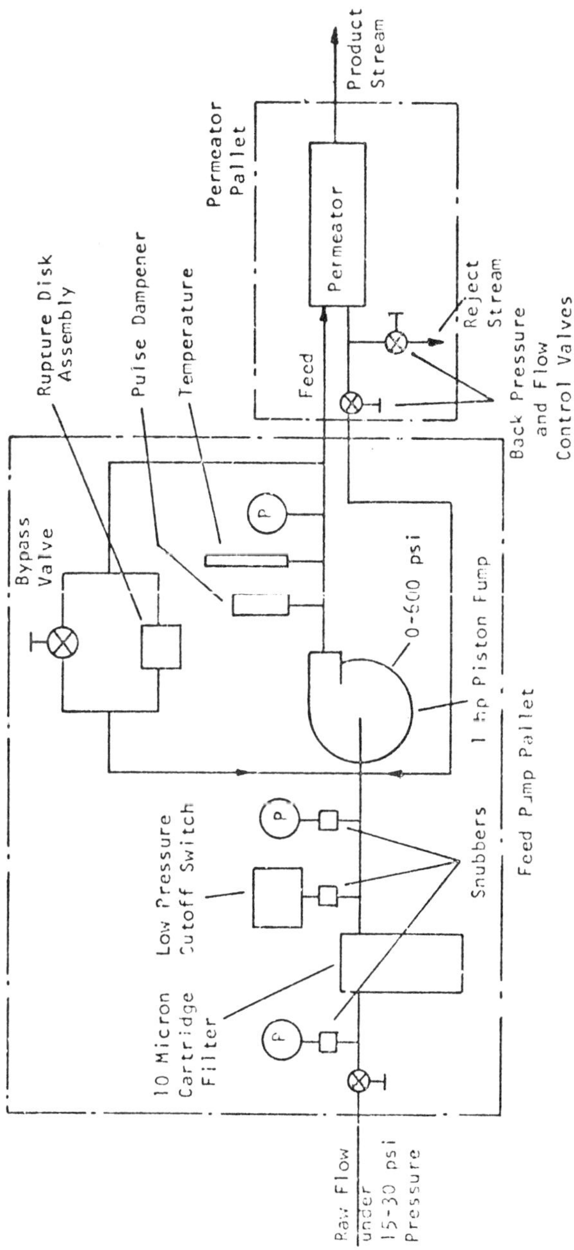

Source: PB 230 756

(h) Investigation of $CaSO_4$ supersaturation via second stage processing of the brines through tubular modules after addition of sodium hexametaphosphate.

Bench scale tests in the field were undertaken on the Proctor #2 mine waters in accordance with the test plans. Two separate visits were made to the test site. The results of the operations on the portable RO unit during these visits are shown in Table 6.20.

TABLE 6.20: RO UNIT OPERATION DATA (ACID MINE WATER SOURCE: PROCTOR #2)

Parameter	Run 1	Run 2	Run 3
Feed flow, gpm	1.474	1.815	1.600
Product flow, gpm	1.071	1.149	1.090
Brine flow, gpm	0.403	0.666	0.510
Temperature, °F	55	57	55
Water Recovery, %	73	63	68
Pressure In, psi	390	400	411
Pressure, Out, psi	371	360	354
Normalized flux, gsfd[1]	2.1	2.2	2.1
Length of Run, hrs	2	6	4
Date	5/23/72	5/24/72	10/4/72

[1]At 400 psi net driving force neglecting osmotic pressure and at 68°F. Membrane area used: 950 square feet.

Source: PB 230 756

All runs were of short duration. The water recovery during these runs varied between 63 and 73%. The water quality data for the raw, product and brine samples are shown in Table 6.21. Reasonable ion balances were obtained. In essentially all cases there was a slight excess of anions present (i.e., SO_4).

Since the above analyses do not include sodium and potassium, these ions could make up the apparent deficiency in cations. In all cases, the sum of the ions was significantly lower than the TDS measured by evaporation. This may be a characteristic of acid mine waters since the same has been observed at other AMD sites. In general, the sum of the ions was equal to approximately 80 to 85% of the measured TDS.

The safe recoveries based on calcium sulfate fouling were calculated to be in the ranges of 50 to 55% for typical composition of the Proctor #2 mine waters. However, some dilution of these mine waters was noticed during May, 1972 because of heavy rains in the Hollywood area. The calculated safe recovery at that time was approximately 70%. The actual recoveries in the three runs conducted at the site, varied between 63 and 73%.

TABLE 6.21: WATER QUALITY DATA FOR RO OPERATIONS

Analysis[1]	Run 1 Feed	Run 1 Product	Run 1 Brine	Run 2 Feed	Run 2 Product	Run 2 Brine	Run 3 Feed	Run 3 Product	Run 3 Brine
pH, units	2.5	3.2	2.25	2.7	3.6	2.4	2.8	3.6	2.5
Iron (III)	120	3.0	389	127	2.0	329	66	2.5	197
Iron (II)	134	5.0	431	133	3.0	341	298	11.5	987
Acidity	--	--	4002	--	--	3300	1484	57	4910
Calcium	110	6.0	355	109	3.0	231	180	14	515
Magnesium	54	3.0	165	53	1.4	140	75	4.0	230
Manganese	5.0	0.3	17	5	0.2	13.8	7.1	0.35	23.2
Aluminum	74	2.0	265	73	1.0	181	88	11	282
Silica	30	1.0	130	30	1.0	86	31.5	3.0	97
Sulfate	1570	78	4350	1635	22	4200	1934	85	5803
TDS	2497	140	7389	2535	79	6707	2100[2]	188[2]	7500[2]
Sum Cations, mg/l	29.6	1.1	97.7	29.7	0.6	76.0	39.4	2.8	122.7
Sum Anions, mg/l	32.7	1.6	101.0	34.0	0.5	87.4	40.3	1.8	128.8
Sum Ions, (TDS)	2097	98	6572	2165	3.4	5572	2648	128	8037

Note: All results expressed in mg/l unless otherwise noted.

[1]All metal analyses conducted by atomic absorption.
[2]Measured on a portable conductivity meter. All other TDS analyses by evaporation.

Source: PB 230 756

The 68% recovery used in Run 3 was in fair excess of the 50 to 55% predicted equilibrium limit and produced a supersaturated brine without addition of any antiprecipitants such as sodium hexametaphosphate (SHMP). Attempts to saturate the brines further via processing through a Calgon Havens tubular module with addition of 50 mg/l SHMP were not successful on a bench scale basis.

Since the brines were already supersaturated, $CaSO_4$ scaling problems were immediately apparent. Precipitated calcium sulfate clogged the process lines of the tubular module and the pump safety rupture disc was blown. These problems became worse because of the recycling of the waters from a supersaturated feed inventory.

The high iron content and the low pH of the supersaturated feed might have rendered the trace amounts of SHMP ineffective in inhibiting the rate of $CaSO_4$ precipitation. However, it appears that supersaturated conditions for $CaSO_4$ can prevail if short brine residence times are insured throughout the operation and shutdown of the RO systems (as shown by the production of supersaturated brine through the hollow fiber unit). The scaling problems encountered when the supersaturated feed inventory was used with the tubular assembly point to the importance of maintaining short residence times and avoiding recycle.

Membrane Fouling Evaluations

Some fouling of the RO membranes was noticed during the short brine production RO runs. It was indicated that such fouling may have been either due to ferric hydroxide precipitation (pH 2.8) or due to $CaSO_4$ scaling. In order to assess the true cause of this fouling, the portable RO unit was flushed with a 4 weight percent solution of sodium hydrosulfite for 2.5 hours.

This cleaning restored the pressure drop across the membrane from 60 psi to 35 psi. A significant amount of precipitated iron on the membranes was flushed out by dissolving it into the cleaning solution. However, no attempt was made to analyze the iron concentrations at this time because of some intermediate flushing with the tap water at the site which contained a significant amount of iron (3.5 mg/l) at a pH of 6.3.

The clean membrane was then tested on Proctor #2 waters along with a new cartridge filter. The recovery was kept below 43% to insure against $CaSO_4$ fouling. The headloss across the membranes steadily built up to a value of 67 psi from 35 psi in only 21 hours of operation. The loss of product flux (normalized flow rates) during the corresponding period was approximately 8%. Since the system was operated well below the $CaSO_4$ scaling range the cause of membrane fouling can be attributed to ferric hydroxide precipitation at the RO membranes.

The feed water pH ranged between 2.8 and 2.9. As long as the brine pH at the low recovery operation was 2.7 or higher, a drop in normalized flux rates was noticed. However, during two readings at 4 hours and 9 hours, when the brine pH dropped below 2.7, the decline of flux rates was reversed. The same observation was confirmed during the test period of 21.5 to 24.0 hours, when the brine pH was lowered to 2.6 by recycling some brine into the feed water.

The recycling of the brine to the feed resulted in a significant increase in the recovery ratio, but the ΔP across the membranes as well as the flux rates during this period were remarkably steady. The check on the flux decline rates during the 21.5 to 24.0 hour period is even more significant when it is noted that the recovery ratio during this period was much higher than the allowable recovery range (50 to 55%) for controlling the $CaSO_4$ scaling. The membranes were not expected to be fouled significantly during the brief period of operation at high recovery.

After the 24 hour operation period, the membranes were flushed with a 4 weight percent solution of sodium hydrosulfite for 2 hours. Analysis of the flushing solution indicated that the total iron concentration increased from 0 to 390 mg/l. Next, the membranes were flushed with a 2 weight percent solution of ammoniated citric acid at a pH of 4.0 for 2 hours.

The total iron concentration of the flushing solution increased from 0 to 40 mg/l before stabilizing. The pressure drop across the membranes was completely restored from 64 psi to 35 psi during the membrane cleaning operation for iron removal.

Next the membranes were flushed with a 2 weight percent solution of citric acid buffered to pH 8 with ammonia in an attempt to remove any $CaSO_4$ scaling. The flushing was continued for 6 hours with this solution. The analysis of the flushing solution showed that total hardness increased from 20 to 107 mg/l as $CaCO_3$ while the total iron concentration increased from 0 to 5 mg/l during this period. No further change in the headloss across the membranes was noticed.

These data show that even though there was some deposition of calcium on the membranes, the main cause of the membrane fouling during these tests was due to ferric hydroxide precipitation when the brine pH was 2.7 or higher. It was also demonstrated from these tests that, if the brine pH can be maintained below 2.7, successful control of the ferric iron fouling can be achieved.

For the acid mine waters in question, i.e., Proctor #2, the lowering of the feed pH to a level of 2.6 will probably well insure against iron fouling of the RO membranes. This can be achieved by adding an acid (H_2SO_4 or HCl). The amount of acid needed to achieve this is very small (10 to 20 mg/l H_2SO_4) for Proctor #2 waters and cost of such addition will be very nominal.

However, the acid costs to bring other AMD waters containing high amounts of ferric iron to desired pH levels may be excessive. In such cases, consideration should be given to the use of reducing agents for the reduction of ferric to ferrous stage. For example, approximately 10 mg/l sodium hydrosulfite or 80 mg/l sodium sulfite may serve both as an oxygen scavenger and as a bactericide. With sodium sulfite, cobalt chloride (0.01 to 0.1 mg/l) is conventionally employed as a catalyst.

Neutralization Tests

The brines produced from Run 2 were neutralized in field with four different alkaline agents, viz lime, limestone, soda ash and sodium hydroxide. As expected, lime was found to be the quickest and the most efficient alkaline agent while limestone was the slowest and least efficient alkaline agent. It should be noted that these neutralization tests were conducted after the ferrous iron in the brine was oxidized to the ferric state by adding an excess amount of hydrogen peroxide based on the results of neutralization literature review.

Additional tests were continued on the brine from Run 2 to develop preliminary neutralization data. Variation of pH with time at different levels of limestone additions was determined. Although the pH was higher than 4.5 for all three samples, appreciable amounts of iron remained in the effluent along with most other ions.

The clarity of the settled effluent was very poor for the first two samples containing 4,000 and 6,000 mg/l of limestone while the clarity of the sample containing 8,000 mg/l of limestone was relatively better (the settling rate for the bulk of the precipitated matter was good in all three cases). The clarity of the sedimented effluent improved significantly with increased sedimentation time.

Two runs (Run 2 and Run 3) were conducted to determine the quality of the neutralized brines with a combination of limestone and lime at near neutral (7.6) and high (10.8) pH levels. In all cases the ferrous iron in the brines was oxidized to the ferric state via H_2O_2 addition prior to neutralization tests. The pulverized limestone dosage used for neutralization was equivalent to twice the acidity of the RO brines.

The limestone was kept in suspension by aerating the brine slurry for 15 minutes. This limestone dosage and mixing time was sufficient to raise the brine pH to approximately 4.5 to 5.0. Lime was added to raise the pH to desired levels. The neutralized slurries were then allowed to settle for various intervals between 1 hour to overnight settling. From the effluent quality it could be seen that a significant increase in the calcium level resulted due to limestone-lime neutralization.

This increase in calcium levels was much higher in Run 2 tests than Run 3 tests. However, greater reliance should be kept on the test results from Run 3, where a multitude of tests were conducted in field compared to a single preliminary test in Run 2. The increase in calcium level at both low and high pH levels were similar in spite of the different lime dosages required for neutralization.

Significant reductions in the sulfate levels were recorded in the neutralized effluents. While only a partial removal of various metals except aluminum was achieved at the low pH neutralization level, the effluent was visually free of all cations except calcium at the high neutralization pH of 10.8. Also the effluent clarity at the high pH was significantly better compared to the clarity at lower pH (as shown by the turbidity results).

However, the sludge volume at the high pH level was higher compared to the sludge volume at lower pH. Settling velocities were generally greater than 3 to 4 feet per hour and are considered to be quite good. The effect of the addition of SHMP in the RO brines was found to be completely insignificant with regard to neutralized effluent quality and solid-liquid separation characteristics.

From the results of the above neutralization tests, it is evident that the treatment to a pH level of 10.8 can provide a quality of effluent that would be suitable for processing through reverse osmosis after softening treatment. The elimination of all magnesium at the higher pH level would help to save on chemical costs because of the use of a cheaper chemical—lime, vs more expensive softening chemicals such as soda ash. Moreover, the use of the softening step would inherently require the raising of the pH level of the solution for facilitating the precipitation of calcium hardness. Therefore, the neutralization of the first stage RO brines should be conducted at a pH level of approximately 10.8.

Softening Tests

Four different ways of softening were evaluated for the neutralized and settled brines. The primary objective in these tests was to reduce the amount of calcium in the treated effluent in order to prepare this water for a second stage of RO processing.

Sodium Aluminate Softening: Softening with sodium aluminate is achieved at a high pH level of 12.0. It is expected to reduce both the calcium and sulfate levels in the treated brines. These tests were conducted in two ways. Initially, sodium aluminate was added to the neutralized and settled effluent at pH 10.8 along with necessary lime to raise the pH to 12.

Later, sodium aluminate was added along with the necessary amount of lime to the limestone treated brine slurry so that the resultant slurry pH was 12.0. The sodium aluminate treated slurries were then allowed to settle for periods between 2 hours and overnight settling.

The results indicate that significant reductions in the calcium content of the treated effluents can be achieved at a dosage of 128 mg/l sodium aluminate per 1 pound per 1,000 gallons $CaCO_3$ total hardness. The reduction in sulfate was appreciable only when sodium aluminate was applied to the neutralized and settled effluents.

The sulfate concentration in the effluent was not reduced to the levels anticipated (600 mg/l target vs 1,350 mg/l achieved). However, this may have been an effect of the long periods of contact during which the mother liquor was in contact with the sludge. It may also be noted that the solid-liquid separation characteristics were not very encouraging. Considerably voluminous sludges and turbid supernates were experienced from the sedimentation-clarification treatment.

Soda-Ash Softening: Softening with soda-ash was evaluated for various combinations of limestone and lime treatment. Soda ash dosages were varied between 2,000 and 4,000 mg/l. This chemical was added to the brines in two ways. Either it was added to the limestone and lime treated effluents or it was added along with lime.

The soda ash treated slurries were then allowed to settle for periods between one hour and overnight settling. The results indicate that for an effective reduction of calcium in the softened brine, separate clarification stages are necessary for neutralization and softening. Soda ash dosage required for optimum softening results lies between 2,000 and 2,500 mg/l.

The solid-liquid separation characteristics for the separate softening step was extremely good. Sludge volumes were less than 2% and settled well in less than 1 hour. This sludge is nothing but precipitated $CaCO_3$ and should be recycled back to the neutralization step for additional savings in the chemical cost as well as towards sludge disposal costs. Also, softening with soda ash was evaluated for limestone treated brines. Again, separate clarification stages for neutralization and softening showed better results.

However, the calcium reduction in the softened effluent was appreciably lower when compared to the results of softened effluents as high pH. Moreover, operation at lower pH resulted in significantly low reduction of magnesium ions. The solid-liquid separation characteristics were also not encouraging with hazy effluents and high sludge volumes.

Carbon Dioxide Softening: The injection of carbon dioxide gas to the neutralized and clarified brines at high pH (10.8) was done in an attempt to precipitate some $CaCO_3$ without incorporating additional cations. It was found that the pH of the brine changed rapidly when CO_2 was bubbled through this water.

Less than 200 mg/l of CO_2 was sufficient to bring the brine pH to 7.0. Because of the lowering of the pH, the insoluble $CaCO_3$ is converted to the more soluble form of $Ca(HCO_3)_2$ and no appreciable precipitation occurs. However, slight haziness of the brines was observed.

The analysis of the resultant effluents at pH of 7.2 and 8.4 after overnight sedimentation did not show any significant reduction in calcium content. Therefore, the carbonation-neutralization treatment which appears to be a difficult softening step at best, introduces a separate contacting and clarification step to an already complicated process schematic.

Ion Exchange Softening: The soda ash softened brine effluent was passed through a commercial cation exchange resin (Rohm & Haas IR-120) to determine the extent of calcium removal in a limestone-lime-soda ash-ion exchange system from the AMD waste brine. The treated effluents contained less than 1 mg/l of calcium and magnesium respectively. The ion exchange system was operated for more than 30 hours. The effluent quality mentioned above was maintained on a sustained basis during this operational period.

Seeding, Sludge Recycling and Precoat Vacuum Filtration Tests

$CaSO_4$ Seeding: $CaSO_4$ crystals were added to the AMD brines during the neutralization test in order to simulate effective precipitation of $CaSO_4$ from the supersaturated brines. These tests proved to be inconclusive for the batch tests performed during this bench scale investigation. No clear advantage of the addition of $CaSO_4$ crystals could be established based on the treated effluent quality.

Also the use of seeding techniques to avoid calcium sulfate membrane scaling when processing supersaturated brines in tubular RO units (on a once-through basis) could not be effectively evaluated in a batch type operation.

Sludge Recycling: Bench scale sludge recycle tests were simulated by mixing the settled sludge in varying proportions with the fresh brines along with the neutralizing chemicals. The resultant slurries were then allowed to settle for periods between 1 hour and overnight settling. The sludge recycle ratios were varied between 0 and 50%. It was indicated that sludge recycling has definite merit in reducing the overall sludge volumes and other settling characteristics. Also, some improvement in the effluent quality was indicated with regard to calcium and sulfate concentrations. The optimum ratio of sludge recycle was indicated to be between 10 and 20%.

Precoat Vacuum Filtration of Sludge: Cursory bench scale experiments were performed on neutralized brine sludge with a 0.1 square foot surface area precoat filter leaf apparatus to evaluate the filterability and handling characteristics of these sludges. The results obtained from these tests (for total cycle time ranging between 2.5 and 5 minutes) were generally good. Excellent solids loading rates of 0.14 to 0.37 lb/sq ft were attained without any chemical conditioning.

The solids loading rates were almost doubled when a polymer (Atlasep IAI) was utilized. Excellent cake solids values ranging between 19 and 25% were obtained without polymer and up to 28% were obtained with 2.5 mg/l polymer conditioning. The cake formed with and without chemical conditioning was stable, did not crack and smooth cuts were obtained indicating the feasibility of dewatering of these sludges by precoat vacuum filtration.

The filtrates were visually clear and volumes ranged between 130 to 300 ml in 2.5 to 5 minutes of total cycle time testing. Based on the solids loading achieved in these tests, it is indicated that the cost of precoat filtration of these sludges would be in the range of \$8 to \$12 per ton solids without any chemical conditioning and approximately \$5 to \$7 per ton solids with chemical conditioning.

Phase II Flowsheet

According to PB 230 756 a two stage reverse osmosis treatment of the acid mine waters is warranted if the calcium and sulfate concentrations in the raw water reach limiting values when processed via reverse osmosis at moderate recoveries.

First Stage RO Treatment: The addition of a sand filtration step is desirable to reduce the load on the finer filters (10 micron) whenever the presence of the particulate materials necessitates such inclusions. The addition of acid to maintain a brine pH of 2.6 or a reducing agent prior to RO processing may be necessary to avoid any precipitation of ferric iron. The addition of acid would eliminate the need for UV light treatment. However flexibility should be provided in the design of the system so that this treatment may be included, if found necessary at a later date.

The recovery through the first stage RO processing should be maintained below the $CaSO_4$ limiting values (Ksp value: 25 to 35 x 10^{-5}) or below 50% for the Proctor #2 waters at the Hollywood, Pennsylvania site. Two low pressure RO units, each of 5,000 gpd nominal capacity, of Monsanto hollow fiber and GESCO spiral wound membranes could be considered for use during first stage RO processing.

Second Stage RO Treatment: It would be necessary to neutralize and soften the reject from the first stage RO processing prior to its treatment via a second stage of RO purification. Based on the results of the latest neutralization technology and the results of the neutralization tests discussed earlier in this report, it was concluded that for the acid mine waters under consideration (Proctor #2), biochemical oxidation and limestone-lime treatment to a pH of 10.8 would be the most suitable method of neutralization.

The biochemical oxidation of ferrous iron can be accomplished with a synthetic media type trickling filter made out of galvanized iron at a very nominal cost. The selection of the limestone-lime combination treatment is expected to provide not only savings in the chemical costs but also superior solid-liquid separation characteristics as well as constituent removals that would be favorable for economical postsoftening of the first stage brines. The recommended chemical softening is soda ash.

The soda ash softened effluents are expected to contain approximately 5 to 10 mg/l calcium and 1,700 to 2,000 mg/l SO_4 typically. The equilibrium recoveries for the second stage RO unit based on the above calcium and sulfate levels would be in the range of 87 to 92% without $CaSO_4$ fouling problems. However, acid addition would be necessary to the softened brines in order to eliminate potential carbonate fouling of the RO membranes at high recoveries and hence the same is recommended for Phase II operations.

The addition of small quantities (10 to 20 mg/l) of sodium hexametaphosphate, an antiprecipitant and dispersant, can help to increase the RO recoveries further to approximately 95% levels. However, it is recommended that the recoveries through the second stage RO unit not be allowed to exceed 90% since the osmotic pressure of the resultant stage two brines at higher recoveries would be extremely high (>200 psi) and would affect the water outputs significantly.

Also, a combination of 50% recovery through the first stage RO unit and 90% through second stage RO unit would provide an overall recovery level of 95% meeting the desired project objectives. The addition of the SHMP is, however, recommended during the second stage RO operations since this additive also helps to control iron fouling associated with any trace iron which may originate from equipment, industrial grade acid or sand filtration. The SHMP should be made up fresh on a weekly basis to avoid natural depolymerization of the stock solution.

The data of the softened supernate clarity indicates a turbidity range of 2 to 6 FTU. It is expected that the sand filtration and membrane protective 10 micron cartridge filtration would reduce the turbidity to below 1 FTU levels, making the treated effluents suitable for processing through hollow fine fiber B-9 membranes.

The second stage RO unit should consist of approximately 6,000 gpd (nominal) capacity Du Pont HFF B-9 membranes to be operated at a pressure of 400 psi. The polyamide base membranes are expected to be resistant to the alkaline feeds. The RO unit can consist either of a 1 x 1 array of full size and half size B-9 modules or 2 x 1 half size B-9 modules. The recommended treatment schematic for Stage 1 and Stage 2 processings is shown in Figure 6.14.

FIGURE 6.14: RECOMMENDED SCHEMATIC FOR PHASE II OPERATIONS

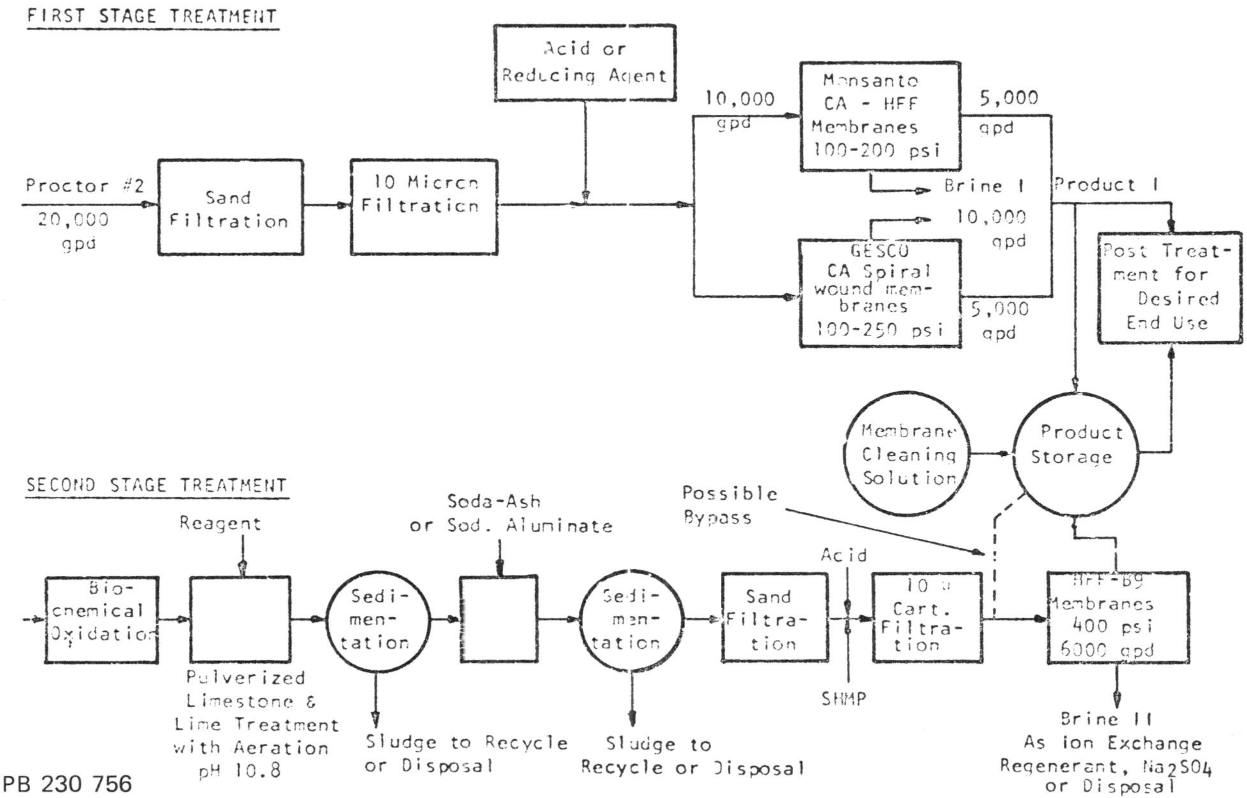

Source: PB 230 756

SOIL AS MEDIUM FOR RENOVATION OF AMD

The material in this chapter was excerpted from PB 228 868 (see Bibliography, p 390).

This study was initiated to determine the feasibility of using soils as a medium for the renovation of acid mine drainage. The detailed objectives were:

(1) To determine the ability of Rayne silt loam and Guernsey silty clay loam soils to renovate acid mine water.

(2) To determine the effect of various soil properties on the reclamation of acid mine water.

MATERIALS AND METHODS

The acid mine drainage water used in the three phases of this study was obtained from the Proctor 2 source at the Pennsylvania State University mine drainage treatment facility at Hollywood, Pennsylvania.

All soil samples used in phase 1 and 2 of this investigation were obtained from the Pennsylvania State University Soil Characterization Laboratory. Bulk samples of Rayne and Guernsey soils used in phase 3 (column study) were obtained at sites in Somerset (Rayne) and Washington (Guernsey) counties. These sites were previously sampled for characterization analyses by the Soil Characterization Laboratory. The Rayne and Guernsey soils as well as most of the other soil samples used were chosen to represent a wide spectrum of soil properties as well as being soil materials common to the coal producing areas of Pennsylvania.

Acid Mine Water Equilibration with Rayne Silt Loam

Samples (<2 mm) from the Ap, B23 and C1 horizons of a Rayne silt loam soil (obtained from the Soil Characterization Laboratory) were used in this phase of the study. The soil samples were added to acid mine water in 100 ml centrifuge tubes, stoppered, shaken for a prescribed time, centrifuged and the supernatant filtered. 60 ml of acid mine water was used in all treatments while the amount of soil varied. The amount of soil used was 60 g (1:1 ratio), 12 g (1:5 ratio) and 1.2 g (1:50 ratio). The soil material after centrifugation and decantation of the supernatant liquid was resuspended with enough distilled water to give a 1:1 soil to water ratio for the 60 g and 12 g samples and a 1:2 soil to water ratio for the 1.2 g samples and the pH was determined on the suspensions.

After filtering, conductivity, Fe^{++}, Fe^{+++}, total iron, total acidity and pH were determined on the filtrate. Single soil samples of 60 g (1:1) and 1.2 g (1:50) were shaken for 5 and 30 minute equilibrations. Duplicate soil samples of 60 g (1:1), 12 g (1:5), and 1.2 g (1:50) were shaken for 2 hour, 8 hour, and 32 hour equilibrations.

In addition duplicate 60 g (1:1), 12 g (1:5) and 1.2 g (1:50) samples of the B23 and C2 horizons were equilibrated for 128 hours. Control samples of soil and 0.005 M $CaSO_4$ were carried through all equilibrations except for 1.2 g (1:50 ratio) samples. Also control samples of acid mine water with no soil added were carried though the equilibrations and analyzed.

Soil-Acid Mine Water Equilibration and Correlation Analysis

Twenty-one samples from seven Pennsylvania soils were selected. These samples came from both surface and subsurface horizons, and represented a wide range of pH values and organic matter, clay, and clay mineral contents. Single samples of 50 g (oven dry basis) of each soil material were added to 250 ml of acid mine water (1:5 soil to water ratio) in 300 ml polyethylene centrifuge bottles.

The soil-water samples were shaken for a 32 hour equilibration period, centrifuged, filtered, and the filtrate analyzed for pH, total acidity, conductivity, total iron, Fe^{++}, and Fe^{+++}. A 100 ml sample of the filtrate was collected for analysis of K, Ca, Mg, Mn, Zn, Cu, Fe, Al, and Na by atomic absorption spectrophotometer. The 100 ml samples were stabilized by adding 3 drops of chloroform and 2 ml of concentrated HCl. The soil samples in the centrifuge bottles were air dried; 10 grams was removed, suspended in enough distilled water to give a 1:1 ratio and the pH of the suspension determined.

The remaining soil in the centrifuge bottle was resuspended with fresh acid mine water, shaken for an additional 32 hours and analyzed. This procedure was repeated a total of 3 times. Control samples of acid mine water with no soil were also carried through these equilibrations. Regression and analysis of variance analyses were made on the data obtained from the equilibration of the 21 soil horizons with acid mine water. The analyses were run on an IBM 370-165 computer using the Pennsylvania State University computer library programs ANOVES for analysis of variance and UPREG for regression analysis.

Percolation of Acid Mine Water Through Rayne Silt Loam and Guernsey Silty Clay Loam Soils

Percolation cylinders were used in this part of the study. They were constructed in the following manner: to the bottom of a 4 foot x 4½ inch Plexiglas cylinder a fiberglass screen mesh was attached with a 4½ inch automobile hose clamp. The cylinder was then placed in a 6 inch diameter polyethylene funnel and clamped by 4½ inch auto hose clamps to a plywood backboard. The funnel was placed in an Erlenmyer flask with a rubber stopper to prevent evaporation. The tops of the cylinders were covered with Pyrex plates.

Duplicate Rayne silt loam and Guernsey silty clay loam soil profiles were reconstructed in the Plexiglas cylinders. The soil material was collected in the field by horizons, brought to the laboratory, air dried, and sieved through a 4.7 mm sieve. The columns were loaded by placing a one inch layer of 2 mm glass beads in the bottom of the cylinder; next the soil horizon material of the upper 40 inches of each soil profile was added in small increments and the cylinder shaken until the material stopped settling, and lastly a 1 inch layer of 2 mm glass beads was placed on the top of the Ap horizon.

The soil columns were then wetted with 10 inches of acid mine drainage water. Acid mine water was then added to each soil column at the rate of 5 inches (1,000 ml) every seven days, and the effluent collected from the bottom of the columns. Infiltration rates, percolation rates and the volume of effluent collected were measured.

The effluent was analyzed for pH, total acidity, conductivity, total iron, ferrous iron and ferric iron by the methods previously described. Water samples were saved, stabilized and K, Ca, Mg, Mn, Cu, Zn, Fe, Al and Na content were determined. The experiments were carried on for 20 weeks.

RESULTS AND DISCUSSION

Acid Mine Water Equilibration with Rayne Silt Loam

The addition of the three soil materials to the acid mine water and its equilibration resulted in an initial increase followed by continuous increases, no change, or a decrease in the pH of the water. The data show that the greatest increase in pH occurred in the first 5 minutes of equilibration. This increase was exhibited by all soil material, (Ap, B23, and C1) and soil-water ratio treatments.

After the initial increase the Ap treatments showed a gradual pH increase with time while the B23 and C1 treatments remained essentially constant with the exception of the B23 128 hour determination. The B23 128 hour determination showed a decrease in pH for all treatments. This decrease is apparently not a result of an interaction between the soil and acid water, but an artifact resulting from a machine malfunction. These data indicate that the initial pH change is very rapid and subsequent changes occur more slowly.

The kind of soil material (horizon) equilibrated with the acid mine water had an appreciable effect on the change in pH of the water. The reaction of the Ap material with the acid water caused a greater increase in pH of the water than the acid water reaction with the B23 and C1 soil material for the 1:1 and 1:5 soil-water treatments.

For the 1:1 and 1:5 treatments both the B23 and C1 materials increased the pH of the acid mine water about the same degree. For the 1:50 ratio all three soil material treatments increased the pH of the acid mine water to approximately the same pH for the first 8 hours of equilibration; after which the Ap treatment increased the acid water pH slightly above that of the B23 and C1 treatments for the 32 hour equilibration.

The soil to water ratio also had an appreciable effect on the pH increase of the acid mine water. The increases in pH were greater when larger amounts of soil material were equilibrated with the acid mine water. The 1:1 soil-water ratio gave the greatest increase in pH while the 1:5 treatment gave less of an increase and the 1:50 treatment gave the least change. This was true for all three soil materials.

Samples of 0.005 M $CaSO_4$ equilibrated with the soil materials showed appreciable increases in pH for the 1:1 and 1:50 Ap treatments while the $CaSO_4$ solution treated with B23 and C1 materials showed essentially no change. Control samples of acid mine water, 0.005 M $CaSO_4$ and distilled water showed little or no changes in pH throughout the study.

The greater pH changes of the 1:1 and 1:5 Ap treatments when compared to the B23 and C1 treatments are most likely due to a higher pH, higher percent base saturation, and higher organic content of the Ap material (Table 7.1). These factors apparently affect the change in acid mine water pH, but the contribution of each cannot be determined from these data.

TABLE 7.1: RAYNE SILT LOAM Ap, B23, AND C1 SOIL PROPERTIES

Horizon	pH	Al me/100g	Base Saturation	Organic Carbon	Free Iron Oxide	Kaolinite
			- - - - - - - - - - -	- - - %	- - - - - - -	- - - - -
Ap (56-2-1)*	6.0	0.3	44.6	1.83	2.4	25
B23 (56-2-4)*	4.8	6.5	12.8	.05	-	40
C1 (56-2-7)*	4.8	5.0	5.2	.03	3.5	40

*Penn State Soil Characterization Number

Source: PB 228 868

The changes in pH noted for the 1:50 soil-water ratio for all three soil treatments were very similar. This would seem to indicate that at a ratio of 1:50 the soil system was overloaded and the soil properties of pH, organic matter content, and percent base saturation were not significant in affecting the pH status of the acid mine water.

The increases in pH noted for the 0.005 M $CaSO_4$ 1:1 and 1:50 Ap treatments were unexpected and may possibly be due to one or more of the following factors: (1) neutralization of coarse limestone particles in the Ap material, (2) reduction of sulfate to sulfide and the production of hydrogen sulfide, or

(3) reduction of sulfate to sulfide and its adsorption on clay minerals with the resulting release of OH's from the soil minerals. This increase in pH with time was also noted in the soil pH data after equilibration (Table 7.2).

TABLE 7.2: pH OF RAYNE Ap RESUSPENDED SOIL MATERIAL EQUILIBRATED WITH 0.005 M $CaSO_4$

Soil:Water Ratio	pH*					
	Control	5 min.	30 min.	2 hr.	8 hr.	32 hr.
1:1	6.1	6.1	5.9	6.2	6.1	7.0
1:50	6.1	5.8	5.8	6.0	6.1	6.6

*Determined on a 1:1 (for the 1:1) and 1:2 (for the 1:50) soil to distilled water suspension

Source: PB 228 868

The Ap treatments were the only ones which exhibited this phenomenon. Thus the factors given above seem reasonable although it cannot be proven which factor or combination of factors caused the response noted.

Total acidity was determined and expressed as ppm $CaCO_3$ equivalent for all water samples in the study except the distilled water control samples. Changes and trends similar to those noted in the pH data were evident, and were expected because of the close relationship between pH and total acidity. The major change noted was a decrease in acidity, the bulk of which occurred within the first 5 minutes of equilibration. After the initial change there was little significant change with additional time.

The type of soil material equilibrated with the acid water greatly influenced the amount of reduction in total acidity noted. The 1:1 and 1:5 Ap treatments were more effective in reducing the acidity than the 1:1 and 1:5 treatments of the B23 and C1 material. The type of material made no difference at the 1:50 ratio. Apparently at the 1:50 ratio the soil system was completely overloaded.

The soil-water ratio also influenced the amount of total acidity reduction. The 1:1 soil-water ratio gave the greatest reduction in total acidity, while the 1:5 treatment reduced acidity somewhat, and the 1:50 had little or no effect.

Organic matter possesses several types of functional groups (carboxyl, phenolic, and amino) which are capable of adsorbing or desorbing hydrogen ions. Hydrogen ions can also be removed from solution by inorganic soil colloids. Percent base saturation is a measure of soil saturation with respect to basic ions, while pH is a measure of the hydrogen ion concentration in solution. The higher the pH and percent base saturation of a soil the more ability it has to remove hydrogen ions from solution. This means the more the organic matter, the higher the pH, and the higher the percent base saturation, the larger the quantity of hydrogen ions that can be removed from solution. These relationships are evident in the total acidity equilibration data. The Ap horizon has the highest organic carbon content, pH, and percent base saturation and it removed the most acidity from solution. The B23 and C1 materials had lower pH's, lower organic carbon contents, and lower percent base saturation values and removed less acidity than the Ap treatment. Although there were differences in these three parameters between the B23 and C1 materials these differences were apparently not great enough to cause significant differences in the acid neutralizaton capability of these two materials.

The filtered supernatant samples were also analyzed for conductivity. The first five minutes of equilibration lowered the conductivity of the various treatments the largest amount, after which there was little or no change with additional equilibration time.

The soil-water ratio also influenced the decrease in conductivity noted. The 1:1 soil-water ratio lowered the conductivity by the largest amount for all three soil materials, while the 1:5 ratio reduced conductivity somewhat and the 1:50 ratio did not significantly affect it.

Sulfate is the major anion of the acid mine water and its retention may explain the conductivity trends observed. Sulfate retention is dependent on pH, the amount of iron and aluminum oxides, the amount of exchangeable aluminum, and the type of clay minerals present. In soils with high levels of aluminum and iron, a low pH causes adsorption of sulfate because of the presence of the multivalent ionic species which attract and adsorb sulfate more readily than monovalent cations. This apparently explains the trends noted in the conductivity data for the 1:1 soil-water treatments. The B23 and C1 soil treatments had the lower conductivity values apparently because they had lower pH's, higher amounts of free iron oxides, more exhangeable aluminum, and they contained more kaolinite than the Ap soil material.

Total, ferrous, and ferric iron were determined on all water samples except the distilled water control samples.

Equilibration of the Ap, B23, and C1 soil horizon materials with the acid mine water resulted in a large initial reduction in iron content of the acid water for the 1:1 treatments, a moderate initial reduction for the 1:5 treatment and little or no change for the 1:50 treatment. The initial change for all soil treatments was followed by a slight decrease or no change for the first 32 hours of equilibration. The 1:5 and 1:50 B23 and C1 treatments showed large decreases in iron content between the 32 and 128 hour equilibration times. Control samples of untreated acid mine water showed little change in iron content during the first 32 hours of equilibration, while between the 32 and 128 hour equilibration times they showed a significant decrease in iron content.

The kind of soil material used in the equilibrations influenced the quantity of iron removed from the acid mine water. The Ap soil material removed the largest quantity of iron when compared to the other soil materials for the 1:1 and 1:5 treatments; the C1 material removed the next largest quantity of iron and the B23 the smallest amount. In general there was no difference between the various soil materials in their ability to remove iron at the 1:50 treatment.

The soil-water ratio was the third factor influencing the quantity of iron removed from the acid mine water. The greatest reduction in iron content was noted with the 1:1 ratio for all three soil material treatments; the 1:5 treatment removed the next largest quantity of iron and the 1:50 treatment removed the smallest quantity of iron.

The amount of iron removed from solution by a soil material is dependent upon the pH and E_h (oxidation potential) of the soil-water system. Since pH was the only one of these parameters measured on the acid water, evaluation of the data will be based primarily on the pH data in relation to the iron content of the water.

Ferrous iron is the most abundant type of iron in the acid mine water. Below pH 8 ferrous iron must be oxidized to the ferric state and precipitated as $Fe(OH)_3$ in order to remove it from solution. As the pH decreases below 8 and as the redox potential lowers to a more reducing environment, ferrous ions become more stable and will remain in solution. Thus the greater the increase in pH caused by the soil treatment and the more oxidizing the conditions the larger the quantity of iron that will be removed from solution.

The changes and trends noted in the reduction of iron content in the acid water indicate that for the first 32 hours of equilibration the change in pH upon equilibration with the soil was the major factor causing the decrease in the iron content of the acid water. This change in pH is governed primarily by the original pH of the soil and the amount of soil equilibrated with the acid water.

The large decreases in iron content noted for the 1:50 B23 and C1 treatments after 32 hours of equilibration is apparently not related to the interaction of the acid water with the soil material. This decrease was noted in the control samples as well as the treatments. Apparently enough oxygen is available in the test tubes to trigger oxidation of the ferrous ion, although it must be a slow process. This conclusion is substantiated by the observation that at the end of the 128 hour equilibration time, the inside walls of the test tubes in which the soils were being shaken had an accumulation of brownish iron precipitate coating them.

Percolation of Acid Mine Water Through Rayne Silt Loam and Guernsey Silty Clay Loam Soils

The acid mine water showed an increase in pH after it percolated through the Rayne and Guernsey soils. The pH of the acid water applied ranged from 2.6 to 2.8 while the pH of effluent water (water that passed through columns) initially and throughout the study was approximately 4.0 for the Rayne soil and 8.2 for the Guernsey soil. The high effluent pH (8.2) for the Guernsey treatment is due primarily to the neutralization of the acid in the water by limestone material in the Guernsey subsoil (Table 7.3).

This conclusion is somewhat substantiated by the fact that $CaCO_3$ in equilibrium with water produces a solution pH of approximately 8. The Rayne soil is not calcareous although its surface horizon has a near neutral pH (Table 7.3). This pH level is due to the application of agricultural lime to the soil during farming operations. The increase in pH of the acid water resulting from the Rayne treatments is probably due to the interaction of the acidity in the water with the ions on the organic and inorganic exchange complex as well as some hydrolysis of soil minerals. Some coarse undissolved limestone particles in the surface

TABLE 7.3: SOIL PROPERTIES OF RAYNE SILT LOAM AND GUERNSEY SILTY CLAY LOAM

Horizon	Depth	pH	Al	Ca	Mg	Na	K	Org. Carb.	Base Sat.	$CaCO_3$ Equil.	Clay Minerals: Kaol	Ill	Verm	Mont	Chl	Int	Qtz
	Inches		meq/100g					%									
								Rayne									
Ap (56-2-1)*	0-9	6.1	0.3	9.0	0.3	0.06	0.34	1.83	44.6	1.4	25	15	35	0	0	15	10
B21 (56-2-2)*	9-17	5.4	1.2	4.7	0.1	0.06	0.19	0.37	21.7	0.0	30	25	30	0	0	5	10
B22t (56-2-3)*	17-21	4.9	5.9	2.7	0.1	0.04	0.21	0.18	18.2	0.0	45	40	15	0	0	0	0
B23 (56-2-4)*	26-38	4.8	6.5	1.9	0.1	0.04	0.21	0.05	12.8	0.0	40	45	15	0	0	0	0
B24 (56-2-5)*	38-51	4.8	6.4	1.3	0.1	0.05	0.35	0.05	10.3	0.0	40	45	15	0	0	0	0
								Guernsey									
Ap (63-6-1)*	0-9	6.2	0.1	11.4	0.7	0.2	0.3	1.84	59.2	1.5	25	35	25	0	0	15	0
B21t (63-6-2)*	9-18	4.9	3.8	9.5	0.9	0.2	0.5	0.14	39.4	0.0	25	30	15	5	0	25	0
B22t (63-6-3)*	18-23	4.2	3.5	12.7	1.1	0.2	0.4	0.11	50.7	0.0	25	30	10	5	0	30	0
B23t (63-6-4)*	23-27	5.5	0.0	23.8	1.4	0.2	0.4	0.12	71.9	2.0	20	30	15	10	0	25	0
B24gt (63-6-5)*	27-33	6.7	0.0	26.0	1.1	0.2	0.3	0.18	100.0	5.0	20	30	10	5	0	35	0
C1 (63-6-6)*	33-40	7.4	0.0	30.8	0.7	0.2	0.2	0.13	100.0	8.8	20	45	15	0	0	20	0

*Penn State University Soil Characterization Laboratory Number.

Source: PB 228 868

horizon may also have contributed to the neutralization of some acidity in the water. It appears from this data that an equilibrium condition has been established between the acid mine water and the soil, the pH level of which appears to be a function of the soil. Thus soils may possess certain levels of acid mine water pH equilibrium which are dependent on the characteristics of the soil.

The data for total acidity, conductivity, Mn, Zn, Cu, Al, Na, Mg, K, and Ca content of the effluent water for the Rayne treatment showed an initial decrease in the concentration of these constituents. Following the initial decrease there was an increase which leveled off or decreased depending upon the parameter measured. The initial parts of this trend most likely can be attributed to the amount of contact time that occurred between the soil material and the acid water. In this study the amount of time it took the acid water to percolate through the columns (percolation time) is considered the best estimate of this parameter, and these data are presented in Table 7.4.

For the Rayne soil this data indicates a progressive increase in percolation time

TABLE 7.4: MEAN PERCOLATION TIME

Total Inches of Applied Water	Rayne		Guernsey	
	Volume Collected	Percolation Time	Volume Collected	Percolation Time
	ml	Hours	ml	Hours
5	960	32	945	39
10	935	46	920	94
15	940	51	950	94
20	950	52	950	93
25	950	70	950	92
30	940	69	950	96
35	960	70	965	92
40	940	70	950	94
45	-	71	-	-
50	950	70	960	94
55	935	70	950	95
100	950	69	960	93

Source: PB 228 868

to the 5th application of water (25 inches); after which the percolation time was constant. The percolation time for the Guernsey soil was constant after the 1st application of acid water. The reason for these changes in percolation time was, the soil material slumped upon the addition of the water causing sealing of some of the water conducting pores and channel ways in the material. This slumping was more extensive in the Rayne soil and extended over a longer time period. This slumping and its effect on the percolation rate is part of the reason for the trend noted.

After the initial changes an equilibrium between the soil and the various constituents in the water was established and there was little change noted in most parameters for the duration of the study.

Of the changes that did occur after equilibrium was established some are due to changes in quality of the acid mine water applied. These changes in quality were largely a result of rainfall dilution of the various constituents of the acid water. Other changes noted will be discussed in the following sections.

Both the Rayne and Guernsey soils caused a decrease in the total acidity of the acid water. The Guernsey effluent water was alkaline throughout the study, while the effluent water collected from the Rayne soil was acid. The equilibrium level in the Rayne soil was reached after 30 inches of acid water was applied while it was reached immediately with the Guernsey soil. Both soils maintained these approximate equilibrium levels throughout the study.

The only change noted in the equilibrium levels was a slight lowering of the level for the Rayne effluent during the last half of the study. This change was due to a change in the quality of the acid mine water applied. The Guernsey's effluent was alkaline because the acid mine water was neutralized by $CaCO_3$ in the soil. The high percent base saturation of the Guernsey soil may also be a contributing factor to the decrease in the acidity of the acid water. The Rayne soil is acid and must rely primarily on cation exchange and hydrolysis reactions to neutralize the acidity. Some coarse undissolved limestone particles in the surface horizon of the Rayne soil may also have contributed to the neutralization of some of the acidity.

Conductivity of the acid mine water was reduced significantly for the first few applications of water for both the Rayne and Guernsey soils. After an initial decrease for the Rayne soil the conductivity of the effluent from both soils rose gradually and leveled off after 35 inches of acid mine water had been applied. After this point there was only a slight difference in conductivity between the effluent and the acid water applied.

The Rayne soil material treatment reduced the conductivity of the effluent more than the Guernsey soil treatment for almost all applicatons of acid mine water. This response is due primarily to differences in soil properties between these soils, and the effect these differences have on sulfate retention. Sulfate is the major anion in acid mine water and its retention greatly affects conductivity.

In general sulfate retention is enhanced by low soil pH, high amounts of aluminum and iron, and the presence of kaolinite clay. The Rayne soil has the lowest soil pHs, and the largest amount of aluminum and kaolinite clay of these two soils (Table 7.3). Thus it would be expected that the Rayne soil treatment would lower the conductivity of the acid mine water more than the Guernsey soil treatment. Although there was an initial reduction in conductivity the data indicate that neither soil was effective in reducing conductivity after 35 inches of water was applied.

Apparently the soils become saturated or nearly saturated with sulfate at this point and can no longer remove any from solution.

Iron content of the Rayne and Guernsey effluent was determined by both colorimetric and atomic absorption methods. The total iron content of the acid mine water applied ranged from approximately 150 to 300 ppm, while the effluent showed no iron in solution for all water samples even after the addition of 100 inches of acid mine water. The amount of iron removed from solution by a soil material is dependent upon the pH and E_h of the system. The dominant ion species of the acid water is ferrous ion.

Apparently both of these soils were well aerated enough to oxidize the ferrous iron to the ferric state which is within the pH range of these two effluents (Rayne pH 4, Guernsey pH 8) and precipitated out as $Fe(OH)_3$.

The manganese content of the acid mine water was reduced by both the Rayne and Guernsey soil treatments. The Guernsey removed all of the manganese from the acid water as it passed through the soil, while the Rayne was much less effective, particularly in the latter stages of the study. The removal of manganese from solution is dependent upon pH, E_h and the availability of carbonate. The higher the pH and the more oxidizing the system the greater will be the tendency for manganese to precipitate from solution. In the case of the Guernsey soil, the manganese is being precipitated as MnO_2, a hydroxide, or possibly a carbonate as well as some being adsorbed on the exchange complex. The manganese is probably being removed from solution in the Rayne soil by adsorption on the exchange complex. This would explain the Rayne's reduced ability to remove manganese from solution after 50 inches of acid mine water had been applied. Apparently the exchange complex at this point was very close to its saturation point with respect to manganese.

The zinc content of the acid water was reduced by the Rayne and Guernsey soil treatments. The Guernsey soil removed all of the zinc after the first application while the Rayne effluent showed a trend after the initial changes of a gradual decrease followed by an apparent state of equilibrium. Zinc retention in soils is dependent upon pH, and type and amount of exchange complex. Zinc applied to alkaline soils will precipitate as a carbonate or hydroxide and apparently this is what is happening as the acid water passes through the Guernsey soil. Zinc is relatively soluble at pH values of the Rayne effluent.

Apparently the retention noted for the Rayne soil is a reflection of zinc being adsorbed on the exchange complex. The trend of decreasing zinc content of the effluent after the initial changes occurred is different than that noted for other effluent constituents, and no explanation is obvious.

Copper content of the Guernsey effluent was lower than that of the acid water applied while that of the Rayne effluent was higher. The Guernsey soil treatment decreased the copper content of the acid water 100% for the first 10 inches of acid water applied; after which a very low concentration was found in the effluent water until 55 inches of acid mine water were applied.

At this point, the copper content decreased to zero when the copper content of the acid water applied decreased to approximately zero. Copper content of a solution is pH dependent and at high pHs it will precipitate as an oxide, carbonate or hydroxide. Apparently this is what is happening in the Guernsey soil. In the case of the Rayne soil because of its low pH, no copper is being retained and in fact the acid water is extracting copper from the soil material.

Aluminum content of the acid mine water which percolated through the Rayne and Guernsey soils was decreased. The Guernsey soil reduced the aluminum content 100% for all applicatons of acid mine water while the Rayne soil showed a significant decrease throughout the study. In the case of the Rayne treatment an equilibrium condition was established after the initial change which varied slightly due to a change in quality of the acid water during the later stages of the study.

Aluminum solubility and adsorption is dependent primarily on soil pH. Aluminum at low pHs (<4.5) is very soluble, consequently in the Rayne treatment aluminum reduction is primarily a function of adsorption on the exchange complex. At higher pHs such as those of the Guernsey soil aluminum precipitates as a hydroxide, and this is probably the major mechanism for the complete removal of aluminum from the Guernsey effluent.

The sodium content of the effluent from the two soil treatments varied appreciably during the study. The Guernsey effluent showed a sodium content higher than that of the acid water applied for all but three applications, while the Rayne soil generally reduced the sodium content of the acid water until approximately

60 inches of acid mine water had been applied. After this point the Rayne effluent showed a higher sodium content than the acid water applied. The high sodium content of the effluent water indicates that sodium is being extracted from the exchange complex of the soils. Sodium is not a strong competitor for exchange sites, and forms no insoluble compounds in soils, thus it is understandable that it is being removed from the soils by the acid water percolating through it.

Potassium content of the acid water was reduced considerably by the Guernsey soil treatments for all acid water applications, while the Rayne treatments reduced the potassium content of the acid water only slightly and only in the first half of the study. During the second half of the study the acid water applied had a lower potassium content than the Rayne effluent. Potassium retention is dependent upon the number of cation exchange sites available, in particular the number on illite and vermiculite minerals.

These minerals have a strong affinity for potassium and will fix large quantities of this element. The Rayne and Guernsey soil have approximately the same amount of illite and vermiculite, thus both should remove approximately the same amount of potassium from solution, but because of a lower pH, the Rayne was unable to adsorb as much potassium as the Guernsey. The pH effect is probably a result of more pH dependent sites available for exchange at higher pHs or possibly it may be due to less blocking of the base exchange sites with aluminum or iron polymers at higher pH levels.

Magnesium content of the acid mine water is reduced by both the Rayne and Guernsey soil treatments. Both soils show a trend of decreasing ability to reduce the magnesium content of the acid water up to 50 inches of applied acid water. After this point the trend continued for the Rayne while an equilibrium level was established for the Guernsey soil.

Since almost all magnesium salts are soluble, magnesium retention in the soil is a function of its adsorption on the exchange complex. Apparently the Guernsey soil's higher total exchange capacity is responsible for the differences noted between these two soils in magnesium retention.

Calcium content of the Rayne and Guernsey effluent was greater than that of the applied acid water. Both soil effluents showed a trend of increasing concentration of calcium up to approximately 40 inches of applied water, after which an equilibrium level was established. The increases in calcium content of the water for both soils is the result of hydrogen ions as well as other ions displacing calcium from the exchange complex. Calcium is the major basic ion on the exchange complex (Table 7.3). Thus this is a feasible explanation of the calcium content data observed. The high content of calcium in the Guernsey effluent is also due to the dissolution of $CaCO_3$ in the subsoil.

Acid Mine Water Equilibration with Samples from Seven Pennsylvania Soils

The data obtained were statistically analyzed by multiple linear regression in an attempt to relate the amount of renovation noted to certain of these soil properties. Only explained variance of 60% or greater and major increases in explained variations within a treatment will be discussed.

The differences noted in acid water pH after equilibration with the various soils are a result of differences in the properties of these soils and the effects these differences have on the hydrogen ion concentration of the mine water. In the first equilibration 87% of the variation in acid mine water pH could be explained by soil pH, and cation exchange capacity (CEC) of the soil material. In the second equilibration 91% of the variation was accounted for by $CaCO_3$ equivalent, CEC, soil pH, and clay content of the soil material, while in the third equilibration 87% of the variation was attributed to $CaCO_3$ equivalent, CEC, and clay content.

All soil properties except clay were positively correlated with acid mine water pH, indicating that as the amount of these soil properties increased so did the acid water pH. This could be expected because with high soil pHs, and larger CEC and $CaCO_3$ reserves, the pH of the water should increase because of greater ability of the soil material to adsorb and neutralize the acidity of the acid water. The increase in importance of $CaCO_3$ in treatments 2 and 3 indicates that the soil's exchange complex is becoming saturated with acidity and it can no longer do a sufficient job in changing the hydrogen ion concentration of the water.

These trends were also noticed in the two previous studies in which soils with high soil pH and $CaCO_3$ equivalents increased the acid water pH considerably more than soils with low soil pH and no $CaCO_3$. The negative correlation of acid water pH to clay content was unexpected. Apparently the hydrogen ions of the acid water are extracting aluminum from the minerals in the clay mineral and the aluminum is helping to maintain a low pH.

Percent base saturation, CEC, exchangeable aluminum, and $CaCO_3$ equivalent accounted for 89% of the variation in total acidity in the first equilibration treatment. In the second treatment CEC, soil pH, clay content, and $CaCO_3$, exchangeable acidity, and percent base saturation accounted for 94% of the variation. Percent base saturation, CEC, soil pH, and $CaCO_3$ equivalent were negatively correlated while total acidity, exchangeable aluminum, clay content, and exchangeable acidity were positively correlated with total acidity of the acid water.

The negative correlation shows that when the soil, equilibrated with acid mine water has a high percent base saturation, a high CEC, a high soil pH and $CaCO_3$ equivalent, the acid water total acidity will be lower than if the soil properties listed had small values. This is reasonable because high soil pH, high percent base saturation, and high $CaCO_3$ equivalent allow for neutralization of acidity and the exchange of hydrogen and other ions for basic cations on the exchange complex.

The quantity of exchangeable aluminum, clay, and exchangeable acidity also influenced the total acidity of the acid mine water. When these soil properties have high values the total acidity of the acid water will also be higher than if the value of these properties were low.

The relationships noted for exchangeable aluminum and exchange acidity were expected but not the one noted for clay content. It is believed that aluminum is probably being released from the clay material by weathering and contributing to the total acidity of the acid water. The first two phases of this study also showed similar trends. The soils with the higher pHs, higher percent base saturation, and $CaCO_3$ content had the greatest effect in lowering the total acidity of the acid mine water.

The differences noted in soil pH after the first equilibration with the acid mine water are attributed to initial soil pH, CEC, and $CaCO_3$ equivalent which accounted for 89% of the variation, while during the second treatment 89% of the variation was accounted for by soil pH, CEC, $CaCO_3$ equivalent, and clay content.

In the third equilibration 83% of the variation was accounted for by $CaCO_3$ equivalent and CEC. All soil properties with the exception of clay content were positively correlated thus indicating that when these soil properties have high values, the soil pH after equilibration will be higher. This could be expected because initially higher soil pHs, higher CEC and larger $CaCO_3$ equivalent will all contribute to higher soil pHs after the addition of acid mine water.

The differences noted in the conductivity of the acid water could only be explained at a satisfactory level by soil properties in the first equilibration. The soil properties which affected conductivity were CEC, $CaCO_3$, clay content, and exchangeable acidity which accounted for 82% of the variation. Calcium carbonate equivalent and CEC were positively correlated, while clay and exchangeable acidity were negatively correlated with the conductivity of the acid water.

This would be expected since sulfate is the dominant anion in the acid water system and its retention by soils is dependent upon low soil pH, and the amount of kaolinite clay (low CEC) present for adsorption. If a soil material has a high pH (low acidity) or $CaCO_3$ equivalent or the clay has a high CEC (indicative of low kaolinite content), sulfate will not be adsorbed readily. The lack of much explained variation in the second and third treatments results from a saturation of the adsorption complex for sulfate after the first equilibration.

The first two phases of this study also showed the same trends, particularly the column study which indicated that after approximately 35 inches of acid water had been applied the soil could no longer adsorb any more sulfate and the acid water being added had approximately the same conductivity as the effluent water coming out of the bottom of the columns.

The differences noted in the total iron content of the water cannot be explained adequately by a variation in soil properties in the first equilibration. In the second equilibration 74% of the variation in total iron content can be attributed to CEC, silt, and organic matter content while in the third equilibration CEC and percent base saturation accounted for 66% of the variation.

All these soil properties were negatively correlated indicating that as the value of these soil properties increases total iron decreases in concentration in the acid mine water. These data suggest that iron is being adsorbed in larger quantities when the soil material has a higher CEC, and a higher organic matter, percent base saturation and silt content. Oxidation of the iron by atmospheric oxygen could not be controlled during these equilibrations. The lack of control over this factor undoubtedly accounts for the inability to account for more of the variability in total iron content as a result of these treatments.

The ferrous iron content of the acid water in the second equilibration showed the only meaningful data. These data indicate that CEC, silt, and organic matter account for 69% of the differences in ferrous iron content of the acid water. The lack of adequate accountability in the variation of ferrous iron by soil properties is due to poor control of oxidation of the iron by atmospheric oxygen.

The ferric iron content of the acid mine water in the second equilibration was the only equilibration where any meaningful level of variation could be explained. During this equilibration 62% of the variation in ferric iron content was explained by the cation exchange capacity of the soils. As was previously mentioned this is due to lack of control over the oxidation of the iron by atmospheric oxygen.

A slight increase in potassium content of the acid mine water after equilibration was noted for most soil samples. This increase could not be related to any of the soil properties by the statistical analysis.

The soil properties related to the 88% of the variation in calcium content of the acid water for the first treatment were percent base saturation, CEC, clay content and $CaCO_3$ equivalent. In the second equilibration CEC, soil pH, $CaCO_3$ equivalent and clay content accounted for 91% of the variation in calcium content of the water while in the third equilibration 89% of the variation was attributed to percent base saturation, $CaCO_3$ equivalent, CEC, and exchangeable acidity. The calcium content of the water after equilibration was generally higher than that of the water added. This difference is due to calcium being replaced on the exchange complex by hydrogen and other ions and the dissolution of $CaCO_3$ in the soil material.

Thus it is reasonable that the amount of calcium on the exchange complex (base saturation), the amount of clay present, the CEC and $CaCO_3$ equivalent are important factors influencing the calcium content of the treated acid water. These relationships were also noted in the column study. The magnesium content of the acid mine water showed no meaningful relationship to the soil properties.

In the first treatment 71% of the variation in zinc content of the acid water was accounted for by CEC and exchangeable acidity of the soils, while for the second equilibration CEC, exchangeable acidity, percent base saturation, and $CaCO_3$ equivalent accounted for 85% of the variation. There was no meaningful accounting for the variation in the third equilibration. Zinc retention in soils is dependent upon soil pH and type and amount of exchange complex. Soils with high percent base saturations, high $CaCO_3$ equivalent, low amounts of exchangeable acidity and high CEC will remove zinc from the water much more readily than will a soil material which is acid.

In the copper analysis the second equilibration was the only treatment where a meaningful level of explained variation could be found and it accounted for 72% of the variation in copper content of the acid water. The 72% of variation was accounted for by the CEC, soil pH, silt, exchangeable aluminum and organic matter content of the soil material. Copper is retained in soils when the soil pH is high and there are exchange sites available for adsorption.

Thus CEC, soil pH, and organic matter with its cation exchange ability are important in copper removal from the water. Exchangeable aluminum explains some of the variation. This relationship is apparently related to the acidity associated with high levels of exchangeable aluminum.

The differences in the manganese content of the acid mine water could not be related to any soil properties to provide a meaningful accounting for the variation. Total iron of the acid mine water as determined by atomic absorption showed no meaningful relationship between soil properties and the iron content of the acid water for the first equilibration while the second and third equilibrations showed meaningful relationships. Organic matter, silt content and CEC of a soil material accounted for 64% of the variation in the second equilibration while CEC accounted for 61% of the variation in the third equilibration.

These data indicate that soils with high organic matter contents, high CEC and silt content will adsorb more iron. This agrees with the colorimetric iron determinations as far as what soil properties affect iron concentration in the acid water.

The first equilibration treatment showed a very low level of explained variation in aluminum concentrations of the treated acid water, while the second equilibration accounted for 88% of the variation with CEC, soil pH, and exchangeable aluminum. The third equilibration had 62% of the variation in aluminum content of the acid water accounted for by CEC and exchangeable acidity. Soil pH, and CEC were negatively correlated while acidity and exchangeable aluminum were positively correlated. These relationships are expected.

The higher the CEC the more sites that are available for exchange reactions with aluminum, and the higher the pH (lower acidity) the greater will be the tendency for aluminum to polymerize and be adsorbed by the soil material.

The variation in sodium content of the acid mine water could only be accounted for in the first equilibration. Clay content, silt content, soil pH, exchangeable aluminum, and percent base saturation accounted for 79% of the variation in sodium content of the acid water.

Sodium is not a strong competitor for exchange sites, thus at lower pH values and lower clay contents there are fewer sites available and more competition for these sites, particularly with aluminum. At high pH values there are more sites available for exchange reactions and apparently sodium can compete more effectively at higher pHs because there is a significant negative correlation between the sodium content of the acid water and percent base saturation.

Although there are relationships between the soil properties and sodium content of the acid water, the major effect according to this study and the previous column study is that sodium is not being adsorbed but like calcium is being released from the soil material.

COMBINED TREATMENT OF MUNICIPAL WASTEWATER AND AMD

The material in this chapter is excerpted from PB 222 936 (see Bibliography, p 390).

Phosphorus removal is becoming recognized as a function of wastewater treatment. This trend favors the increasing adoption of chemical methods of wastewater treatment. Chemical coagulation removes phosphorus and also fine suspended and colloidal solids from wastewater to a greater extent than conventional processes.

Among the chemicals suitable for the coagulation of wastewater are some of the salts of ferrous iron. Acid mine drainage is a natural source of ferrous iron which may serve as a coagulant. Both wastewater and AMD contain pollutants that may be removed if the two are treated together under suitable conditions. This is known as combined treatment.

The principal pollutants in AMD are acidity and dissolved iron, both of which would be removed in combined treatment. Some acidity would be neutralized by the alkalinity in wastewater. The alkalinity of wastewater typically exceeds the alkalinity of water before use by 81 mg $CaCO_3$/l, which is the amount added to the water during use. Further, lime would be added in the combined treatment process so as to neutralize any residual acidity, in order to attain the pH necessary for the coagulation of wastewater solids with ferrous iron.

Combined treatment appears to be especially promising for Pennsylvania, which has widespread and severe AMD pollution problems. In the research described in this report the technical and economic feasibility of combined treatment at various sites in Pennsylvania was investigated.

EXPERIMENTAL

Sampling Program: During the sampling program, from January to April 1972, eleven pairs of samples of wastewater and AMD were collected and processed in a laboratory scale combined treatment plant. Three sampling sites that appeared most favorable for combined treatment had been selected and three or four sample pairs were collected from each site, on different days and at different times of the day. In this way spatial variation of significant factors between sites, as well as time variation at particular sites was taken into account. There was, however, some seasonal variation in the composition and flow of AMD that could not be fully characterized in this study, due to the relatively short duration of the sampling program.

Sampling sites are designated by the letters A, B and C located as follows:

Site A — Central Pennsylvania; wastewater from Indiana Borough raw wastewater outfall; AMD from Imperial Keystone coal refuse pile (near Clymer);

Site B — Eastern Pennsylvania; wastewater from Wyoming Valley Sanitary Authority wastewater treatment plant; AMD from the largest pipe outfall from the Delaware Pumps (near Plains);

Site C — Western Pennsylvania; wastewater from Bethel Park wastewater treatment plant;

C-1 — AMD from Chartier's Creek major source 6035 (near Bridgeville);

C-2 — AMD from Chartier's Creek major source 6034 (near Bridgeville).

Laboratory Scale Treatment Plant: Wastewater and AMD were mixed and treated by flocculation and sedimentation in a jar test apparatus. Two runs were conducted for each sample pair, the first run to investigate the effect of varying the pH of treatment and the second run to investigate the effect of varying the ratio of AMD to wastewater.

The jar test procedure was as follows:

(1) To a 200 ml sample of raw wastewater add a small measured amount of AMD, adjust the pH to 8, then stir for 3 minutes at approximately 60 rpm using a magnetic stirrer. Repeat the addition of AMD, adjustment of pH, and stirring until a floc is visible at the end of the stirring period after the stirrer is switched off.
(2) To 1,000 ml samples of raw wastewater in each of six 1,500 ml beakers add AMD at the dose rate previously found necessary for the formation of a visible floc.
(3) Adjust the pH values in the mixtures to different values according to an integer or half-integer gradation encompassing pH 8.
(4) Flocculate the mixtures simultaneously for 20 minutes using the jar test apparatus set to approximately 60 rpm.
(5) Remove the stirrer blades from the liquid, allow the flocculant suspension to settle for 60 minutes, then decant the supernatant.
(6) Determine turbidity, total phosphorus, and ferrous iron on the supernatant. The other determinations required on the supernatant are completed at a later stage. (After it was established that pH 8 was persistently optimal for removal of phosphorus

and iron, steps 2 to 6 were omitted.)

(7) To 1,000 ml samples of raw wastewater in each of six 1,500 ml beakers add AMD at 25, 50, 75, 100, 150 and 200% of the dose rate previously found necessary to produce a visible floc.

(8) Adjust the pH of each mixture to the optimum value determined in step 6.

RESULTS OF LABORATORY SCALE TREATMENT

Analyses of AMD and of raw wastewater on each sampling occasion (run) appear in Table 8.1. In that table the following code was used:

(A) = Run number
(B) = Sampling site
(C) = Collection date (1972)
(D) = Time of wastewater collection
(E) = pH, AMD
(F) = Acidity (mg $CaCO_3$/l)
(G) = Ferrous iron (mg Fe/l)
(H) = Total iron (mg Fe/l)
(I) = Aluminum (mg Al/l)
(J) = Sulfate (mg SO_4/l)
(K) = Flow (mgd)
(L) = pH, Sewage
(M) = Alkalinity (mg $CaCO_3$/l)
(N) = Total phosphorus (mg P/l)
(O) = BOD (mg/l)
(P) = TOC (mg C/l)
(Q) = Min. Wastewater:AMD mixing ratio for visible floc

The results of laboratory scale treatment were plotted and the graphs interpreted as follows:

(1) Turbidity vs pH — Low effluent turbidity was most consistently observed at or above pH 8. At pH 9 the effluent was green in color.

(2) Phosphorus vs pH — Treatment at pH 8 gave the best removal of phosphorus. At pH 8 the removal of phosphorus increased with the molar ratio of ferrous iron to total phosphorus, with comparatively little further removal of phosphorus when the molar ratio was higher than 2.

(3) BOD vs pH — Results appear not to indicate any strong effect of treatment pH on removal of BOD.

(4) Ferrous iron vs pH — At least 99% of ferrous iron was removed by treatment at pH 8 and above.

(5) Total iron vs pH — pH 8 or above was optimal for the removal of total iron.

(6) Aluminum vs pH — pH 8 appears to be optimal for the removal of aluminum.

(7) Turbidity vs ferrous iron dose — Generally most of the removal of turbidity is attained by 10 to 20 mg Fe(II)/l. However, in Runs 8 and 9 turbidity increased with iron dose because AMD

from that site (C-1) contained suspended ferric hydroxide which does not settle. Incidentally, the treatment of wastewater with chlorinated copperas at a dose rate of 10 to 20 mg Fe(II)/l is reported to allow 80 to 90% of suspended solids to be removed in settlement.

(8) Phosphorus vs ferrous iron dose — Most of the removal of phosphorus was generally attained when the molarity of ferrous iron was twice the molarity of phosphorus. At this molar ratio of 2 the reduction in phosphorus ranged from 68 to 98%, with a median of 94%.

(9) BOD vs ferrous iron dose — Where the dose of ferrous iron was 10 mg Fe(II)/l the reduction in BOD ranged from 30 to 93% with a median of 72%, while at 20 mg Fe(II)/l the range of BOD removal was 50 to 92%, with a median of 78%. This compares closely with the reported performance of chlorinated copperas, which allows 70 to 80% removal of BOD at a dose rate of 10 to 20 mg Fe(II)/l.

(10) TOC vs ferrous iron dose — At 10 mg Fe(II)/l about 60% of TOC was removed.

(11) Effluent alkalinity vs acidity dose — For Runs 1 to 7 the results appear to approximate the theoretical. With regard to Runs 8 to 11, it is not clear why the alkalinity required for pH correction greatly exceeded the acidity introduced with the AMD.

(12) Ferrous iron survival vs ferrous iron dose — Ferrous iron was never detected in the effluent from treatment at pH 8, but was not removed at pH 5.

(13) Total iron survival vs total iron dose — Total iron represents ferric iron since ferrous iron is absent. Ferric iron is substantially, but not completely removed in treatment, except at pH 5.

(14) Sulfate vs sulfate dose — Results for all runs except Run 3 were similar indicating that sulfate is conserved through treatment. Results for Run 3 show that it is incorrect to assume that no sulfate was originally present in the wastewater. Concrete in contact with water containing as little as 240 mg SO_4/l may be corroded.

Overall Evaluation of Optimum pH: Performance of the treatment is at or near optimal in all respects examined, at pH 8. At pH 8 there is no greater variation in performance between runs using samples from different sites than between runs using samples collected at different times from a particular site.

Overall Evaluation of Optimum AMD:Wastewater Mixing Ratio: For the removal of both turbidity and BOD (or TOC) an AMD dose equivalent to 10 to 20 mg Fe(II)/l achieved results comparable to those attributed to the chemical coagulation of wastewater by the same dose of ferrous salt. Most of the phosphorus removal was attained at a molar ratio of ferrous iron to total phosphorus of 2. The variation in process performance between runs for which samples had been taken from different sites was significantly greater than the variation in performance between runs for which samples had been taken from a single site, only where the AMD had been collected from site C-2, where much of the iron is in the ferric form.

TABLE 8.1: ANALYTICAL AND EXPERIMENTAL DATA FOR SEWAGE AND AMD SAMPLES

					AMD						Sewage					
(A)	(B)	(C)	(D)	(E)	(F)	(G)	(H)	(I)	(J)	(K)	(L)	(M)	(N)	(O)	(P)	(Q)
1	A	Jan. 13	10:30 am	3.9	3,020	1,010	1,010	--------	3,650	0.08	-----	--------	7.8	220	----	0.025
2	A	Jan. 25	5:30 am	3.8	1,468	465	465	235	1,850	0.21	7.4	--------	0.6	56	----	0.025
3	A	Feb. 7	9:30 pm	3.2	10,010	3,271	3,295	1,440	22,000	-------	7.2	47.7	10.6	288	----	0.0075
4	A	Apr. 9	10:00 am	3.3	3,722	953	1,171	695	4,200	0.13	7.5	102.3	4.2	165	22	0.020
5	B	Feb. 15	7:15 am	5.4	566	303	303	10	2,800	-------	7.0	43.8	3.2	250	----	0.075
6	B	Mar. 5	9:30 am	5.5	618	303	308	6	2,600	-------	6.7	29.6	1.4	152	----	0.025
7	B	Mar. 29	1:00 pm	5.4	559	297	314	9	2,600	-------	7.0	102	8.7	223	----	0.125
8	C-1	Feb. 27	10:30 pm	6.2	38	94	112	--------	2,100	0.12	7.5	224	6.3	1,160	----	0.375
9	C-1	Mar. 11	1:00 pm	6.0	27	27	88	--------	2,200	0.45	7.1	166	10.7	--------	----	0.375
10	C-2	Mar. 21	9:45 am	6.5	33	143	144	12	1,800	-------	7.4	163	5.7	456	----	0.188
11	C-2	Apr. 9	1:00 pm	6.2	67	114	119	--------	1,900	-------	7.4	154.9	5.9	463	20	0.200

Source: PB 222 936

SUMMARY AND CONCLUSIONS

Chemical methods of wastewater treatment are likely to be used increasingly, in view of the recognition of phosphorus removal as one of the functions of wastewater treatment. In certain cases AMD may not only be an economical source of chemical coagulant, but also diversion of AMD to combined treatment with wastewater would contribute towards the abatement of pollution due to AMD.

Three locations were selected in Pennsylvania where sources of AMD near wastewater treatment plants appeared suited to combined treatment. At each location three or four pairs of samples of wastewater and AMD were collected on different dates.

For each sample pair, laboratory scale treatment by flocculation and sedimentation was conducted, under a range of conditions of pH, and of AMD to wastewater mixing ratio. The effluent, as well as raw wastewater or AMD, were analyzed for turbidity, total phosphorus, BOD, total organic carbon, ferrous iron, total iron, aluminum, sulfate, and acidity or alkalinity.

The treatment process was at or near optimal performance at pH 8 in reducing total phosphorus, ferrous iron and turbidity. At pH 8 the optimum mixing ratio of AMD to wastewater for removal of phosphorus was such that the molar ratio of ferrous iron (in AMD) to total phosphorus (in wastewater) was about 2, when the median removal of total phosphorus was 95%. The optimal mixing ratio for removal of BOD was such that the dose of ferrous iron was 10 to 20 mg Fe/l wastewater, this ferrous iron dose was also considered optimal for reduction in turbidity. Ferrous iron was almost never detectable in the effluent from treatment at pH 8. Sulfate from AMD may be sufficiently concentrated in the mixed AMD and wastewater that the need for cement-rich concrete of sulfate-resisting cement should be investigated.

Combined treatment of wastewater and AMD was ineffective where the AMD was oxidized, with a significant part of the iron in the ferric form.

The economic feasibility of combined treatment was investigated. In the absence of a realistic measure of monetary benefits from combined treatment, benefits were measured as alternative costs of the equivalent chemical treatment. On this basis, combined treatment could only be economically feasible where it costs less than chemical treatment.

Examination of the cost relationships showed that there is a maximum distance for which it is economical to pump AMD to a wastewater treatment plant for combined treatment, beyond which chemical treatment is less expensive than combined treatment.

UNDERGROUND DISPOSAL OF COAL MINE WASTES

The material in this chapter is excerpted from PB 245 130 (see Bibliography, p 390).

Coal mine waste, already a serious problem due to that accumulated from past coal mining operations, presents a serious environmental and social stress, especially on Appalachia. With increased coal production, waste disposal will become an even greater problem.

Approximately 25% of the coal extracted from underground mines in the United States is rejected as waste and disposed on the surface near coal preparation plants. The amount that has accumulated over the past 200 years is considerable. In the eastern coal fields alone there are at least 3,000 to 5,000 sizable active and abandoned waste piles and impoundments which cumulatively contain over three billion tons of refuse.

These waste piles produce acute social as well as environmental problems. They have been the direct cause of the death of mine workers and local inhabitants, and the source of severe local air pollution and stream and groundwater pollution, as well as having had regional, economic, psychological and sociological impacts. Proper surface or underground disposal can minimize such problems at future sites but inactive piles continue to present a long-term negative impact to the region and its people.

Underground disposal, or backfilling, of coal mine waste has been clearly demonstrated to be technically feasible in active longwall mining operations in Europe and in abandoned room and pillar mines in the United States. However, the practice has declined drastically in Europe, being considered a drawback in mechanized, high production-rate mining; although a recently developed lateral discharge pneumatic backfilling system appears to overcome some of the earlier disadvantages.

Most U.S. mines use the room and pillar mining system, but the longwall mining system, for which present backfilling methods were designed, is becoming more widely adopted in the drive to obtain higher production but currently accounts for less than 5% of underground coal mine production.

A safe, operational system of concurrent backfilling and extraction at operating room and pillar faces has not yet been developed, and would require substantial development efforts and operational changes.

Backfilling of abandoned mines or abandoned sections of operating room and pillar mines through boreholes from the surface appears technically less difficult than concurrent backfilling and extraction in active sections of the mine. Such systems could involve some surface neighborhood inconveniences and also would be limited to those situations where fill transportation distances from the preparation plant to the mined out areas are not excessive.

The health and safety of underground miners could be affected by underground disposal systems, particularly those where concurrent mining and backfilling take place in close proximity. In particular, the increased congestion at the face, the increased dust and noise of pneumatic systems, and, in hydraulic systems, the hazard of barricade ruptures, would present safety hazards.

In general the economic costs of underground disposal exceed by far the normal costs of surface disposal. Only if the physical conditions at a mine site for environmentally acceptable surface disposal were so adverse or the social benefits of underground disposal so great would underground disposal become economically competitive with surface disposal for modern, high production rate coal mines.

METHODS OF BACKFILLING AND APPLICABILITY

Methods

Pneumatic Backfilling: Standard, Older System — The classical (European) method of pneumatic backfilling on a longwall coal face is shown diagrammatically in Figure 9.1. The waste is transported along the tailgate haulageway by rail or conveyor belt and introduced into the stower where it is entrained in a flow of high pressure air. It is projected along the tailgate in special abrasion-resistant pipes, forced around a bend in order to travel parallel to the face, and ejected immediately behind the working face into the area that has been mined out. A brattice cloth (heavy sacking material) progressively extended along a row of wooden props, is erected to contain the waste within the backfilled area.

Backfilling is interrupted as each pipe-length of backfilling is completed, and a section of pipe (2 to 3 meters) is uncoupled and placed in position for the next stowing sequence. Then stowing resumes. This cycle is repeated until the backfill has been placed along the entire length of the working face. Then the coal is mined to a depth of 1 to 2 meters, conveyed along the face, discharged onto a conveyor belt in the main haulageways, and transported out of the mine.

Pack walls, constructed by hand of large waste rock broken down from the roof to increase head room in the main haulageways help support the roof above the main haulageways and serve to contain the backfilled waste. Waste crushed to a maximum size of one-third the pipe's inside diameter can be carried several hundreds of meters in pipes, but 150 to 200 meters is considered the desirable maximum. Under suitable air pressure and seam height conditions waste can be projected about 50 meters from the nozzle. The air pressure is up to 4 kg/cm^2 (60 psi).

This method is best suited for the three 8-hour shift daily cycle, in which coal mining is carried out on one shift, stowing the haulageways advancing on the nest and general equipment maintenance on the next, or on the four 6-hour cycle in which coal mining and stowing are carried out on alternate shifts. It has not proved well suited for the continuous (24 hours) coal mining longwall method in use today in Europe or in the United States.

FIGURE 9.1: CONVENTIONAL PNEUMATIC STOWING LAYOUT

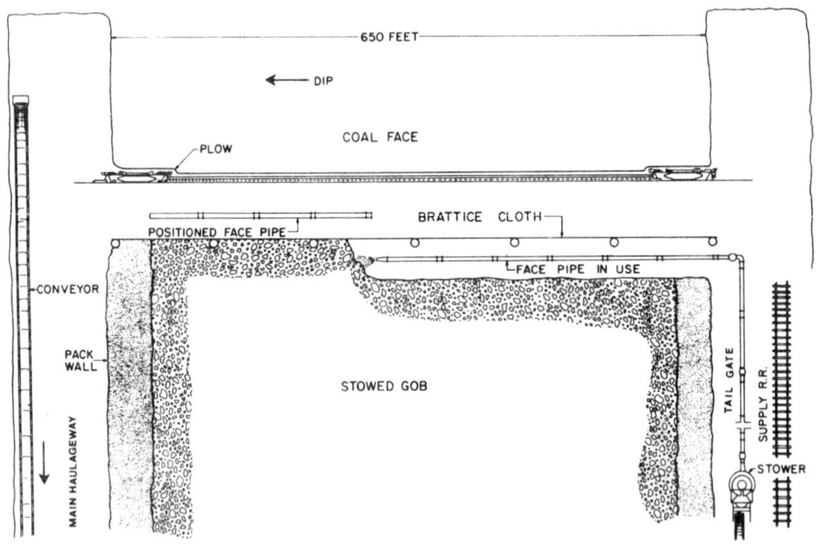

Source: PB 245 130

Pneumatic Backfilling: Lateral Discharge System — The rise in labor and capital costs in European coal mining have resulted in highly mechanized mining systems and increased outputs which are dependent on rapid and continuous advance of the longwall face. This development, the shift to a 24 hour continuous mining cycle, and the relative inefficiencies and cost of backfilling led to a substantial decline in backfilling throughout the 1960s.

However, more efficient stowing machine designs developed over the past several years in Germany lower the cost of handling the stowing materials and are capable of keeping pace with continuous longwall miners. The incorporation of regularly-spaced, side-discharge deflectors into the face pipes and overall mechanization of the stowing operation have made pneumatic backfilling compatible for working concurrently on rapidly advancing mechanized longwall faces (Figure 9.2).

The waste is discharged laterally into the stowage area having been deflected from the pipe via a side-discharge unit. The side-discharge unit can be rapidly

FIGURE 9.2: LATERAL PNEUMATIC STOWING LAYOUT

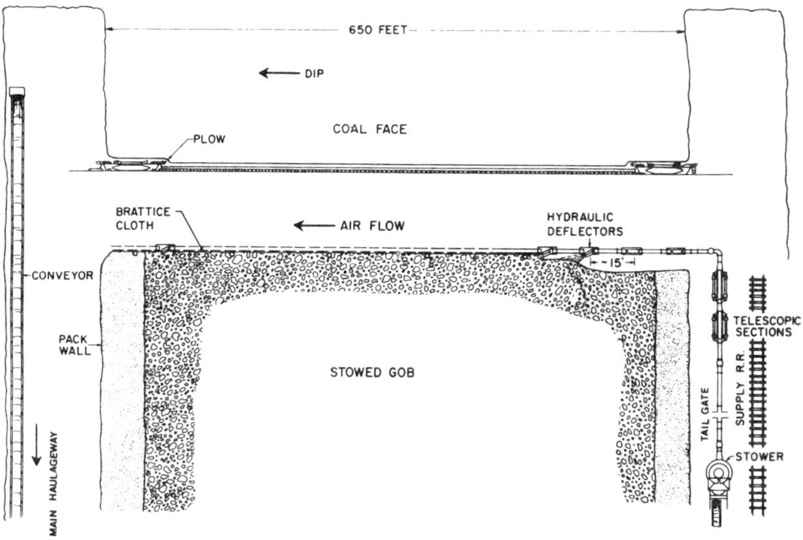

Source: PB 245 130

opened and closed hydraulically, without interruption of the waste delivery. The stowage area is filled progressively along the face by first opening the furthermost deflector to fill the first section to be backfilled; then closing this deflector, and simultaneously opening the next. The face pipes are attached to the rear of a (self-advancing) mechanized face support system and are advanced with the face, eliminating the tedious pipe dismantling and reassembly of the traditional system. The brattice is usually replaced by heavy rubber belt guards attached to the base of the rear support. Telescopic sections of pipe in the tailgate road permit the system to be advanced continuously with the longwall face.

The side-discharge system can reduce the cost of pneumatic stowing to about 50% of that of the traditional system, but experience with it to date has been too limited to obtain reliable cost figures. Mechanized side-discharge backfilling, although still in the experimental stage, is currently being used in several European mines.

Hydraulic Backfilling: Hydraulic backfilling of waste in coal mines is not used as extensively as pneumatic methods in Germany and Poland but it is used more extensively than pneumatic methods in Belgium and France. Hydraulic stowing has been used in active U.S. coal mines. In a typical installation in Poland crushed washery waste is placed at the surface into a large, 50 meter deep underground bunker.

When needed, the waste is flushed from the bunker using hydraulic monitors (high pressure water jets) and conducted several hundreds of meters down a vertical pipe to abrasion resistant pipes. These carry the slurry along inclined ventilation and service haulageways to the longwall (retreat) coal faces. The material is backfilled at a depth of approximately 800 meters using the hydraulic pressure produced by the full hydraulic head of the 800 meter high pipe system.

Brattice cloths are erected along the face as with traditional pneumatic stowing, and lateral discharge pipes are used to direct the waste-laden water behind the cloth. The removal and relocation of pipes is very similar to that described for the traditional pneumatic system. Stowing is carried out up the face to allow water to drain away into a sump in the downward dipping tailgate ahead of the face. The water in the waste drains quickly so that within twenty minutes of stowing the waste appears as dry as pneumatically stowed waste. No coal mining operations are permitted at the face during hydraulic stowing, because of the possibility that the brattice cloth will burst, and the water and waste would flow onto the face.

The potential hazard of waste and slurry flowing onto the face area during hydraulic backfilling suggests that the method is less likely to be used in conjunction with fast-moving mechanized longwall faces than pneumatic backfilling and not likely to be used at active room and pillar faces.

Hydraulic backfilling is most frequently used in steeply dipping seams. There is some evidence that it can be used in gently dipping conditions, but water handling problems appear to prohibit its use in flat-lying seams such as those mined in the United States. The density of packing of the waste is superior to that obtained in pneumatic stowing, so mine roof disturbance and subsidence are more effectively minimized. The hydraulic method is preferred in Poland for use in the mining of thick coal seams (6 meters or more in thickness) where the coal is extracted in three stages, of approximately two meters each.

Although there are similarly thick seams and even thicker seams than these in the western United States, no such thick seams currently are being mined by underground methods. In Europe, these thick seams are mined in stages beginning with the uppermost seam or from the lowermost. In Poland the lowermost two meters is extracted and then backfilled hydraulically. Using the top of the previously backfilled material as the floor of the second mining the second two meters are then extracted in a similar fashion, at a distance (100 meters or more) behind the bottom working face and backfilled in preparation for the third, final level. After the third extraction, the mine roof is allowed to cave. In this way, it has proved possible to limit to tolerable amounts the surface subsidence and associated damage above the mine.

Mechanical Backfilling: Mechanical backfilling is rarely used for underground backfilling in coal mines. In this backfilling method, coal waste is mechanically slung into a void. It has not proven effective in reducing subsidence because the packing density and the percentage of the void that is filled are considerably less than that of hydraulic or pneumatic systems. It could prove more promising as a method of waste disposal in certain applications. For example, in conjunction with a partial extraction system, coal waste could be moved from a borehole injection site to an area to be backfilled using mobile scoops and ram cars.

Hand Packing: Hand packing used to be a system popular in older coal mining methods in which strip packs were constructed manually by building a box-like structure of dry walls and filling them using rock available from local (controlled) roof falls. No waste was brought into the mine for the packs, so the system is of little interest as a possible method for disposing of mine waste. The packs served to support the roof over the caved area and localize roof falls along the caved area parallel to the face. Hard wood or steel roof supports on the pack side of the face area served to limit roof collapse to the caved area, thus preventing it from advancing into the area at the working face. The system did not prove very effective for reduction of subsidence.

Strip packing is a relatively slow operation and although extensively used in the 1950s is no longer practiced to any extent. It gave way to complete caving, in which the roof is allowed to collapse fully behind mechanized roof supports. Caving eliminates delays in face advance that may occur with backfilling and thus is ideally suited for the high advance rates associated with mechanized longwall mining. However, caving does result in greater surface subsidence than any packing or stowing system.

Controlled Flushing (Hydraulic): The term flushing is generally used to characterize systems for introducing backfill materials into voids in abandoned underground mines. In hydraulic flushing, a slurry of crushed coal mine waste and water is transported from the ground surface through boreholes and pipelines to the point of disposal in the mine. The water is then drained off and pumped back out of the mine leaving the waste material in the mine voids. The slurries generally contain 30 to 50% solids. The volume of water varies with pipe size, waste material size, and the desired pumping rate. Hydraulic flushing can be used to backfill flooded or dry mines.

Of the various types of hydraulic flushing, controlled flushing has been the most effective in backfilling abandoned mines. Controlled flushing employs men in the mine to place pipes and to control the distribution of the injected waste. It can be used in mined out portions of active mines, and in those abandoned mines where it is safe for personnel to install pipelines and perform the flushing operation.

In controlled flushing, a pipe, generally six inches in diameter, is installed in a shaft or borehole from the surface to the mine level where it is connected by a large radius pipe elbow to other pipes, either metal or plastic, which convey slurry to the areas to be backfilled. When a mine shaft is not open or safe, a large borehole (44 inch diameter or more) can be drilled into safe areas of an inactive mine to permit the entry of men and equipment. Wooden bulkheads and drain boxes installed downslope from the flushed area impound the waste slurry and permit the water to drain off.

Water pressure developed in the vertical section of pipe is sufficient to force the slurry through the pipelines in the mine. The distances to which the slurry can be distributed underground (under the pressure head in the vertical pipeline) varies with the depth to the workings, the frictional resistance of the pipe, the water-solid ratio of the slurry and the slope or dip of the mine. Generally 300 to 600 feet of horizontal distribution can be obtained for each 100 feet of vertical drop. Controlled backfilling should begin at the far limits of the area to be filled and progress to the point of final egress to the ground surface. Lower sections of a mine can be readily filled, but higher areas sometimes require the construction of bulkheads to control the slurry.

Blind Flushing (Hydraulic): Blind flushing employs no men underground and is used when it is impossible or dangerous to enter a mine because the mine workings are water-filled or have collapsed. In blind flushing, boreholes are drilled into the mine on a selected grid pattern, and the crushed waste is flushed from a hopper installed at the top of the borehole. The slurry flows into the mine with little control of backfill distribution. Placement of backfill continues until the void is filled and no additional slurry can enter the borehole.

An alternative method of blind flushing has been developed in which large volumes of slurry are transported into the mine at high velocity. The process was designed to backfill flooded mines using sand or crushed mine refuse. The material is placed in suspension in a mixing plant and pumped into the mined out areas through a closed pipeline system. The underground mine workings fill as solids settle out in response to changes in flow velocity.

Peripheral to the area of turbulent flow at the base of the injection borehole, a doughnut-shaped mound of solids builds up toward the mine roof. The slurry velocity increases with the resulting restriction of available channel space and the suspended material is transported over the mound and deposited in quiet water on the outer slopes. Thus the mound builds outward.

When one channel becomes blocked, another opens at the place of least resistance, and the process continues until a large area is filled. This technique has achieved a wide distribution of a large quantity of backfill from one borehole. At Scranton, Pennsylvania, in 1972 the maximum amount of refuse implaced through a single borehole was 204,000 tons, some of which carried more than 500 feet from the hole. The average slurry was 17% by weight solids and the maximum amount injected was 61,000 cubic yards in two 8-hour shifts. A total of five holes were drilled, two of which were used to inject two-thirds of the material. Conventional blind flushing techniques would have involved 175 to 200 boreholes to achieve the same large areal distribution of backfill.

Hydraulic backfill requires large amounts of water. This may pose problems in some areas. Water for transporting fill into abandoned mines has been obtained from flooded sections of the mine or from dewatering activities in active mines. Water drained from the backfill or collected at a sump in the mine can be recycled into the backfilling system. Hydraulic backfilling of mines located above the water table may produce problems of increased stream pollution or slope stability. Hydraulic flushing into acid mine pools may exacerbate short-term water pollution problems.

Pneumatic Flushing: Experience with pneumatic flushing in the western Pennsylvania coal fields has been limited to the pneumatic placement of fly ash. All but one of these demonstration projects has been blind injection. In the only controlled flushing project access was possible to a mine beneath a school addition. A cement block wall was built at the entrance to the mine room underlying the school addition and fly ash was injected until the room was filled.

In pneumatic blind flushing, six inch diameter boreholes are drilled on a grid spacing of 20 to 30 feet. The fly ash is transported in pneumatic tank trucks which blow the fly ash into the mine through the boreholes at approximately 25 tons per hour. Photographs of mine backfilling have shown that the fly ash builds up in the mine void at a relatively flat slope (angle of repose is approximately 10°). The spread of the material is a function of the dimensions of the underground void. In large rooms ash spreads on the order of 30 to 40 feet from the borehole filling the voids with reasonably good roof contact. The properties of fly ash are quite different from those of coal waste, but a pneumatic flushing system could be adapted to handle coal mine waste.

Transportation of Backfill Material

The preceding discussion assumes that the backfill material is available at the hopper entry to the stowing or flushing equipment. However, delivery of the waste from the coal treatment plant or mine dump may pose problems.

Transportation systems in coal mines are designed to move coal, men, and supplies. The systems use various combinations of equipment depending on seam height, mining method used, and capacity required. The coal transportation system is designed for continuous movement of coal (and associated waste) from the face to the surface treatment plant. It may consist of locomotives hauling trains of mine cars for miles to the preparation plant for rapid unloading into a bunker using drop-bottom cars or rotary dump cars. It may consist of a series of conveyors, starting with a chain conveyor on the mining face itself, or with shuttle cars, or scoops at the face, dumping the coal from a room into a surge feeder for delivery onto a secondary belt conveyor. Several such conveyors feed a main or trunk conveyor delivering coal to the surface.

Where a rail system is in use, trains returning from the surface sometimes are used to transport supplies. Conceivably, a system could be set up to transport mine waste for backfilling provided these cars could be delivered to the stowing machine hopper, unloaded, and returned to a coal filling point without interruption of coal production. Such scheduling would overload present transport systems and pose insurmountable problems in high production, mechanized mines in terms of synchronization of the backfilling and coal mining operations. A breakdown or delay in either would slow down the other and result in a loss in mine productivity.

If rail transport by rotary dump cars or drop-bottom cars is chosen to move the waste, dumping facilities will have to be excavated and installed underground near the stowing machine. In most U.S. mines this would involve the development of considerable additional head room at the dumping station and special feeder conveyors for transferring the waste to the stowing machine. With each move of the stowing machine new dumping stations and the conveyor systems would have to be relocated. In Europe the provision of waste dumping stations in tailgate haulageways is not a serious problem since head room in traditional European longwall mines differs from that in the United States because of at least two important mining practices:

(1) Single entries rather than multiple entries are driven. Longwall advance rather than longwall retreat is still the more prevalent mining system. The entries are usually enlarged and maintained at a height of 8 to 10 feet to provide sufficient cross-sectional area for adequate ventilation at the face and to allow for access

to the face even when convergence of the roof occurs with compaction of the caved and backfilled material.

(2) Mine cars used for pneumatic stowing in European mines are generally smaller than mine cars now used in the United States.

Because United States longwall systems do not enlarge the entries as much as in Europe, and because longwall retreat is preferred to longwall advance allowing major convergence of the roof behind the retreating face when it is of less consequence than in advance mining, the head room in U.S. mines is invariably inadequate for rotary dumping. An entirely new system of rapid dumping and temporary storing of mine waste for stowing would have to be developed in U.S. mines. Special waste storage bunkers, with chain unloading of rail cars for some low head room situations have been developed in Europe and conceivably could be adapted for use in the United States mines.

Conveyor systems for coal transportation are popular in high production mines and have been suggested for transportation of refuse to pneumatic stowers. Although efficient for transportation, they are designed to be run in one direction only. Not only do the secondary conveyors overhang the trunk conveyors at transfer points, the pulleys supporting the belts are oriented so as to centralize the belt as it runs forward. This action is adverse when running in reverse.

In current operations the need for supplies in room and pillar mining (roof bolts, supports, etc.) and for mechanized longwall mining, can be met by relatively infrequent delivery during a working shift. Introduction of waste disposal would require the establishment of a continuous supply system; separate from coal transportation where this is done by conveyor.

Direct hydraulic or pneumatic transport of the mine waste in pipes from a surface bunker appears more feasible. The Miekowicz mine in Poland uses such a system in its hydraulic stowing operation. Pneumatic conveyance of backfill through vertical pipes is used to transport wastes several thousand feet in metal mines. Although a potential problem, blockage of system has not been a serious practical problem. Research conducted on the hydraulic and pneumatic transportation of coal to the preparation plant may resolve some of the potential problems of transportation into the mine for waste disposal. The use of booster pumps to maintain velocity of the waste in long horizontal stretches of mine roadway may be necessary.

The introduction of a waste disposal system into an operating mine will require a backfill transportation system suited for the particular conditions of that mine. In shallow mines, it might be practical to introduce waste into the mine through boreholes from the surface to a point near the area to be backfilled. Additional boreholes would be drilled as the disposal area became filled or as the mine face advanced. Such a system could encounter problems in populated areas or in hilly terrain, where in both cases, surface accessibility for boring could be limited and where surface transportation of the waste using trucks (or pipelines) on a continuous (24 hour) schedule would be undesirable.

Technical Benefits and Disadvantages

The principal benefits claimed for backfilling are reduced ground subsidence, disposal of waste, and reduction of ventilation leakage on longwall advance faces. The principal disadvantages claimed against backfilling are increased dust, fire and gas hazards, increased noise, more men underground, and roof control problems.

The volume of the surface subsidence trough resulting from complete extraction systems of mining is directly related to the real distribution and volume of the mine voids. Backfilling in partially extracted areas by lending lateral support to pillars can prevent subsidence.

Reduction in subsidence is of major concern when mining under developed areas or under rivers and bodies of water (either surface or groundwater aquifers). Backfilling in complete extraction and longwall systems will reduce subsidence, but cannot prevent it completely since the fill material is subsequently compacted by the weight of the overlying rock. Pneumatic stowing, the most widely used system in coal mines, reduces the maximum amount of vertical surface subsidence to about 50% of the extracted seam thickness without backfilling. Room and pillar mining in the United States has resulted in vertical subsidence of approximately 75% of the extracted seam thickness. Surface strains (the principal cause of damage to buildings and other installations) will be reduced correspondingly, but not necessarily enough to prevent damage.

Backfilling also reduces the disruption of strata in the vicinity of the mine workings, a feature that is of particular significance when mining multiple seams in close (vertical) proximity. It is often difficult to extract coal from a seam that has been disrupted by the effects of caving of underlying mined out seams. The method of multiple layer mining of thick seams in Poland is an extreme example of backfilling being used advantageously to maximize coal recovery from thick seams and multiple seams.

It is claimed in Europe that backfilling has improved ventilation of longwall advance faces (almost all U.S. longwall faces are retreat, not advance). Leakage or short-circuiting of air from the intake haulageway (usually the main entry or entries) to the return haulageway (the tailgate entry) via the voids in the caved area can be a serious problem in advancing longwall faces, especially as the haulageways become long (Figure 9.1). Backfilling may be very effective way of reducing this leakage and insuring that an adequate supply of fresh air reaches the coal face.

There are several technical disadvantages of backfilling. Blockage of the pipe is an unresolved technical problem of pneumatic stowing systems. Care must be taken to avoid an excess of fines or clayey material from the waste as these tend to cause blockages in the stowing pipe system. Mechanized coal mining equipment produces a greater proportion of finer particles than older mining methods.

Backfilling represents a major addition to the steps involved in the underground mining process by requiring a waste transportation and supply system to the backfilling equipment. If backfilling must be synchronized so as to keep pace with the other coal face operations, the possibility of coal production loss due to pipe blockage, equipment breakdown, or lack of adequate waste supply could be considerable.

The high cost of backfilling was a principal reason for the decline of backfilling in Europe. Backfilling reached its peak in Europe during the late 1950s when the demand for coal was very high and it was necessary to mine as much of the available coal resource as possible while at the same time minimizing subsidence damage. This implied not only longwall mining of the coal seams, but also the extraction of contiguous seams. In order to restrict surface subsidence and facilitate multiple seam mining some form of stowing was deemed necessary.

During the 1960s the demand for coal declined. Economic and social circumstances have dictated increased mechanization and a shift to a continuous 24 hour mining cycle. Less economical operations have been abandoned. Current mining practices include such economical measures as:

(1) Mining only the most profitable of the coal seams in a series (thereby reducing the total volume of voids contributing to subsidence but also reducing the resource recovery);
(2) Permitting caving instead of stowing; except in a few isolated situations;
(3) Introducing mechanized longwall systems in conjunction with the caving;
(4) Where feasible, designing surface installations to accommodate the resulting surface subsidence;
(5) Where not feasible, paying for subsidence damage rather than lowering production rates or institute a less profitable mining system involving backfilling.

Modifications of Mining Methods for Backfilling

Experience has been gained with pneumatic and hydraulic backfilling behind longwall faces, and backfilling in abandoned mines. In room and pillar operations extensive changes in mining and transport methods would be necessary to allow the use of backfilling. Some of these had been practiced when mining methods were considerably less mechanized than today. Others are possible mining techniques where the technology is partially or fully developed. Still others reflect drawing board concepts based upon technology which is not currently available. These include techniques applicable to mining methods other than room and pillar such as in situ extraction. Some of these methods would circumvent the waste disposal problem altogether.

Creation of Partially Extracted Waste Disposal Areas: In a mining operation where only the wastes created as a result of current operations at a particular mine need to be disposed of, it is probable that wastes will account for less than 25% of the volume of extracted material. Under this circumstance it may be technically and economically desirable to dispose of all the wastes in some particular section of the mine.

An example of this would be an operation in which several mine panels would be developed by pillaring and the rooms left open as a repository for wastes produced from the coal mine. The wastes could be transported to the disposal area by rail or conveyor and pneumatically or mechanically packed into place. Such a system would have the advantage of not slowing coal production because of a lack of backfill material. A disadvantage of this scheme, as with all partial recovery mining methods, is that the pillars in the disposal area could not be recovered.

Secondary Recovery of Pillars After Flushing from the Surface: Secondary recovery of pillars after flushing was accomplished in the Anthracite region when mining methods were not mechanized and not regulated by current laws. In 1956, the Bureau of Mines with the cooperation of an anthracite mine near Wilkes-Barre, Pennsylvania, also extracted pillars after flushing. The mines were reentered after flushing, and the pillars extracted by conventional cut, shoot, and loading. The resulting voids were also in turn backfilled by flushing. Some rehandling of the wastes in the original cross-cuts was inevitable. An important feature of the mine was the existence of a thick, competent, sandstone roof overlying the coal.

Aware of this success it is tempting (but generally misleading) to expect its possible application to bituminous room and pillar mines. Sections of the mine would be partially extracted, then barricaded and filled using hydraulic flushing. Shallow mines could be filled from the surface; in other cases the fill could be introduced at seam high points and water removed at low points. Once backfilled and solidified, the pillars within the section would be mined. The mine layout could be arranged to minimize rehandling of backfill, for example by a system of rectangular pillars with minimum width cross-cuts.

From a materials handling point of view, this backfilling scheme has the advantage of not interfering with routine daily mining operations and of using separate crews for backfilling. However, the system has several drawbacks. When first introduced, the waste material simply fills (and not completely) the rooms and carries none of the weight of the overlying rock, this being supported by the pillars.

The ability of backfill to support an applied load develops progressively as it is compressed. Extraction of the pillars will throw increasing proportions of the overburden weight onto the fill. The overburden load due to the weight of overlying rock remains constant. Initially the entire load is carried by the pillars. As pillars are removed the roof sags and converges onto the fill. A sizable amount of the total convergence must occur before the fill carried a sizable portion of the load.

Shifting of the load gives rise to shear stresses and substantial deflections in the roof layers. Unless these are competent and able to sustain the stresses, poor roof and hazardous working conditions are likely to develop. The amount of roof deflection could be as much as 20% or more of the pillar height for hydraulically placed fill and appreciably more for pneumatically placed fill.

Flushing of the secondary mining voids, where required, would require separate boreholes, significantly raising the overall costs, probably to a prohibitive level except perhaps in a very shallow mine under level terrain. If roof conditions are such that a significant fraction of the pillars must be left unmined, it would become even less economical.

Rehandling of a percentage of the material that had been initially flushed into the mine would be inevitable particularly with multiple entry mining systems and obligatory cross-cuts between them. As much as 40 to 50% of the fill material would be rehandled.

Backfilling in Room and Pillar Operations with Complete Recovery as Part of the Mining Cycle: Room and pillar mining generally involves two stages: mining of rooms and extraction of pillars. In the second stage it is usual to allow the roof to cave behind the workings. A stepped pattern of extraction is developed to promote caving at some distance behind the pillar extraction line, thus providing some protection to the miner.

As in longwall retreat mining, the voids created during pillar extraction are behind the mining region, and it seems logical to suggest that backfilling into these voids could be accomplished without adverse effects. Indeed, if the roof is competent enough to allow a sizable part of the mined out area to remain open for a long period, it would be possible with existing technology to stow pneumatically from entries or rooms, other than those being mined, projecting the fill into the voids from a supported area. Projection of the waste to distances of the order of 50 feet into a void is considered possible in seams of moderate height (5 feet or more). However, a pneumatic system in a room and pillar operation would be more complicated than in the comparable longwall retreat system.

In most operating U.S. mines, the roof is relatively weak and caves very quickly after each segment of a pillar is extracted. Where this occurs it would be virtually impossible to use current, pneumatic backfilling systems. Modified undeveloped systems in which the stower is operated close to the continuous mining machine would increase congestion at the face, reduce maneuverability, and increase ventilation and safety hazards.

The system has several overwhelming disadvantages. Not only would it only be applicable under a typical U.S. mining condition, be awkward to maneuver and potentially increase safety hazards, it is not based on technology that is currently available or able to be modified. Even if practical, it would be a very expensive operation due to high equipment costs and extensive labor costs incurred by frequent moves.

Foam Encapsulation of Fill Material: The use of foams in underground mines was first investigated in 1954 by the U.S. Bureau of Mines in an attempt to apply various expandable plastics to sealing broken ground and limit access of air to fires in abandoned workings. From this first investigation many uses have been found for foams underground and considerable information has been developed that indicates more uses are possible.

The term foam covers a wide variety of materials from detergent or soap foams, protein, rubber, and an extensive variety of resin base foams to the foams of cellular inorganics, cement, gypsum, anhydrite, and silicate. Of these, detergent and polyurethane foams have been used in underground mines. Cellular concrete is being considered as a material for building walls in mines (packs or stoppings) but no underground demonstration work has been conducted. It is normally more fluid than that used in concrete construction and can be pumped over horizontal distances of 1,000 feet. Anhydrite (with setting accelerators) is used as a mixture with mine waste for construction of high strength mine packs in underground coal mines in Germany.

The utilization of foams in returning coal mine wastes underground offers several advantages. In pneumatic stowing, foams could be used to suppress dust and to encapsulate the stowed material thereby preventing spontaneous combustion and acid mine drainage. Foam or cellular concrete with coal waste as the aggregate would be stronger and less compressible than untreated waste and might provide enough overburden support to enable safe secondary recovery of coal pillars in relatively shallow mines.

A disadvantage of foam encapsulation of waste is that it represents a substantial added cost to the disposal process. In the case of cellular concrete it probably would not be economical to consider making a mix strong enough to support the overburden. Even using foam concretes there still would not be sufficient waste material to backfill an entire mine since waste is only 25% of the mined volume.

Hydraulic Mining Systems: It is reasonable to assume that there will be in the near future prototype systems for underground hydraulic mining and handling of coal. If this assumption becomes reality, there will be underground mines designed to handle volumes of water as part of the coal production cycle. Hydraulic transport does not imply loose water in the mining area nor would hydraulic cutting of coal necessarily result in large quantities of additional water at the mine face because the high pressure jets utilize comparatively little water. Hydraulic backfilling, on the other hand, requires relatively large quantities of water. Although such water could be recycled into the system, its drainage from the backfill may not be continuous enough to guarantee the flow necessary for a continuous hydraulic system.

Moreover, the problems of controlling water in nearly horizontal seams would be very difficult to overcome. Therefore, it cannot be assumed that hydraulic backfill can be added to hydraulic mining systems without significant redesign of the system. However, with the development of systems that handle water underground hydraulic backfilling would be more feasible. It is impossible to postulate equipment and manpower needs for such a backfill system until the mining part of the hydraulic system is developed.

Underground Coal Washing: Inasmuch as coal mining and backfilling are primarily materials handling problems, any change that decreases the amount of rehandling should be considered. One suggested change is to locate preparation plants in the mine. Waste could then be replaced in the mine voids by backfilling techniques without long transportation hauls. It is unlikely that this could be accomplished safely because of hazards from methane. Also, it is impractical because of the impermanence of coal mine openings. A variation of this idea would be to design a continuous mining machine that contains the primary cleaning circuit within it. If it were possible to remove some percentage of the waste material during retreat mining, then it could be backfilled immediately. The size and weight of present cleaning units make this plan strictly conjectural.

Automated Mining Systems: Significant research has been conducted in automating mining machines in order to reduce the exposure of miners to hazardous face conditions. The most successful projects, for the short term, do not remove operators from the mine. They are removed, however, from the hazardous face area (usually accepted as being with 25 feet from the face of coal being cut or supported) and placed behind an instrument that remotely controls the mining machine. The operator's own senses are used to perceive conditions and their commands are relayed to the machine by wire or radio control.

More complicated schemes of automation have been proposed. These involve remote sensing of mining conditions and long-distance remote control machines. Although the operators would be removed from the coal face by these schemes, the increase in complexity of machinery would result in greater exposure of maintenance personnel to the more dangerous face situations. As problems of sensor reliability and durability are overcome, it is conceivable that automated mining could be coupled with nonexplosive atmospheres to reduce personnel accidents. In principle, backfilling equipment could be incorporated into an automated mining system, but might tend to make the already complicated machinery even more cumbersome.

In Situ Extraction by Gasification and Solvent Extraction: Techniques for in situ gasification of coal have been receiving increased attention during recent years. In situ gasification involves converting coal into a combustible gas by partially burning the coal while passing a stream of air and steam through the coal seam. The steam reacts with the coal and creates a gas that then is forced to a series of collection wells. In situ gasification would have the advantages and disadvantages of normal gasification.

It would produce relatively clean fuel and theoretically make low-grade coals and lignites economical deposits. In situ extraction would have the further advantage of doing away with the mining process itself and hence with the problem of disposal of mine wastes. This process may have severe disadvantages such as subsidence, pollution of groundwater, and aquifer interruption where the coal bed itself or overlying strata are major water aquifers. It is difficult to postulate whether these disadvantages would be more or less pronounced than in conventional mining.

Inasmuch as some of the proposed coal desulfurization methods involve the use of organic solvents to dissolve selectively the hydrocarbon fuel, there is reason to investigate the use of these solvents for in situ extraction of coal. The major problems of such a scheme are solvent cost and recovery, and, as with any solution-mining scheme (salt, or sulfur for example), control of solvent leakage to the groundwater and subsidence of the resulting cavity. A major advantage is the elimination of the handling of coal mine wastes and pollutants. Such a scheme would eliminate the problems of both surface and underground disposal.

ECONOMIC FEASIBILITY

The costs of handling mine waste in an environmentally acceptable manner must be considered part of the normal expense of production for active coal mines. Economically, it does not appear that the cost of underground stowage is necessarily prohibitive in all mines, although underground disposal is considerably more expensive than current methods of surface disposal. The economic feasibility of waste disposal underground must be evaluated for each mine operation in the light of the competitive position it holds in the industry and the competitive position of coal in the energy market.

The cost of underground disposal will vary widely from mine to mine and in some cases may be prohibitive. Advances in waste disposal technology may significantly reduce the costs of returning coal mine wastes underground in future mines, but underground disposal may be expected to be more expensive than current methods of surface disposal. If wastes are to be placed underground, economic and legal incentives will no doubt have to be provided or current surface disposal methods will have to be further regulated or made less economically attractive.

Data are available for the costs of backfilling for subsidence control in the United States and in Europe, but figures for backfilling strictly for waste disposal can only be estimated. Some, but not all of the costs for subsidence control are directly transferable to costs for disposal. Experience in Europe has shown that backfilling is expensive and that there are less expensive ways of limiting subsidence damage than by means of underground disposal of waste. However, the unfavorable economics of backfilling for subsidence control in Europe does not mean that backfilling for disposal will be economically prohibitive in the United States. It has been estimated that at minimum, under unusually favorable mining conditions, and average conditions of waste production that it would cost at least one to four dollars per ton of coal to stow wastes underground.

If the return of coal mine waste underground should be required of all active operations, the added costs to underground mines in competition with surface operations could force closure of some mines. However, increased energy prices suggest that industry-wide improvements in technology can be afforded at this time although some operators would be hurt more than others.

REFUSE RECLAMATION BY VEGETATION

The material in this chapter is excerpted from PB 226 725 (see Bibliography, p 390).

Coal mine waste and refuse accumulations resulting from mining and preparation plants are increasing each year. It is estimated that over 100 million tons of refuse is produced from bituminous coal mining operations and nearly a billion cubic yards of coal mine wastes have despoiled the landscape in the anthracite region of Pennsylvania.

Every reclamation project is an individual problem. Therefore, the projects described may not be identical to projects and problems that may be encountered elsewhere. Likewise, the methods and costs derived may not be the same for any other similar project. PB 226 725 (1973) is intended to serve as a guide to, and to stimulate activity in, reclamation work as it applies to mine refuse piles.

FACTORS AFFECTING COSTS OF DISPOSAL AND RECLAMATION

Capital Investment

The capital investment costs shown in Table 10.1 cover all costs to industry for refuse disposal and reclamation and include road preparation, loading and transportation equipment for refuse disposal, and equipment used in soil covering and planting.

Some of the equipment may have been used previously in the mining operation but, to simplify costs, the investments shown include the cost of new equipment necessary to do only reclamation work. Capital expenditures for waste disposal equipment, in most cases, are estimated on the basis of replacement costs for equipment in use by the companies. The size of equipment used is not always described by company information.

TABLE 10.1: CAPITAL COST OF INDUSTRY RECLAMATION PROJECTS

Company	Mine location	Mine name and type	Capital investment[1]
Bethlehem Mines Corp...	Barrackville, W. Va.	Idamay mine No. 44 (underground).	$268,000
Do...............	Ellsworth, Pa.......	Mine 60[2] (underground).	375,000
Beth-Elkhorn Corp......	Jenkins, Ky.........	Hendrix No. 22 (underground).	50,000
Do...............	.........do.........	No. 27 mine (underground).	50,000
Eastern Associated Coal Corp.	Rachel, W. Va.......	Joanne mine (underground).	[3]100,000
U.S. Steel Corp........	Uniontown, Pa.......	Ginger Hill (underground).	NA

NA Not available.
[1]Includes land costs and preparation, roads, loading and transporting equipment, and planting equipment.
[2]Abandoned dumps; refuse from Mines 51 and 58 being disposed at Mine 60 dump by aerial tram.
[3]Estimate.

Source: PB 226 725

Refuse Source and Disposal Sites

Source: Coal refuse consists of waste coal, slate, carbonaceous and pyritic shales, and clay associated with a coal seam. These minerals are separated from the coal during the cleaning and preparation process.

Preparation plant refuse generally consists of fractions ranging in size up to one-quarter of an inch. The coarse waste is sent directly to the refuse dump. The fine wastes are pumped as a slurry to a settling pond where the suspended solids settle.

In 1969, over 435 million tons of raw bituminous coal were processed in cleaning plants; this resulted in more than 100 million tons of coal refuse. An average of 23% waste resulted from the cleaning process.

Disposal Sites: Prior to the passage of reclamation regulations, disposal sites were selected primarily on the basis of size, commensurate with expected refuse. Currently, consideration is given to size and those topographic features that will permit the construction of a refuse disposal site that is amenable to reclamation and near the source of the waste.

In mountainous regions, valley filling with coal refuse is common practice. This method of disposal has several environmental advantages. The refuse can be layered and compacted across the valley floor where two sides are naturally sealed, and the upper and lower ends of the edges can be terraced and sealed.

Reclamation of refuse dumps on flat or level land requires more work and, therefore, tends to be more costly. More machine-time and man-hours are needed to terrace a pile on all sides, more soil is needed to cover the terraced pile, and more vegetation is needed to plant a larger surface area. A pile that is exposed on all sides tends to allow more drainage of water, more siltation to surrounding areas and streams, and more opportunity for spontaneous combustion and burning.

Active and abandoned strip mines can be used for the disposal of both slurried refuse and coarse refuse from preparation plants. Strip mine pits have a distinct advantage over other surface disposal. Refuse is covered with material from the adjacent spoil pile, which makes compacting and terracing unnecessary.

Most coal companies, with existing refuse accumulations, continue to pile refuse on top while reclaiming the edges. Generally, reclaimed piles were terraced, compacted, and covered with soil. Additional improvements may include settling ponds where acid waters may accumulate; this water will then be neutralized and released to a nearby stream.

Refuse Disposal and Preparation

Transportation: Transportation of coal refuse by all methods is shown in Table 10.2. Most refuse is disposed of by trucks. Truck size, although not always mentioned in the company reports, depends upon quantity of refuse to be disposed, distances from mine to waste dump, and number of shifts worked per day.

Transportation by aerial tram is used only at one mine included in the study. Conveyor belts can be used for transporting refuse from the preparation plant to a loading hopper, from there it can be trucked to a refuse dump. The conveyor system would provide constant removal of refuse from preparation plants. Slurry material from the preparation plants can be transported by pipeline to settling ponds.

Spreading and Compacting: Preparation of the refuse dump starts with spreading the waste material. Spreading consists of smoothing the top surface to the slope desired to conform with the surrounding countryside. On flat reclamation sites, spreading is accomplished by flattening the top and leaving the sides to slope at the angle of repose that will reduce landslides or erosion. If the angle of repose of the banks is too great, it may be necessary to form terraces. For the same reason, terraces are used to prevent water run-off. At some reclamation projects it was necessary to make the top slightly concave to hold rainwater to obtain better growth of plants.

At some dump sites, refuse disposal is a continuous process. The waste is spread as it is dumped. As the pile becomes higher, the sides are compacted and planted at specified intervals. Bulldozers are primarily used to spread and compact refuse. However all types of hauling equipment are used to some extent to compact by traveling over previously deposited material. Compacting the waste is necessary to hold the material in place and to prevent air and water pollution of the surrounding area. Compaction also provides a better base for the soil cover. Spreading and compacting are two of the least costly operations at the dump site. Costs of these methods range from 3 to 20 cents per ton, with the average cost being 11 cents per ton.

TABLE 10.2: REFUSE DISPOSAL

Company	Mine location	Mine name and type	Refuse produced, tons per day	Mode of transportation	Distance to disposal site (one way), miles	Refuse disposal cost/ton/mile
Bethlehem Mines Corp.	Barrackville, W. Va.	Idamay mine No. 44 (underground).	1,100	Truck..........	1.0	$0.28
Do..............	Ellsworth, Pa.	Mine 60[1] (underground).	2,800	Truck and aerial tram.	1.0 .5	.17 .10
Beth-Elkhorn Corp....	Jenkins, Ky...	Hendrix No. 22 (underground).	765	Truck..........	1.0	.13
Do..............	......do......	No. 27 mine (underground).	1,170	Conveyor belt and truck.	1.0	.13
Eastern Associated Coal Corp.	Rachel, W. Va.	Joanne mine (underground).	600	Truck..........	1.0	.16
U.S. Steel Corp......	Uniontown, Pa.	Ginger Hill (underground).	4,500	Underground rail mine cars and carryall.	4.0	[2].09

[1]Abandoned dumps; refuse from Mines 51 and 58 being disposed at Mine 60 dump by aerial tram.
[2]Direct cost only.

Source: PB 226 725

Preparation for Planting

Soil Cover: Soil applied to coal refuse dumps is the most costly part of the reclamation operations. Most soil cover can be obtained from nearby farm land, previously stripped and restored land, or from road scrapings and ditches. Some of the cover might be considered topsoil, but most is clay or subsoil.

Trucks or carryalls haul the soil material from the source to the waste dumps. Cost of hauling depends upon the availability of soil, distance carried, and size of equipment used. Some of the reported handling costs are not reported but are aggregated in Table 10.3 under heading "Soil cover, cost per acre." However, costs shown cannot be correlated with any degree of accuracy to reflect any conclusion as to average cost.

Trucks can be used for transporting soil when reclamation of dumps is done periodically. Trucks may normally be used by mining companies in several facets of mining operation; therefore, purchase of additional equipment for reclamation work would be unnecessary. When reclamation of a refuse area is a continuous operation, equipment such as carryalls and bulldozers may be the least costly method of doing the job.

Spreading and Compaction of Soil: Trucks hauling soil are capable of spreading the material to some extent and at the same time compact the soil by repeated travel over previously dumped and spread material. However, general practice is to use bulldozers for spreading after the material is dumped. The dumped material is usually spread in layers 8 to 12 inches thick and compacted by repeatedly passing over the spread material. Carryalls and scraper pans are also used for spreading and compacting soil.

Planting: The compacted surface is cultivated and soil conditioners added to obtain good germination of seeds. After seeding, the surface is recompacted lightly to prevent the seed from blowing or washing away.

Certain varieties of grasses and legumes used in vegetating the soil cover have a good survival rate. It was generally concluded that, with proper soil conditioning, grass seed in combination with legumes gave the best survival results. In a previous investigation by the Bureau of Mines, it was determined that the quantity of grass seed planted should be three to four times greater in weight than legumes. Legumes, which are slow to become established, will take over as ground cover within 2 to 3 years.

Table 10.3 indicates that Kentucky fescue grass and a legume, usually *Serecia lespedeza,* was used more often than other combinations and that satisfactory results were obtained. A few companies reported obtaining good results using perennial and annual rye, orchard grass, and love grass. Other legumes that are and could be used include crown vetch and bird's-foot trefoil.

Soil conditioners generally included quantities of lime, depending upon the acidity of the soil, and paper mulch. Various fertilizer formulas were used, depending upon fertility of the cover material. Mulches were used either after planting or in connection with planting to protect the seed and soil from drying.

TABLE 10.3: SOIL COVERING AND PLANTING

Company	Mine location	Mine name and type	Soil cover, tons per acre	Soil cover, cost per acre[1]	Vegetation planted	Planting cost per acre[2]
Bethlehem Mines Corp.	Barrackville, W. Va.	Idamay mine No. 44 (underground).	3,300	$833	Grasses, black locust, white pine.	$165
Do...........	Ellsworth, Pa.	Mine 6C[3] (underground).	21,750	[4]10,900	White pine, perennial rye, Kentucky fescue, Serecia lespedeza.	[5]442
Beth-Elkhorn Corp.	Jenkins, Ky...	Hendrix No. 22 (underground).	4,350	1,300	White pine, annual rye, Kentucky fescue, Serecia lespedeza.	50
Do...........	......do......	No. 27 mine (underground).	-	-	White pine, short leaf pine, loblolly pine.	[6]60
Eastern Associated Coal Corp.	Rachel, W. Va.	Joanne mine (underground).	13,235	1,323	Grasses, locust, white pine.	[7]148
U.S. Steel Corp...	Uniontown, Pa.	Ginger Hill (underground).	3,263	665	Crown vetch, black locust, pine trees, perennial rye grass, Kentucky fescue.	[8]455

[1] Includes soil, transportation, spreading, and compacting.
[2] Costs not footnoted are those for planting grasses and trees on soil cover not conditioned.
[3] Abandoned dumps; refuse from Mines 51 and 58 being disposed at Mine 60 dump by aerial tram.
[4] Privately contracted at 50 cents per ton.
[5] Includes cost of lime, paper mulch, fertilizer, grass seed applied by hydroseeding, and trees planted by hand. Weighted average cost for 1966, 1969, and 1970-71 projects.
[6] Soil cover not used. Trees and grasses planted directly on refuse not previously conditioned.
[7] Includes cost of lime, mulch, fertilizer, seed, and trees.
[8] Includes cost of soil conditioning, hydroseeding, and tree planting. Direct cost only.

Source: PB 226 725

Tests made by the Bureau of Mines indicate that Norway spruce, black locust, Japanese larch, pitch pine, white pine, and jack pine trees, in descending order, have the best survival rate. Norway spruce trees are relatively expensive, so most companies used black locust and some variety of pine, generally white pine. Some companies reported that planting trees and seeding was done on an experimental basis and that when failures occurred, additional trials were made, using new methods of soil conditioning and new types of vegetation.

Generally, it was found that a single planting and seeding directly on soil cover without benefit of conditioners was the least costly. Four projects listed in Table 10.3 used this method and the cost of planting was about $60 per acre. This method of planting resulted in a low survival rate.

Most companies planted trees and seeds on soil treated with conditioners. Plantings by hand are greater in cost than those planted by machine. In some cases, hand planting was a necessity—especially for trees planted on steep slopes. The range in costs reported by companies was from $148 to $455 per acre (Table 10.3). Some companies reported using a combination of hand and machine planting. The most popular planting machine was one in which a mixture of water, mulch, fertilizer, and seed was sprayed over the soil.

INDUSTRY RECLAMATION PROJECTS

Bethlehem Mines Corp.

Idamay Mine No. 44, Barrackville, W. Va.: The reclaimed refuse dump at the abandoned Idamay Mine No. 44 covers an estimated 75 acres. The dump and adjacent slurry ponds contain about 8 million tons of coal refuse, which amounts to 16% of the 50 million tons of coal produced during the life of the mine. During the last 10 years of the mine operation, about 1,100 tons of preparation plant refuse was produced daily. The refuse was hauled about one-half mile to the dump site in 10-ton trucks where it was leveled and compacted with a front-end loader. The total estimated cost to haul, spread, and compact was 31 cents per ton of refuse.

Each of the 30-foot-high terraces formed was covered on the top and edges with an average of about 1.5 feet of clay soil. About 250,000 tons of soil cover was dug and hauled from company property at a cost of 28 cents per ton.

The reclaimed refuse site has become an attractive development area. Foundations are being constructed on an adjacent reclaimed slurry pond for the erection of a plastic manufacturing and fabricating plant. This reclaimed area is also under development as a site for a housing project. Further commercial and housing development for the entire reclaimed 75 acres is expected in the future.

Mine 60, Ellsworth, Pa.: The Bethlehem Mines Corp. has three refuse dumps at this operation. One is located 1 mile from active Mine 60; the other two are one-half mile each from inactive Mines 51 and 58. The disposal site at Mine 60 is located in a valley covering an area of 123 acres. This dump is 128 feet high on the east side and 163 feet high on the west side. Slopes of these sides of the dump are maintained at a 2 to 1 ratio to insure stability.

The top surface of the refuse pile is graded to prevent water run-off and slope washing. A band of vegetation is cleared from the base of the dump for fire protection.

Disposal from Mine 51 starts from a bin at the cleaning plant by way of a continuous bucket system to a transfer area; from there the refuse moves by means of a balanced two-tower aerial tram system to a dump site. Mine 58 refuse is disposed of in a similar manner except that the refuse moves from the transfer area to a dump site by means of a single-tower aerial tram system. Refuse from Mine 60 and the preparation plant continues to be disposed of by truck to the active dumping area. As the slopes increase in height, they are vegetated.

Starting in 1966, 24 acres of refuse without soil cover were treated with 1,000 pounds of hydrated lime. Each acre of the area was hydroseeded with 75 pounds of perennial rye (grass), 25 pounds of Kentucky fescue (grass), 500 pounds of fertilizer, and 1,200 pounds of combined paper and water used as a mulch. Subsequently, 1,200 pine trees were planted on 6-foot centers. The growth and survival rate of the plantings was discouraging. Planting cost of the 24 acres was $385 per acre.

In 1967, a program was undertaken to determine the sulfur content and temperature of the coal refuse surface on which the plantings were made and to evaluate the findings relative to their detrimental effects on plant growth. Summer temperatures of the refuse surface were as high as 114°F. This high temperature was one of the causes of poor growth.

In 1968, experimental work was done by replanting certain areas with a variety of grasses and trees to determine those with the best growth and survival rate. None had a satisfactory growth or survival rate when planted directly on coal refuse.

In 1969, planting was done on 5 acres of the northwest face of the refuse bank, which was covered to a depth of 10 feet of soil obtained from the surrounding hillside and valley floor. Three scrapers of 30-yard capacity placed the soil in 1-foot layers. The soil was compacted by the scrapers with the aid of a bulldozer and a sheep's-foot roller. A clay soil wall was built 5 feet higher than the actual top edge of the dump in the form of a dike, which served as a guide for the refuse dump trucks and to prevent rainwater from running over the sides. Approximately 21,750 tons of soil were required to cover 1 acre of refuse.

The cost of the soil was estimated at 50 cents per ton, based on the availability of the soil, length of haul, and equipment used. Benches were provided at several heights to allow heavy equipment to maintain the slopes and for operation of the seeding equipment. The benches provided additional stability to prevent landslides.

Each acre was hydroseeded using 200 pounds of fertilizer, 20 pounds of Kentucky fescue, 20 pounds of perennial rye grass, 10 pounds of *Serecia lespedeza* (legume), and 1,000 pounds of paperwater mulch. Pine trees were also planted using a density of 1,000 trees per acre. The total cost of the planting was about $500 per acre. The results of planting were good.

In 1970 and 1971, plantings similar to those done in 1969 were completed on

3 additional acres of the northwest face and on certain of the acres where the previous planting survival rate was poor. Planting cost during this period amounted to about $800 per acre.

Beth-Elkhorn Corp.

Hendrix No. 22, Jenkins, Ky.: Hendrix No. 22 produces about 4,200 tons of raw coal per day. The reject rate from the coal preparation plant is 18% and results in about 765 tons of coal refuse per day. The refuse disposal area, occupying about 40 acres, is located in a valley 1 mile away from the mine. Refuse is hauled by trucks which usually operate three shifts per day. The trucks dump the refuse, which is then spread by a bulldozer and compacted by both vehicles.

The south face of the refuse bank was started at a 15% grade, which will be progressively increased as the face is raised. Transportation of the coal refuse to the dump is reported to cost about 13 cents per ton-mile. Spreading and compacting are done for about 16 cents per ton.

Located in back of the refuse area, a new pond was constructed to collect and hold surface water. This prevents the natural drainage water from coming in contact with either the slurry or the refuse, preventing the discharge of acid water. An overflow culvert, which discharges the water from the pond into Rockhouse Creek, was built along the east side of the valley to bypass the slurry pond and refuse bank.

As areas of the refuse bank are inactivated, they are covered with clay and seeded. Kentucky fescue, annual rye (grass), *Serecia lespedeza*, and native white pine trees have been planted. A high survival rate for the grasses was achieved; tree growth, however, was poor. The cost of the soil cover was about 30 cents per ton. Planting costs amounted to $50 per acre.

In 1968, an inactive, 85-foot-high refuse bank, which was located behind the preparation plant was leveled, resulting in a flat area of about 4 acres. Refuse had been dumped on the pile for approximately 15 years at an average rate of 700 tons per day. Two bulldozers were employed intermittently over a period of several months to level the pile. Soil cover, which is extremely limited in the area, was obtained from various sources, such as roadways and material from cleaning roadway drainage ditches.

It is estimated that it took about 1 year to obtain sufficient soil material to cover the area. The area was planted with *Serecia lespedeza* and other vegetation without the aid of lime or fertilizer. The survival rate was about 80%. Due to the intermittent nature of the work, exact cost-data could not be obtained; however, the estimated reclamation costs were about the same as those for Mine 22 described above.

No. 27 Mine, Jenkins, Ky.: This site contains an active preparation plant and refuse dump, both started in 1943. Approximately 4,600 tons per day of raw coal is cleaned, which produces about 1,200 tons of refuse. The reclamation area covers 26 acres of ground and contains between 15 and 20 million tons of waste. The height of the refuse dump is estimated to be about 400 feet. The procedure used is to transport the refuse on a 26-inch belt that is 880 feet long and on a 16% slope from the preparation plant to a loading hopper. Refuse is then transferred to a 30-ton truck that dumps the refuse in the active section of the disposal area where it is spread and compacted by a bulldozer and by truck travel. Transportation of the coal refuse is estimated to cost 13 cents per ton-mile. Spreading and compacting of refuse is estimated to cost an additional 13 cents per ton.

Because of the rugged topography of eastern Kentucky, flat land is very scarce and, therefore, valuable. Most refuse is placed in valleys. Soil in the valley is estimated to be less than 1 foot in thickness, which makes the availability of soil inadequate. Therefore, planting is usually done directly on the refuse. The environment of the refuse in many cases results in a very low survival rate of the trees or grasses. It is necessary in some cases to make many plantings to establish growth on the abandoned portions of the bank.

The front face of the Jenkins refuse bank was planted eight times using various species of pine trees including short leaf pine, white pine, and loblolly pine. The number of trees planted each time varied between 1,000 and 5,000. It is estimated that about 20 acres have been planted. The planting was usually done by a contractor and was estimated to cost between $50 and $60 per acre. The trees varied in price but were estimated to cost between $10 and $20 per 1,000, depending on species. It is estimated that tree growth was established on 40% of the face slope.

To eliminate siltation on the local stream, portions of the face slopes were terraced. Water run-off is now collected in a holding pond. The active part of the refuse bank is being held 20 feet back from the existing front face slope to provide a bench for access to equipment and to maintain the stability of the refuse bank.

Eastern Associated Coal Corp.

Joanne Mine, Rachel, W. Va.: A reclaimed refuse dump that is 50 feet in height and covers an area of 6.8 acres was constructed in compacted layers 20 feet thick, and each layer was covered with 12 inches of clay soil. The compacting operation was done by the loaded trucks hauling the refuse to be dumped and by bulldozers leveling the refuse after it was dumped.

The slope area of the refuse dump was covered with 6 feet of clay removed from an adjacent mountain side. It required some 90,000 tons of clay soil to cover the slope area. The dig-and-haul cost of the soil was 10 cents per ton. The slope area was planted with grasses, 4,000 locust trees, and 4,000 white pine trees and subsequently fertilized and limed. The total cost per acre for soil cover, treating, and planting was about $1,470.

United States Steel Corp.

Ginger Hill, Uniontown, Pa.: The United States Steel Corp.'s Ginger Hill mine, opened in the late 1950s, is located near Uniontown, Fayette County, Pa. This mine produces metallurgical coal from the Pittsburgh seam for its Clairton coke works.

Because the preparation plant is located within a populated area, the refuse is transported 3.2 miles by underground rail in specially designed mine cars to the shaft at Ginger Hill. There it is lifted 532 feet to the surface by an automatic skip hoist and discharged into a surge bin, from which it is loaded into carryalls and hauled and spread over the disposal area in layers tightly compacted by the carryall traffic.

Two carryalls are required, working two shifts per day, to dispose of approximately 4,500 tons of refuse produced daily. Prior to the disposal of any refuse, the ground surface area is cleared of all combustible material and vegetation. Topsoil is recovered for later use in planting.

A grader is used to maintain haulage roads at the site and aid in disposal and reclamation work. Slopes of the sides of the refuse area are maintained at inclines that will insure stability. The outer limits of the dump area are maintained at a higher elevation than the dump surface to direct all water run-off toward the center of the disposal area where a pond is located.

To provide erosion control and beautification of the refuse disposal area, a program was initiated several years ago to cover the slopes of the disposal area with clay or topsoil to a depth of about 18 inches and to plant these slopes with stabilizing vegetation. This was considered feasible since the refuse is mildly alkaline.

The soil is transported by the carryalls and is spread over the slopes by a bulldozer and grader. This is usually done each time the refuse area elevation increases about 10 feet. About 7 to 8 acres of new slope area is formed every year.

About 44 acres of slope area have been covered with soil. This is equivalent to about 3 tons of soil per 100 tons of refuse. The direct costs are about $665 per acre.

In addition to controlling run-off water on the refuse dump surface, run-off water from the areas immediately surrounding the disposal site are diverted by ditches to settling ponds on the valley floor. After the silt settles out, the clarified water is checked periodically for acidity, then flows into the local stream.

First attempts at planting in 1966 were not too successful—partially due to an extremely dry summer. Of the 5,000 white pine and 5,000 Austrian pine (8- to 10-inch high) that were hand planted at a rate of 385 trees per acre, only 20% survived. Natural grass and weed growth on the soil-covered areas was satisfactory.

Seeding was first attempted in 1970. As the upper slopes became formed, they were hydroseeded by a contractor. The seeding mixture recommended, which cost about $100 per acre, consisted of the following: 150 lb fertilizer, 100 lb silva fiber mulch, 50 lb perennial rye grass, 7.5 lb crown vetch, 2.5 lb chopped grass roots, 1.5 lb black locust seed and 750 gal water.

This mixture, with slight variations, has been used on about 30 acres of slope area. Results have been good to poor, based on weather conditions after planting. It is estimated that over 90% of the black locust seed germinated, but the crown vetch and rye grass did not survive as well.

In 1971, to permit hydroseeding at more selective times, 800-gallon bantam hydroseeding equipment was purchased for about $5,000. From the experience gained with this equipment, it is apparent that by seeding at more opportune times and by watering and applying additional fertilizer as needed, a much more successful program is possible. A new seeding mixture, which costs about $200 per acre, is used. It consists of the following: 500 lb fertilizer, 200 lb silva fiber mulch, 200 lb hydrated lime, 75 lb crown vetch, 50 lb Kentucky fescue grass and 1,600 gal water.

An additional step taken to improve the appearance of the reclamation site was to screen the area by planting 94 large trees costing about $20 each. Through the implementation of detailed planning to include esthetic, as well as structural requirements, the construction of this refuse disposal area blends with the surrounding area, making it unobtrusive in appearance. The only noticeable indication that it is a refuse disposal area is the incremental layering of refuse at the top before the exposed slope is covered with soil and seeded.

CONTROL OF DRAINAGE FROM COAL MINE MINERAL WASTES

The material in this chapter is excerpted from 14010 DDH (see Bibliography, p 390).

A substantial amount of coal mined in this country undergoes a beneficiation or a cleaning operation. This is done to remove some of the dirt and impurities present in the coal. These impurities form the rejects or unmarketable portion of the coal mining operations and are usually referred to as refuse or gob.

Disposal of the refuse varies with the type of mining operations conducted, i.e., surface or underground. When coal from a surface mine is cleaned, modern practice frequently consists of trucking the refuse back to the strip pits to be buried in the spoil bank under an adequate thickness of overburden material. The land is then graded and planted with a suitable cover of grass, shrubs, or trees.

When a coal cleaning operation is practiced in conjunction with an underground mine, the disposal of refuse becomes a more complex problem. Since strip pits are not normally available to an underground mine, disposal of the larger pieces of refuse, up to 8 inches in diameter, is to the nearest open field or valley. Fine reject material, usually 20 mesh and smaller, is transported in slurry form, by pipeline, to diked enclosures, slurry lagoons, or surface impoundments.

The coarse refuse portion of a coal cleaning operation consists largely of coal intermixed with pyrites, sandstones, clays, and shales of a carbonaceous character. When stored outdoors in piles or heaps and exposed to the elements, chemical reactions take place on the surface of the refuse pile. Rainfall, oxygen in the air, and the pyrite in the refuse provide an ideal environment for the formation of an acidic drainage containing dissolved iron and other compounds which enters the streams and rivers from runoff and seepage through the pile. Additional problems follow in that the clays, shales, and sandstones are continuously decomposed and erosion constantly washes away the silt, exposing new material for oxidation and acid formation. Acid drainage and siltation occur during mining operations, and can continue for decades after operations cease.

Slurry lagoons associated with coal mining operations present a different type of environmental problem. The lagoons contain the fine reject material from a cleaning plant and can analyze as much as 50% coal with the balance ash and some pyrite. Rainfall on these lagoons percolates into the beds, seeps through the dikes, or is returned to the atmosphere via evaporation, with little surface runoff. The dikes are usually well built and compacted from clean earth, but occasionally are built from refuse and covered with a layer of earth. In many instances, a grass cover or trees are planted on the slopes to prevent erosion, or vegetation can develop from volunteer growth. During active operations, a pool of water exists on the surface and only minor problems are experienced involving repairs to a leaking dike. When mining operations cease, maintenance often ceases and the dikes can wash out during heavy rainstorms. In addition, during extended periods of dry weather, blowing winds entrain the surface material and create a dust problem in the vicinity of the site.

Scores of these types of refuse piles and slurry lagoons, from underground and surface mining operations, exist in both the Appalachian and midwestern coal fields. Only a limited number of options are available to effectively handle this problem. Topography tends to make each situation unique. In a large number of instances, the refuse piles have been abandoned.

In some instances, covering the pile with a thick layer of clean earth and planting a vegetative cover has been effective but very expensive. In certain cases, earth cover may not be available or it may be so expensive as to make the covering operation very costly. Chemical treatment of the runoff and seepage, using hydrated lime or limestone, may be an interim measure during active operations but is obviously not the long-term solution since the formation of acid can continue for decades.

In the latter part of 1968, Truax-Traer Coal Company, a Division of Consolidation Coal Company, entered into a cooperative grant with the Federal Water Pollution Control Administration to demonstrate effective and practical means of abating air and water pollution from coal mining refuse piles and slurry lagoons. The intention of this demonstration project was to provide engineering data and design parameters that could be applied to minimize or prevent these types of pollution.

The specific site chosen for this project was an abandoned mining operation, formerly known as the New Kathleen Mine, located in southern Illinois and active from 1943 to 1955. This site consisted of a refuse pile occupying 40 acres and an adjacent slurry lagoon complex consisting of 50 acres.

The project is divided into two phases. Phase I consists of determining the system characteristics and acid formation rate of the refuse pile and testing potential abatement measures for both the refuse pile and slurry lagoons. Phase II will implement specific remedial procedures for the entire demonstration site to be followed by a monitoring program that will evaluate the degree of pollution abatement. This report presents the results of Phase I.

DESCRIPTION OF SITE

The site forms a part of an abandoned coal mining operation that included a coal cleaning plant operated by Union Collieries Company in conjunction with the New Kathleen Mine. This was a slope mine in the Herrin (No. 6) Seam at a depth of approximately 110 ft. Surface drainage from the site is generally to the west, into Walker Creek which flows southwest past the western end of the pile and which has its headwaters approximately ½ mile north of the pile. Flow in Walker Creek often ceases during the dry summer months.

The site contained an irregularly shaped refuse pile approximately 40 acres in area, standing 65 ft at its highest point and containing about 2,000,000 yd^3 of coarse refuse. In addition to the refuse pile, the site contained a complex of six slurry lagoons, standing approximately 15 ft high, essentially flat, and occupying some 50 acres in area.

The lagoons were completely enclosed by earthen dikes and contained the fine coal rejects transported thereto by hydraulic means. At the west end of the slurry lagoons, six small lakes remained from the abandoned mining operations that were used to collect the runoff from the slurry lagoons, and so arranged as to eventually overflow into Walker Creek. The refuse pile and slurry lagoons were separated by a strip of original ground, approximately 1,500 ft wide, that was used for a railroad siding during the active mining operations. The material in the refuse pile was a mixture of shale, clay, and low-grade coal, in which both sulfur ball and large-crystal pyrite forms were included.

The composite material was sufficiently cohesive to stand at an angle of repose nearly vertical, but it was very susceptible to erosion, as reflected in the deep gullies which were formed on the side slopes of the pile. The refuse material throughout the pile appeared to be heterogeneous in its physical characteristics, reflecting the irregular manner in which the pile was formed. Much of the pile was placed by end-dumping from trucks. Further examination revealed a relatively large number of individual seepage points at the base of the refuse pile.

Some were continuous flows while others appeared to be intermittent, indicating either a storage pool of water existed in the pile or parts of the pile were resting on groundwater springs, or both. Portions of the western ⅓ of the pile showed evidence of burning at some time in the past. Both the refuse pile and the slurry lagoons were essentially barren except for the eastern edge of the slurry lagoons, where volunteer growth was established on soil, apparently washed down from the top of the dikes. Vegetation in the form of trees, grass, and shrubs was established on the outer periphery of the slurry lagoon area.

An aerial survey of the site was made from which contour maps were prepared which showed a natural division of the refuse pile into three sections. The western ⅓ features a plateau with a surface elevation of about 470 ft, relative to an original ground elevation of about 430 ft. The central, and highest section of the pile has an elevation of somewhat more than 490 ft at its western edge, and slopes toward the east to an elevation of about 475 ft, where there is an abrupt drop to the eastern ⅓ of the pile.

To develop a better understanding of the pile structure below the surface, a backhoe was used to cut a number of trenches in several sections of the pile to a depth of 8 ft. Examination of the cross section of the pile indicated three distinct layers or zones. The first was the outer mantle of the pile, consisting of a layer 4 to 10 inches thick, from which much of the clay had been washed out by precipitation. From the standpoint of permeability to air and water, this outer mantle appeared to be relatively open. The second zone was comprised of a layer of clayey fines, an inch or more thick, packed tightly by rain action into the refuse immediately beneath the outer mantle that served as a partial barrier to air and water. This layer had discontinuities which provided points of entry for water into the main body of the pile. The third zone was the main body of the pile which showed only little evidence of weathering.

It became apparent at that time that any efforts to obtain representative samples of the solid material in the pile, using classical procedures, would be difficult from a practical standpoint because of the heterogeneous nature of the material and the large variations in particle size observed, ranging from fine silt to pieces 6 to 8 inches in diameter. However, recognizing these limitations, a set of random samples was taken from the top 4 inches of the refuse pile and analyzed for sulfur forms and total sulfur. The results are shown in Table 11.1.

TABLE 11.1: FORMS OF SULFUR ON REFUSE PILE

Sample No.	% Totals	% Sulfates	% Pyritics	% Organics
1	5.51	4.83	0.029	0.65
2	9.12	3.36	4.08	1.68
3	14.05	3.27	10.60	0.10
4	14.02	2.73	9.78	1.51
5	7.77	3.07	3.70	1.00
6	5.35	3.94	0.72	0.69
7	5.82	5.05	0.078	0.69

Soil Type Composition of
Composite of Above Seven Samples

Clay: 34.7%
Carbonaceous material: 37.8%
Shales, other coarse non-carbonaceous material: 27.5%

All analyses reported on a dry weight basis.

Source: 14010 DDH

ACID FORMATION RATE STUDIES

The visual examination of the refuse pile cross sections formed the basis for the hypothesis used in determining the acid formation rate of the pile. It was postulated that acid formation, i.e., the oxidation of pyrite, was confined to a relatively narrow zone at or near the surface with the products of the reaction accumulating in this zone. During periods of precipitation, a portion of these products was flushed out and appeared in the surface runoff while the remainder percolated into the interior, appearing later as seepage at the base of the pile eventually entering Walker Creek.

To determine the acid formation rate from the pile, consideration was given to dividing the refuse pile into six watersheds and monitoring the surface runoff from each watershed. Seepage into and out of the pile, plus the evaporation effect, would then have to be related to the rainfall and surface runoff before an average acid formation rate could be determined. This approach, with certain modifications, was used in the study.

Installation and Operation of Monitoring Stations

The situation at the site was favorable for the determination of a hydrologic water balance around the refuse pile in that the only water input to the area was in the form of precipitation. The pile rests near the headwaters of Walker Creek and is located at an elevation sufficient to prevent either ground or surface water from entering the refuse pile. A water balance around the pile would have consisted of rainfall on the pile, direct runoff, seepage from within the pile, and evaporation. Rainfall was measured by a single continuous bucket recorder type rain gauge. This instrument was located approximately 500 ft east of the pile.

Instead of installing six monitoring stations, sections of the refuse pile were shaped, with a minimum amount of grading, into three watersheds from which runoff and water quality data could be obtained. Flumes were sized and installed to monitor surface runoff from three areas, designated Areas 1, 2 and 6. The 3.0-ft flume in Area 1 had a flume section constructed of stainless steel, while the other two were constructed of Corten steel. The drop boxes of all three flumes were constructed of mild steel. Each flume was equipped with a water level recorder to measure surface runoff.

Severe corrosion was experienced with the Corten flumes and with the drop boxes for all three flumes, in spite of attempts to protect them with asphalt mastics. Although the corrosion problems were brought under control by August, 1969 by the installation of stainless steel liners, some water loss probably occurred at the Area 1 and Area 2 flumes throughout the course of the study.

Siltation in the drop boxes of the flumes and stage recorder float chambers created a continuous maintenance problem. A single storm of sufficient duration could completely fill these units with silt. Manual removal of the silt with shovels after the storm was the most practical way of handling this problem. In December, 1969, a three-point continuous temperature recorder was installed near the flume in Area 6. A continuous record of ambient temperature, temperature 2 inches below the surface, and 20 inches below the surface was obtained.

During all of 1969, water quality samples were taken by hand at approximately 2-minute intervals throughout the duration of a rain storm. Samples were analyzed at the on-site laboratory for pH, acidity, total alkalinity, ferrous and total iron, and sulfate. The acidity values, together with the runoff data obtained at the flumes, were used to calculate the mass flow of acid flowing from each of the monitored areas. While this procedure was very accurate, it required considerable amounts of manpower, and storms occurring during unattended periods could not be effectively monitored.

Experiments conducted with a laboratory conductivity meter showed a correlation between total acidity and conductivity for any given flow point. However, separate correlations for each flow point were necessary. Based on these correlations, portable battery-operated recording conductivity meters were installed at the three monitoring flumes in January, 1970. Meter readings were correlated regularly with acidity analyses to provide a basis for the estimation of acidity concentrations from the conductivity record during a period of storm runoff. All storms monitored in 1970 used conductivity measurements as a basis for estimating the acidity values in the runoff.

The battery-operated conductivity meters did not function satisfactorily. Leak-proof batteries leaked and the response of the instrument was sluggish in cold weather. In January, 1970, the conductivity meter installed in Area 6 was converted to operate on 110 v AC and this proved to be very satisfactory. The other two meters continued to operate on batteries for the duration of the data collection period.

The final item in the hydrologic water balance was evaporation. In this system, evaporation occurs from the pile surface at a decreasing rate for several days after rainfall and continuously from the pile surface adjacent to the springs. Although the evaporation phenomenon was recognized as a variable, it was not related to the overall hydrologic balance. An assumption was made, based on the physical characteristics observed on the pile, that the concentration of acid products in the direct runoff was similar to the concentration in the water infiltrating into the pile. The difference in concentrations observed in the base flows was attributed to evaporation losses rather than to any further significant oxidation within the pile.

Ground Water Survey: This survey attempted to identify the nature of the water seepage at the base of the refuse pile (hereafter referred to as base flows). The bucket-stopwatch technique was used to measure these flows. Fourteen observation wells were drilled to the original ground level below the pile and each cased with 1-inch diameter PVC plastic pipe. Water level measurements of the pile storage pool were determined regularly by lowering an electrical probe down the tube.

Simulated Rainfall Test Plots: To allow the collection of supplementary acid formation data under closely controlled conditions and to guard against the possibility of a sustained drought during the data collection period, a small (0.109 acre) simulated rainfall test plot was installed on a representative section of the refuse pile.

Uncontaminated water from a nearby groundwater source was applied via irrigation sprinklers at desired metered rates. The direct runoff was collected in the ditching system and measured by a flume at the outlet. Throughout the simulation runs, water quality samples were manually collected. Acid formation rates from this test plot, together with those obtained during periods of natural precipitation, were used in estimating the average acid formation rate from the refuse pile.

Hypothesis of Acid Formation

The following fundamental hypothesis was used to allow the calculation of an average acid formation rate for the refuse pile:

(1) The oxidation of pyrite is primarily confined to a relatively narrow zone at or near the surface of the pile with the products of the reaction accumulating in this zone to be flushed out during periods of

precipitation and appearing in the runoff, and

(2) The acid load from the refuse pile is directly proportional to the acid load from the surface runoff and inversely proportional to the ratio of total storm runoff to the total rainfall, seepage at the base of the pile being disregarded.

This hypothesis can then be expressed mathematically using the following relationship:

$$P = \frac{\Sigma R}{A \times \Sigma t \times f}$$

where

P = Average acid formation rate, lb/acre/day
ΣR = Total weight of acidity in runoff from all storms on record for a given drainage area, in lb acidity as $CaCO_3$
A = Surface drainage area in acres
Σt = Total period of acid formation corresponding to all storms on record, in days
f = Ratio of total storm runoff volume to total rainfall volume for all storms on record

Results and Discussion

The acid formation rates, in pounds of acidity, as $CaCO_3$, per acre per day (lb/A/day) from the simulated rainfall test plot (SRTP) and from Areas 1, 2, and 6 are summarized in Table 11.2

TABLE 11.2: SUMMARY OF ACID FORMATION RATES

Area	Acid Formation Rate lbs/A/day	Number of Tests or Storms Monitored
SRTP	120	13
Area 1	79	5
Area 2	193	22
Area 6	305	29

Source: 14010 DDH

Because of topographic similarities of Areas 1 and 2, the acid formation rates were averaged for these two areas, weighted in accordance with the number of storms monitored, to produce a single value representative of the steep slope area typical of the periphery of the refuse pile, thus

Area 1	79 x	5 storms =	395
Area 2	193 x	22 storms =	4,246
		27	4,641

Average for Areas 1 and 2 = 172 lb/A/day

Using the contour map of the refuse pile and visual inspection, the refuse pile was divided into areas considered representative of the test data. This procedure resulted in estimating approximately 13 acres of compacted refuse area typical of the area on which the simulated rainfall tests were conducted.

The steep slope areas were estimated to consist of approximately 14 acres represented by the results from Areas 1 and 2. The uncompacted dumped refuse area was approximately 13 acres represented by the results from Area 6. Applying the acid formation rates to these respective areas, an average acid formation rate for the pile was calculated:

Compacted refuse (SRTP)	13 acres	x 120 =	1,560
Steep slope area (Areas 1 and 2)	14 acres	x 172 =	2,408
Uncompacted refuse (Area 6)	13 acres	x 305 =	3,965
	40		7,933

Average = 198 lb/A/day

Thus the average rate of acid formation for this particular refuse pile was 198 lb of acidity, as $CaCO_3$, per acre per day. Two points should be mentioned regarding the results obtained by this method. The first relates to the temperature effects on acid formation. The data are not considered sufficiently accurate to reflect temperature effects and thus the acid formation rates reported herein can only be considered as average values. Although a continuous record of temperatures was obtained near the Area 6 flume at ambient level, 3 inches below the surface, and 18 inches below the surface, the temperature effects were not related to the average acid formation rate. It is believed that due to the dark surface of the pile, sunlight warms the surface mantle rapidly and acid production probably continues during all but the coldest months of the winter at this location.

The second relates to the fact that seepage flows are tributary to the ponds above the flumes of Areas 1 and 2. Due to the rapid flush of water during a storm, it would appear that seepage water accumulated in the ponds from a previous storm has little effect on the acid weight determined from the runoff data. Due to the lack of sufficient data, however, this point cannot be checked with any precision.

The data obtained from the groundwater survey were inconclusive. Although considerable effort was expended in monitoring the water levels in the observation wells and obtaining base flow and quality data, these data were not amenable to any analysis and no rational correlations were evident.

Supplementary Experiments

The following supplementary experiments were conducted during the course of this study and are briefly summarized below.

Role of Bacteria in Acid Formation: A study was conducted to determine what role, if any, bacterial catalysis may play in the oxidation of pyrite. The initial conclusions indicated sound evidence for the occurrence of significant catalysis of pyrite oxidation only in environments such as are found at the surface of refuse piles and spoil banks. Bacterial catalysis is less likely in underground environments.

Application of Bactericides: During the bacterial catalysis studies, a number of bactericides were applied to the simulated rainfall test plot on the refuse pile to determine if the oxidation of pyrite could be reduced or arrested. Hexionic acid, sodium lauryl sulfate, and linear alkyl benzene sulfonate were applied separately. No significant reduction in the acid formation rate was noted.

Laboratory Acid Formation Rate Studies: In an effort to substantiate the acid formation rate estimated in the field studies, refuse samples were tested in the laboratory for oxygen uptake at 25°C in a Warburg apparatus. An average of nine tests on the random samples taken from the pile mantle indicated an acid formation rate of 107 lb/A/day.

ABATEMENT MEASURES

One of the objectives of this project was to demonstrate water and air pollution abatement measures which would be essentially permanent, require little need for continued maintenance, and would present a pleasing aesthetic appearance. The basic approaches discussed were ways to minimize the movement of air and/or water into the pile by sealing the pile with a suitable cover (organic, mechanical, or chemical), thus reducing or eliminating completely the formation of acid, siltation, erosion, and dust entrainment. A cover should function to varying degrees in one or more of the following ways.

The cover may prevent erosion and thus prevent the continuing exposure of fresh pyrite surfaces. Since oxygen must be continuously supplied to support the pyrite oxidation reaction and since any layer of material separating pyrite from the atmosphere will function as a resistance to diffusion, then any physical stabilization of the pile surface will cause the zone of oxidation to move deeper into the pile and the overlying diffusion barrier will eventually control the rate of pyrite oxidation. The reaction will decrease with time due to this effect, although the decrease may be very slow.

The cover may be sufficiently impermeable to oxygen transport to act as an efficient diffusion barrier. For example, a plastic sheet placed over the refuse may effectively stop all oxygen transport to the pyrite and oxidation will cease.

The cover may be sufficiently impermeable to water movement to decrease or stop water movement into the refuse. If this occurs, then oxidation products will not be flushed away from the oxidation sites and the only movement of acid salts into the interior of the pile will be through seepage generated by the hygroscopic nature of the acid salts themselves. Depending on oxygen availability, pyrite oxidation may continue, but the products will be largely retained at or near the site of oxidation.

The cover may function as an oxygen-consuming layer. A vegetative cover such as grass might build up a sufficiently high concentration of organic matter in the soil to support high rates of aerobic bacterial activity. Such a layer might be effective in removing oxygen from the soil atmosphere before it reaches the zone of pyrite oxidation.

Experimental

Attention was directed largely toward vegetative covers that could be applied using standard agricultural techniques without resorting to new and untried equipment and/or machinery. Such covers would be self-healing and would prevent further exposure of pyrite by erosion. A selection of grasses and legumes were considered to be the best approach in this experimental program. Accordingly, fourteen 0.1 acre test plots were established on the refuse pile outside the limits of drainage Areas 1, 2, and 6. In addition, three test plots were established on the slurry lagoons.

Plots 3, 4, 5, 6, 7, 8, 9, 10, 12, 13, and 14 were equipped with a tile underdrain laid one to two feet below the surface of the gob and the boundaries of the plots were ditched to prevent the flow of surface acid runoff from adjacent areas across the plots. Monitoring of the plots consisted of visual observation of the growth of grasses, chemical analysis of soil samples taken from the refuse layer at the interface of the refuse with the various applied covers, and chemical analysis of subsurface drainage caught in the underdrain.

The grass species selected were recommended by the USDA Soil Conservation Service based on their experience at this locale. In establishing grass covers on test plots using a topsoil cover, samples of soil were submitted to a soil testing laboratory to determine limestone and fertilizer requirements. In the case of test plots planted directly on refuse or on slurry lagoon material without the use of topsoil, a modified soil test was developed and used as a guide after it became apparent that the standard soil test was not applicable to these types of material. The treatment procedures used in establishing the experimental covers were as follows.

Plot 1: Established June, 1969. Agricultural limestone, 48 x 100 mesh, applied at 40 T/A (tons/acre) rototilled into the refuse to a depth of 8". Commercial fertilizer 6-24-24, applied at 1,500 lb/A. Seeded with a mixture of Kentucky fescue (37½% by weight) and perennial rye (62½%) at 80 lb/A. Straw mulch spread on surface at 1½ T/A.

Plot 2: Established July, 1969. Agricultural limestone, 48 x 100 mesh, applied at 40 T/A, rototilled into the refuse simultaneously with 1 T/A straw to a depth of 8". Commercial fertilizer, 6-24-24, applied at 1,500 lb/A. Seeded with a mixture of Kentucky fescue (37½%) and perennial rye (62½%) at 80 lb/A. Straw mulch spread on surface at 1½ T/A.

Plot 3: Established September, 1969. Agricultural limestone 48 x 100 mesh, applied at 40 T/A disked into refuse to a depth of 8". Commercial fertilizer, 6-24-24, applied at 1,500 lb/A. Seeded with a mixture of Kentucky fescue (37½%) and perennial rye (62½%) at 80 lb/A. Straw mulch spread on surface at 1½ T/A.

Plot 4: Duplicate of Plot 3 with identical results.

Plot 5: Established September, 1969. Agricultural limestone 48 x 100 mesh, applied at 40 T/A, disked into refuse to a depth of 8". Commercial fertilizer 6-24-24, applied at 1,500 lb/A. Seeded with a mixture of Kentucky fescue (37½%) and orchard grass (62½%) at 80 lb/A. Straw mulch spread on surface at 1½ T/A.

Plot 6: Established October, 1969. Black polyethylene membrane, 39 mils thick, was placed on the grated refuse. A 4" thickness of field soil was placed on the membrane. Set out for percolation tests initially. In March, 1970, agricultural limestone, 48 x 100 mesh, applied at 2 T/A, was raked into the soil. Commercial fertilizer, 6-24-24, applied at 1,000 lb/A. Seeded with Kentucky fescue at 50 lb/A. Straw mulch spread on surface at 1½ T/A.

Plot 7: Control plot; no treatment.

Plot 8: Established October, 1969. A 4" thickness of field soil was placed on the refuse. Agricultural limestone, 48 x 100 mesh, at 2 T/A, was hand-raked into the soil. Commercial fertilizer, 6-24-24, applied at 500 lb/A. Seeded with a mixture of Kentucky fescue (57%) and orchard grass (43%) at 70 lb/A. Straw mulch spread on surface at 1½ T/A.

Plot 9: Established October, 1969. A 12" thickness of field soil was placed on the graded refuse. Agricultural limestone, 48 x 100 mesh, at 2 T/A, hand-raked into the soil. Commercial fertilizer, 6-24-24, applied at 500 lb/A. One-half of test plot seeded with Kentucky fescue at 40 lb/A and ½ of plot seeded with orchard grass at 40 lb/A. Straw mulch spread on surface at 1½ T/A.

Plot 10: Established October, 1969. A 24" thickness of field soil was placed on the graded refuse. Agricultural limestone, 48 x 100 mesh, at 2 T/A, hand-raked into the soil. Commercial fertilizer, 6-24-24, applied at 500 lb/A. One-half of test plot seeded with mixture of Kentucky fescue (80%) and lespedeza (20%) at 50 lb/A. The other half of test plot seeded with mixture of orchard grass (80%) and lespedeza (20%) at 50 lb/A. Straw mulch spread on surface at 1½ T/A.

Plot 11: Established in October, 1969.on a 4:1 slope with southern exposure. Agricultural limestone, 48 x 100 mesh, applied at 40 T/A, disked into refuse. Commercial fertilizer, 6-24-24, applied at 1,500 lb/A. Seeded with a mixture of Kentucky fescue (37½%) and perennial rye (62½%) at 80 lb/A. Straw mulch spread on surface at 1½ T/A.

Plot 12: Established October, 1969. Dried sewage sludge at 4" thickness was applied to the surface of the refuse. Seeded with a mixture of Kentucky fescue (50%) and perennial rye (50%) at 20 lb/A.

Plot 13: Established October, 1969. Lime sludge from an oil refinery water treatment plant at 95 T/A was spread on the surface. The material was slightly oily and was not mixed into the refuse. Test plot set out for percolation tests only.

Plot 14: Established November, 1969. A finely ground mixture of lime and limestone, labeled Code L from Mississippi Lime Company, applied at 90 T/A, was disked into refuse to a depth of 6". Test plot was set out for percolation tests only.

Three additional test plots were established during this investigation on the slurry lagoons and are summarized as follows:

Plot 15: Established June, 1969. A series of 24 strips, 200 ft long, 3 ft wide, and on 9 ft centers were staked out on the slurry lagoons. Field soil at 3" thickness was applied to the strips and rototilled to a depth of 3". Agricultural limestone, 48 x 100 mesh, at 2 T/A, was hand-raked into the surface. Commercial fertilizer, 6-24-24, at 500 lb/A was applied. The strips were seeded with a mixture of Kentucky fescue (31%), lespedeza (15%), red clover (8%), and perennial rye (46%) at 65 lb/A. Straw mulch at 1½ T/A was spread on the surface.

Plot 16: Established May, 1970. Agricultural limestone, 48 x 100 mesh, was rototilled into a 0.1 acre test plot of slurry lagoon material to a depth of 8", at 15 T/A. Urea, 45-0-0, at 200 lb/A, triple-super phosphate, 0-46-0, at 300 lb/A, and potash, 0-0-60, at 300 lb/A were applied. The test plot was seeded with a mixture of Kentucky fescue (22%), sudan grass (34%), and oats (44%) at 90 lb/A. Straw mulch at 1½ T/A was spread on the surface. This test plot was completed on an extremely windy day and the mulch was held down with binder twine anchored to stakes to prevent the seed from blowing away.

Plot 17: Established May, 1970. A 10 ft by 10 ft test plot was established to determine the stabilizing effect of a petroleum based product, Coherex. The use of Coherex as a stabilizing agent for mineral wastes has been reported by the Bureau of Mines. A one-gallon sample of Coherex was mixed with 6 gallons of water and the mixture was applied at a rate of one gallon per square yard to the slurry lagoon test plot using a garden sprinkler can.

Results and Discussion

In summary, grass was established on all test plots planted to grass on the refuse pile, with and without the use of topsoil. The grass plots weathered well through one winter season and continued growth was visible during the second season. The control plot, No. 7, remained barren during this entire period. Grass covers established on Plots 1 and 2 produced excellent stands of grass. As visually observed, the vegetative covers on Plots 1 and 2 were significantly more uniform, with denser grass than the grass covers on Plots 3, 4, and 5.

This improvement is attributed to the use of rototilling equipment when applying the limestone on the former plots as contrasted to discs used on the latter. A more intimate mixing of limestone with refuse material is believed to have taken place. Average pH values of the soil on Plots 1 and 2 were pH 4.7 and pH 4.9 respectively for the first six months of 1970. No data were collected for Plots 3, 4, and 5 because of the spotty condition of the vegetation.

Plot 6, using a polyethylene membrane and a 4-inch thickness of topsoil, presented several problems. Spreading of plastic was a very difficult task even during a calm day. Soil was piled on one edge and then spread ahead by a small bulldozer. Great care was taken to protect the plastic but it was difficult to be certain that it was not punctured in the process of spreading the soil. This plot was initially set out without grass cover for percolation tests only. Grass was planted on this plot in Spring 1970 and a good cover of grass was established.

Plot 7, control plot, remained barren during the entire test period. One sample of the refuse material taken in June, 1970 showed a pH 2.3. Plot 8, established with a 4-inch thickness of topsoil, had good grass growth where the soil thickness was maintained. However, the plot was relatively small to be worked with even the smallest farm equipment and refuse was exposed, resulting in a spotty grass cover. This problem was not experienced to the same degree in Plots 9 and 10 that were covered with 12 inches and 24 inches of topsoil. Plot 9 wintered well into the 1970 growing season. The fescue did better than the orchard grass. Plot 10 showed good grass growth on both sides in the Spring 1970. The legumes (lespedeza) appeared in June 1970.

Plot 11 produced a spotty grass cover, apparently due to poor mixing of limestone with the refuse. A disc attachment was used on a small farm tractor. Results were similar to Plots 3, 4, and 5. Plot 12, established with dried sewage sludge in October 1969, produced no vegetation until Spring 1970, when a heavy growth of grass appeared including tomato plants and a number of weed species. Plots 13 and 14 were set out for percolation tests only.

The rows of grass established on the slurry lagoons grew well in 1969 and even better in 1970. Visual observation of the areas between the rows in the Summer 1970 showed grass seed present but in an ungerminated form. Excellent new growth on the strips was evident in the Fall 1970. Grass growth was visible within one week after planting on Plot 16. By mid-Summer 1970, the vegetation was 5 feet tall and continued growth was expected.

Plot 17, treated with Coherex, developed a hard crust on the surface which penentrated approximately 4 inches deep with decreasing hardness. The condition of the test plot was essentially the same six months after application. This type of treatment would probably reduce any air pollution from the slurry lagoons to an insignificant level and would also reduce infiltration rate by increasing runoff.

Refuse samples were taken from the test plots on the refuse pile only three times per month during the first six months of 1970. The samples were taken immediately below the interface between the refuse and the prepared cover, or from the first 4 inches of refuse in plots having no cover. One hundred grams of refuse were mixed with 500 ml distilled water and the resulting supernatant taken for chemical analysis. Due to the effect of the agricultural limestone used in the preparation of some of the plots, pH, acidity, and soluble iron analyses do not reflect the amount of oxidation products in the solid sample. Sulfate, however, was present in proportion to the amounts of pyrite oxidation products and should reflect any change in the oxidation rate, given a sufficiently long period of time.

For all of the plots, there was a considerable amount of scatter in the data and no apparent decrease in sulfate concentration in the refuse with time. However, due to the slow rate of flushing of the refuse by infiltrating water, detecting a definite change in oxidation product concentration in the refuse might require as many as five years, and no conclusion as to pyrite oxidation rate change can be drawn from these data. The variation in average sulfate concentrations found in the samples taken from the plots during the six-month sampling period is believed to be due to differences in the plots before covering and to have little relationship to the effect of the cover.

Samples of subsurface drainage from the test plots were taken once per week during the period February to May, 1970. No flows were detected in the underdrains before or after these dates and several plots showed no flow at any time. Samples were analyzed for pH, acidity, iron, and sulfate. As in the case of the solid refuse samples, there was no trend evident for meaningful sulfate content change with time for any of the test plots. Here, again, it is believed that several seasons might be required to flush the refuse significantly of oxidation products, even if oxidation of pyrite were completely stopped. The chemical analyses were inconclusive in regard to an evaluation of the ability of any of the covers to retard or stop pyrite oxidation.

The most significant result of the experimental abatement measures applied to the test plots was the discovery that a grass cover could be established directly on highly acidic refuse material without the use of a topsoil base. However, because of the short duration of the test, it is not known to what extent direct application of limestone and fertilizer to refuse material can be effective in abating pollution by the vegetative cover technique.

EXTINGUISHING AND REMOVING BURNING COAL REFUSE BANKS

The material in this chapter is excerpted from PB 198 258 (see Bibliography, p 390).

Burning coal refuse banks are a significant local source of air pollution. The smoke and fumes emitted are obnoxious and a hazard to health. The banks occupy land that could be used for industrial, residential, and other purposes and detract from the aesthetics of the environment.

In 1963, a Bureau of Mines survey revealed that 495 coal refuse banks were burning in 15 of the 26 coal-producing states. It is difficult to determine the direct cause of many of these fires, some of which have been burning for a number of years. Some are attributed to (1) man's reckless burning of trash on or near the bank; (2) an abandoned mine fire under the bank that has broken through to the surface; (3) possibly spontaneous combustion and (4) brush, forest and camp fires.

The major air pollutants emitted from burning coal refuse banks include sulfur dioxide, hydrogen sulfide, carbon monoxide, carbon dioxide and suspended particulate matter. During times when an area experiences a temperature inversion, the air pollutant concentration can reach serious levels.

High concentration of these contaminants may have toxic effects on human, animal, and plant life. Property damage is also attributable to these gases. Hydrogen sulfide attacks lead-base paints, and sulfur dioxide which converts to sulfur trioxide when combined with moisture becomes corrosive to many metals. At times this corrosive mist reaches a high level of concentration and impairs visibility.

Many attempts have been made by coal companies, state and federal government agencies to control burning coal refuse banks. Several of the methods which have been employed or have been tested include the following:

Isolation: This method involves the excavation of a trench to natural surface, thereby separating the burning portion from the nonburning refuse. Following isolation, the burning portion of the bank is blanketed with incombustible material.

Blanketing: A covering mantle of fine incombustible clay material several feet in depth is applied on the entire bank and serves to smother the fire by excluding air. This method requires continued maintenance to assure the seals effectiveness.

Grouting: A slurry of water and finely divided incombustible material, such as pulverized limestone or sand, is forced into the burning bank. By filling the voids, air is excluded from the refuse with the slurry providing some cooling action.

Explosives: Explosive charges placed deep into a bank through horizontally drilled boreholes are used to create fissures in the fused material. Water is then applied through the newly formed crevices and the quenched material is loaded out.

Spraying: Continuous application of a fine water spray over the entire refuse bank is utilized in this method. Water must be applied until all evidence of hot spots is eliminated. The effectiveness of this method is principally dependent upon the cooling action of the water, since no material is removed.

Loading-Out: The loading-out method is generally applied to small fires and consists of hydraulically cooling and excavating the burning material. The method found most effective in this demonstration project may be considered a refinement of the load-out method.

DESCRIPTION OF PROJECT SITE

The demonstration project site locally referred to as the Baker Bank is located in the city of Scranton, Pa. The coal refuse bank is situated on the west-central border of the city.

The actual cause of the fire is unknown; however, burning and smoldering material was first detected at the western foot of the bank in 1960. In an earlier effort to isolate and control this fire, the bank was divided by a trench into two sections. This effort proved unsuccessful and the fire eventually spread throughout the entire bank. The Baker Bank is estimated to contain 3.5 million cubic yards of refuse covering approximately 48 acres.

The bank was built by depositing refuse that was separated from the marketable coal in a preparation plant. The refuse material ranged in size from six inches to three sixty-fourths inch. The refuse in the bank consisted of coal, rock, slate, shale, bone (a term applied to material containing thin strata or layers of slate sandwiched between coal), and pyritic compounds.

As a result of years of continuous burning, the bank has been transformed into what appears to be a mountain of ash, predominantly red in color and commonly called red dog. The composition is now loose ash, burning and unburned carbonaceous material and a conglomeration of material which has become fused into slaglike masses with nearly the hardness of concrete.

The Bureau's demonstration project was confined to the east section of the bank, an area of approximately 12 acres. This section of the bank averaged 137 feet in height and contained an estimated 1.1 million cubic yards of material.

Water for quenching the burning refuse material was pumped from a vertical shaft which interconnects several levels of an inundated abandoned coal mine. A vertical, deep-well pump, rated at 4,000 gpm was used on the project. Approximately 900 feet of pipeline was required from the shaft to the project site.

A strip-mine pit, approximately 64 feet in depth, adjacent to the demonstration site was backfilled with the quenched material, compacted and graded to correspond with the surrounding terrain.

PROCEDURE AND RESULTS

Hydraulic Treatment with Bulldozer Haulage

The basic techniques used in this phase of the demonstration consisted essentially of quenching and sluicing the hot material with the available mine water and pushing it by bulldozer into the adjacent strip pit. Operations were conducted on a one-shift basis, 5 days per week. The operating personnel include six men: a pump operator, bulldozer operator and four laborers.

The equipment used for this series of tests consisted of 6-inch water monitors, all monitors were operated at the rated 100 psig, used singly or in batteries of two and three units. A variety of commercial nozzles were used to determine their effectiveness. A bulldozer pushed the cooled refuse from the bank into the adjacent strip pit.

The quenching and haulage rates in the following tests, conducted to determine the optimum quenching and haulage rate, were the estimated quantities derived from observation and comparison:

(1) Two monitors, each equipped with a 2,000-gpm fog nozzle, were placed at the foot of the bank approximately 20 feet apart. Water from each unit was directed over separate areas of the bank. The nozzles were operated at several settings, ranging from a straight stream discharge to the maximum fog pattern.

The straight stream setting was used for quenching and sluicing while the fog pattern was intended to control dust and fumes; however, the straight stream jets proved to be effective in controlling air pollutants. The quenching and haulage rate for this particular test was 40 cubic yards per hour.

(2) The same general procedure, as described above, was used in the second test except one of the 2,000-gpm fog nozzles was replaced by a 2,000-gpm straight stream nozzle. Again, the monitors were directed on separate areas of the bank, independent of each other.

The straight stream nozzle had a greater impact than was observed with the fog nozzle. This resulted in an increase of 25% in the quenching and haulage rate, amounting to 50 cubic yards per hour.

(3) A third monitor was added and located adjacent to the two units used in the previous tests. Two monitors were equipped with 1,000-gpm straight stream nozzles instead of the 2,000-gpm nozzles, and the third unit used a 2,000-gpm fog nozzle. The lower capacity nozzles were not effective in sluicing the refuse material in individual areas; therefore, both monitors were directed on one section of the bank.

The fog nozzle was operated exclusively in the straight stream setting. The quenching and haulage rate did not show any increase while using this arrangement of nozzles even though an extra monitor was installed. This test achieved only 50 cubic yards per hour.

(4) Two monitors, both equipped with 2,000-gpm straight stream nozzles, were tested. In the initial period of the test each monitor was directed on a separate area of the bank and operated independent of the other. A greater yield per hour was achieved, however, when both monitors were directed on the same section, thereby concentrating the total 4,000-gpm on one area. The volume of material quenched and hauled increased by 40%, equivalent to 70 cubic yards per hour.

(5) A single monitor was tested using a 4,000-gpm straight stream nozzle. This arrangement provided the maximum efficiency in stream control, water concentration, and sluicing action. One-hundred-and-twenty cubic yards per hour, the highest rate compared to all preceding tests, was achieved using this high-volume monitor.

(6) In all of the preceding tests the monitor or monitors were placed below the elevation at which the water stream made contact with the bank. In this test one monitor using a 4,000-gpm nozzle was positioned on top of the bank. This required quenching and cutting a ramp over hot portions of the bank to permit moving the monitor and supporting the pipeline.

It was assumed that the monitor located on the top of the bank with the water stream pushing material downward would increase the hourly rate; however, no significant increase in the hourly rate was observed. The quenching action of the hydraulic operation was effective in reducing the temperature of the hot and burning material to below 150°F. The temperature of the refuse was further lowered to ambient air temperatures as it was pushed by the bulldozer for disposal in the strip pit. The temperature of quenched material both at the bank and disposal area was monitored daily.

The results of the various tests indicate that the single, higher capacity monitor was more effective in quenching and moving the burning material than several low-capacity monitors operating together. The amount of material removed by the water was limited to the loose rock dislodged by the hydraulic action of the monitors. The sluicing effect of water on the loose material proved to be of little advantage because of the following factors: (1) the relatively flat shape of the loose material, (2) the infiltration of water into the porous bank and (3) increased frictional loss of the irregular surface of the bank. It was necessary to relocate the monitor frequently to provide sufficient hydraulic force to advance the material.

One of the difficulties encountered during the demonstration was to dislodge and fracture large embedded slaglike masses found throughout the refuse bank. Hydraulic undercutting was tried to dislodge to slag boulders. For example, one slag boulder required 12 to 16 hours of continuous hydraulicking before it was dislodged. The bulldozer was then required to push it off the bank.

A mechanical method of crushing and/or dislodging the slag formation was also tried. A steel-wracking ball was swung repeatedly from a crane against the face and sides of the hardened slag surface as well as dropped a number of times on the same general area. The steel ball had only a minimal effect on the hardened slag.

The atmosphere in the working area of the burning bank occasionally became laden with noxious fumes and during these periods the operating personnel were required to use a breathing apparatus consisting of a rubber-mask face piece, pressure regulator, airflow demand regulator, hose, and a standard 244-cubic-foot compressed air cylinder.

The prevailing wind direction often was a factor to be considered when locating the monitors. Generally, the monitors were directed to spray downwind to protect the operator from the backwash, stream and fumes.

Quenching with Tractor-Scraper Haulage

This phase of the demonstration consisted of quenching the hot coal refuse with the water monitors and a sprinkler system. A bulldozer was used primarily for ripping the quenched material while a tractor-scraper transported, spread, and compacted the extinguished material. Prior to the testing work it was necessary to construct access roads and ramps from the top of the bank to the disposal area to facilitate haulage.

During construction of the ramps hot spots on the surface were encountered. These areas had to be cooled before the construction of the ramp could begin. The bulldozer was also used to prepare the ramps in the strip pit since no quenching was required.

The testing was on a three-shift basis, 5 days per week. On the first shift a quenching and hauling operation was conducted. The operating personnel included five men: a pump operator; bulldozer operator; tractor-scraper operator; and two laborers. On the second and third shifts one laborer was required on each shift to quench an area in preparation for the next day's haulage. It was both advantageous and economical to quench on an around-the-clock basis because quenching could be concentrated on the extremely hot and gaseous areas.

The equipment used during this phase consisted of a 6-inch water monitor and a 4,000-gpm straight stream nozzle. In addition, a fabricated flat-type nozzle was used. It was made from a piece of 4-inch diameter, schedule 40, 10-inch long steel pipe. The discharge end was flattened to form an elongated shape and the opposite end was threaded with standard pipe thread.

Fabricated sprinkler pipes were laid and used during the overnight quenching operation. A series of holes were drilled in 6-inch i.d. by 20-foot-long aluminum irrigation pipes. The holes were five-eighths inch in diameter, spaced on 12-inch centers on opposite sides of the pipe. The lightweight pipe was easily connected by simple hook-type couplers, and a complete setup could be made in less than 1 man-hour.

The original bulldozer was replaced by a larger unit rated at 389 fly-wheel hp, equipped with rear end rippers. This larger machine was used to cut through the fused material and to operate in tandem with the tractor-scraper during its loading operation.

The surface temperature of the bank before quenching varied widely on different areas. The surface temperatures ranged from ambient air temperatures on areas of ash or red dog through 200°F on streaming areas and reached 700°F in gaseous areas. Some areas exposed glowing material and flames through cracks and fissures. The temperature of the quenched material averaged about 150°F which was safe for the earth-moving equipment to handle. Occasionally minor explosions occurred deep below the surface of the bank. This was caused by the formation of water gas when the steam contacted the incandescent carbon.

The operating procedure was as follows: The first-shift operation consisted of using the water monitor for quenching an area approximately 100 feet square. Because of its greater range, a distance of approximately 300 feet, the straight stream nozzle allowed quenching at a safe distance from the hot spots. For close-range quenching, the fabricated flat-type nozzle was effective in supplying a deluge over a wider area. Quenching continued until the refuse has been cooled to a depth of several feet.

Operating on a continuous cycle, the tractor-scraper would scrape a load of cooled material, transport and spread the material in layers averaging 14 inches in thickness in the adjacent disposal site. The temperature of the quenched material which was deposited in the strip pit decreased to ambient air temperature in approximately 10 minutes.

The bulldozer, using its rippers, was effective in furrowing the slag formations. Further crushing was accomplished by the leading edge (blade) of the tractor-scraper as it cut its way through while loading. The combined action of the rippers and scraper blade crushed the fused material and virtually eliminated the slag-boulder problem.

Discussion and Evaluation

The objective of this demonstration was to develop a workable method for extinguishment of burning coal mine refuse banks. The results of this demonstration show that a burning coal refuse bank can be effectively extinguished by cooling with water and hauling the quenched material by earth-moving equipment. The strip pit was completely backfilled and the surface was graded to correspond with the surrounding surface terrain.

To summarize the method, the use of water may be considered the basic component. The water percolates into and wets the material to reduce the heat below the ignition temperature. In this case, supplying water in large volumes was only a part of the total extinguishment process. The water was effective in reducing the temperature of the hot material only to a limited depth. Even though the water infiltrated the bank in the present test, there could be no assurance that every hot spot within the core of the bank was extinguished because a substantial proportion of the water was undoubtedly converted into steam and evaporated.

It is possible that these hot spots are protected and insulated by the hardened and encrusted material. The expansion and contraction due to burning and cooling created cracks and fissures in this protective shell, thus allowing air to enter, supplying sufficient oxygen to permit rekindling the material, and sustaining combustion. The other component in the method was the use of earth-moving equipment to haul the quenched material.

To verify complete extinguishment of the quenched material, a grid pattern of boreholes cased with a 1-inch steel pipe were driven through the entire depth of the quenched material. The internal temperatures of the material were recorded through the boreholes by using thermocouple lead wires and a portable pyrometer indicator.

The borehole temperatures were recorded every month for 6 months following the completion of the project. All readings recorded indicated normal subsurface temperatures relative to the ambient air temperature. These data indicate that the backfilled material was extinguished and has not rekindled.

The method of quenching the material with water and hauling it by tractor-scraper yielded a greater volume at a lower cost than the method of hydraulic treatment and removal by bulldozer. The tractor-scraper eliminated any need for rehandling material since it picked up material from the bank and deposited it directly in the disposal pit.

The slag-like formations were as efficiently broken and hauled as the loose material. Quenching was accomplished at a greater distance, since high impact pressure was not required as in the hydraulicking method. Also, the working area was relatively flat and easier to traverse providing greater safety to personnel.

Complete and positive extinguishment was obtained only by alternate quenching and removal. The method may be figuratively referred to as the quench and carry method.

SEALING COAL REFUSE PILES

The material in this chapter is excerpted from PB 198 427 and PB 198 230 (see Bibliography, p 390).

All methods of coal mining, whether surface or underground, contribute to the formation of acid mine drainage. The refuse materials from coal mines and coal cleaning are major sources of AMD, since the breaking and crushing operations necessary for mining and coal separation provide a tremendous increase in particulate surface area available for oxidation.

Extensive work is under way to develop and demonstrate methods for either preventing the formation of AMD or treating it before its seepage into streams. Prevention is far to be preferred, because it would offer a final solution to the problem rather than require a continuing and perhaps growing treatment cost. Methods considered for prevention include diversion of the surface drainage, elimination of air, passivation of the rock surfaces, deactivation of bacteria associated with AMD formation, cultivation of sulfate-reducing bacteria (within the rock strata) to reverse the action of the sulfate-forming bacteria, and modification of pH (within the rock strata) above that conducive to AMD formation.

SILICATE TREATMENT

Tyco Laboratories, Inc. has been engaged in developing a method which will prevent the generation of AMD by direct treatment of the coal mine refuse. The basis of the method, which has been shown to be feasible on a laboratory scale, is to isolate the AMD generating sites from drain water via treatment of the coal refuse with a solution of sodium silicate or sodium silicate plus sodium aluminate.

Waste coal refuse containing significant amounts of pyrite will, in the presence of air and water, generate a dilute solution of iron sulfate and sulfuric acid. Secondary reactions between these ingredients and the local minerals will add concentrations of aluminum, manganese, calcium, magnesium and sodium to the effluent water.

The overall stoichiometry for the acid generation process is:

$$(1) \quad 4FeS_2 + 15O_2 + 14H_2O \longrightarrow 4Fe(OH)_3 + 8H_2SO_4$$

The mechanism of this reaction is under study. There is some evidence that microorganisms also contribute to pyrite oxidation.

This reaction proceeds in stages, the first involving the formation of soluble ferrous sulfate:

$$(2) \quad 2FeS_2 + 7O_2 + 2H_2O \longrightarrow 2FeSO_4 + 2H_2SO_4$$

Air will then slowly oxidize ferrous iron to ferric:

$$(3) \quad 4FeSO_4 + 2H_2SO_4 + O_2 \longrightarrow 2Fe_2(SO_4)_3 + 2H_2O$$

which, as the acid becomes more dilute and/or the soluble iron content increases, will hydrolyze:

$$(4) \quad Fe_2(SO_4)_3 + 6H_2O \longrightarrow 2Fe(OH)_3 + 3H_2SO_4$$

The quantity of these pollutants in acid mine water is variable, and depends on the particular pyrite-coal source and the history of the water before reaching the analyst. The worst cases have a pH less than 2.5 and soluble iron contents up to 10,000 ppm.

Treatment Approach

Mechanisms: To prevent the generation of acid mine water, it is necessary to interrupt reaction (1) at some point. Tyco's approach has been to prevent access of the pyrite to water via the formation of a silica gel over the potentially active acid generation sites. This treatment is based on the following reaction:

$$(5) \quad Na_2SiO_3 + H_2SO_4 \longrightarrow (SiO_2)_n + Na_2SO_4 + H_2O$$

When a solution of sodium silicate (waterglass) is neutralized below pH 10.7, silicic acid is first formed:

$$(6) \quad SiO_3^{=} + 2H^+ \longrightarrow H_2SiO_3$$

Initially, silicic acid and metasilicic acid (Eq. 6) are monomeric. On aging, a three-dimensional structure of Si—O—Si bonds is formed and the solution sets into a gel. The rates at which these reactions take place depend, in a complex fashion, on pH, temperature and sodium silicate concentration.

A second set of reactions can also be involved, i.e., the precipitation of iron silicate. Most transition metal silicates, including both ferrous and ferric silicate, are extremely insoluble, forming the basic chemical structure of most rock formations. A solution of a soluble silicate, in which the anions exist as monomers, dimers and possibly trimers, reacts rapidly to form insoluble precipitates. The extent to which this takes place is a function of pH and the oxidation state of the iron. Ferrous silicate is apparently more insoluble than ferric silicate. At high pH, e.g., ~ 10, soluble complex ions are also formed. However, under the conditions of AMD generation, i.e., at low pH, the addition of sodium silicate can result in the formation of some insoluble iron silicate mixed in with the gel developed according to reaction (5).

Thus, there are three mechanisms by which treatment of coal mine refuse with silicate can inhibit the generation of acid mine water:

(1) Blocking AMD generation with a water impermeable silica gel,

(2) Precipitation of iron silicate which scavenges soluble iron and also coats the pyrite particles, and

(3) Neutralization of AMD acid, according to reaction (5).

Experiments have indicated that all three mechanisms are operative. When a surface gel is formed, most water is diverted from the pile. On disrupting the gel, the effluent water composition becomes similar to that from the neutralized pile. Insoluble iron silicate is observed on removing samples of refuse from inside the pile.

Experimental Methods: Treatment techniques were evaluated by testing piles of actual coal mine refuse under simulated natural conditions. The test piles each consisted of 200 kg of representative refuse, formed to a cone 3 ft in base diameter and 18 inches high. A measured artificial rainfall (generally 100 liters per day which is about 6 inches of water) was maintained on essentially an 8-hr per day, 5-day work schedule, with percolated and runoff water samples being collected on a daily basis. The wash water samples were analyzed for total iron, ferrous iron, sulfate, pH, calcium carbonate acidity and soluble silica.

The test piles were placed in plastic basins on a tilted table to allow easy collection of the runoff. An overhead sprinkling system was constructed to simulate intermittent rainfall. The plastic basins had two sample points from which the runoff and percolated wash water were collected: one collecting from a 14-inch diameter center area and the other from the outer anulus. The data reported are for composite samples of the two collection points.

The specific refuse material used for evaluation was a washed waste coal refuse originating from the lower Kittanning Seam. The analysis is given in Table 13.1.

TABLE 13.1: ANALYSIS OF BITUMINOUS COAL REFUSE

Size Passed	Size Retained	Wt %	% Pyritic Sulfur (Dry Basis)	% Fe_2O_3 (Ignited Basis)	% Sulfur*	% Ash*	% Moisture*
	1/2 in.	2.5	4.14	14.53	5.41	2.82	0.98
1/2 in.	3/8 in.	10.5	3.12	11.45	3.62	54.11	1.01
3/8 in.	1/4 in.	25.3	1.52	7.73	1.89	58.01	1.08
1/4 in.	8 mesh	46.2	2.70	10.31	3.30	53.45	1.14
8 mesh	30 mesh	13.8	5.24	17.47	6.53	55.25	1.30
30 mesh		1.7	3.32	17.75	4.63	44.14	1.80

* As received

Source: PB 198 427

Details of Refuse Pile Treatments

Silicate Material: The sodium silicate solution used in the laboratory experiments to determine the lag time for gelation was produced by Philadelphia Quartz (type N) and had the following composition: SiO_2, 28.7%; Na_2O, 8.9% and H_2O, 62.4%. The SiO_2 to Na_2O ratio was 3.22. The density was 1.39 g/ml (41°Bé at 68°F).

Neutralized Pile (Pile B): A pile of coal mine refuse was neutralized by treatment with 3 liters of a waterglass solution containing 4% SiO_2. Neutralization was defined as the condition when the effluent solution of the pile during the waterglass treatment had the same pH as the fresh silicate solution. Examination of the pile showed that the particles of coal refuse were coated with a thin layer of some gel-like material. This coating did not block the flow of water through the pile, but did tend to prevent contact of the water with the surface of the coal.

Silica Gel: Surface Treatment (Pile C) — It was intended that one pile of coal mine refuse should have a thin layer of silica gel to form a sealant layer. To accomplish this, it was necessary to determine the conditions under which waterglass would gel. It was known that acidifying the sodium silicate solution would cause gelation, and it was felt that it might be possible to allow the acid in the refuse pile to act as the acidifying material.

First, some tests were made to determine the gelling properties of the type N silicate. Reagent grade sulfuric acid was used to lower the pH. Distilled water was used for diluting the commercial silicate and making up all solutions.

Gel times were determined by adding a dilute solution of H_2SO_4 to sodium silicate solutions (in which the concentration of SiO_2 ranged from 1 to 28.0%), mixing well, and allowing the mixture to stand in 100-ml beakers kept at 25°C until gelling.

The data obtained from these experiments are as follows:

Concentration: 1% SiO_2

pH	9.5	8.0	7.0	4.0
Gel time, sec	5×10^5	2,100	2,580	4×10^5

Concentration: 2% SiO_2

pH	8.5	6.5	6.0	2.5
Gel time, sec	1,800	370	415	5×10^5

Concentration: 3% SiO_2

pH	9.0	8.5	7.5	6.5
Gel time, sec	320	170	90	250

Concentration: 4% SiO_2

pH	9.0	7.5	6.5	1.0
Gel time, sec	90	20	100	8 days

Several high concentrations of SiO_2 solutions (10 to 28.0% SiO_2) were tested for gel time at various pHs. Gel was formed instantly.

Several experiments on small pyritic piles were set up to determine which waterglass should be used in treating the larger piles. In the first experiment, 250 g of the fresh, high-sulfur coal refuse described above were slowly washed with 1,000 ml distilled water. The washing was collected and analyzed. The analysis showed: pH 3.3, acidity 540 ppm as $CaCO_3$, ferrous iron 429 ppm, total iron 1,177 ppm and SO_4 475 ppm. This refuse was neutralized, without gel formation, with 100 ml of waterglass solution containing 3% SiO_2 at pH 11.3.

The pile was slowly washed with 250 ml of distilled water and the effluent collected and analyzed. This analysis showed: pH 5.7, acidity 200 ppm as $CaCO_3$, ferrous iron 13.6 ppm, total iron 181.2 ppm and SO_4 247 ppm.

In the second experiment, gel was formed on the top of and partially inside a 500-g pile of fresh refuse by treating with waterglass solution containing 5% SiO_2 at pH 9.5. The treated pile was left for 1 hour to harden the gel and was then washed slowly with measured amounts of distilled water. The effluent pH was 4.0 after washing with 200 ml distilled water. Acidity was 100 ppm as $CaCO_3$.

To determine the reason that the AMD continued to be produced from this treated pile, it was cut into two halves. When examined, it was noted that gel was formed irregularly and that some pieces of refuse were not coated with gel. Water permeability of the pile was unimpaired.

Based on these experiments, it was decided to form the gel on Pile C using a silicate solution containing 4% SiO_2 and to acidify the solution to pH 9.5 before application. The actual treatment of Pile C was accomplished by passing the solution through the pile several times until gelation occurred. The gel formed was somewhat nonuniform, but the treatment was continued until the entire pile was covered with gel. In order to increase the residence time of the silicate solution in the pile, some of the gel was formed by thickening the silicate solution with Cab-O-Sil, a very fine silica powder. This increase in residence time greatly simplified the formation of a continuous gel across the pile. A total of approximately 4 liters of concentrated waterglass was used.

Alumina/Silica Gel: Surface Treatment (Pile D) — The difficulty in forming the silica gel on the pile prompted a search for better gelation methods. In addition to the thickening procedure described it was found that a strong, continuous gel could be formed rapidly by mixing the sodium silicate with sodium aluminate. In fact, this method worked too well and the gel formed before the solution could be applied to the pile. This situation was remedied by first treating a pile with 1 liter of sodium silicate diluted to a solution containing 4% SiO_2 and then immediately treating the pile with 10 liters of 5% sodium aluminate. A strong gel was formed, primarily on the surface of the pile; very little of the material ran through the pile.

Alumina/Silica Gel: In-Depth Treatment (Pile E) — Gel was formed by slowly spraying the pile with sodium silicate solution (SiO_2/Na_2O, 3.22) containing 4% SiO_2 in order to let the solution go deeply through the pile. This was followed by spraying with sodium aluminate solution having a concentration of 5%. This procedure was repeated until the gel formed a layer inside the pile and covered the surface. The difference between this treatment and that of Pile D was that the silicate solution was allowed to penetrate into the pile before the aluminate was added. Thus, when the aluminate was sprayed on, it could not form a sealant layer of gel on the surface, but had to penetrate into the pile to form a gel. Successive treatment finally formed a blocking layer which caused a fairly uniform layer of gel to be built up.

The pile was immediately disturbed and it was found that the gel indeed had formed in the interior, but did not fill all the spaces between the rock particles. This caused the washing water to pass very slowly through the pile (as compared to the relatively rapid flow of water through the untreated pile).

Results and Discussion

Five test piles were set up during the course of the experimentation and were evaluated as described above. The treatments used are summarized as follows:

Pile	Treatment
A	Control—no treatment
B	Neutralized pile—a minimum amount of sodium silicate solution was permeated throughout the pile to neutralize acid via reaction (5), while not forming a continuous gel
C	Silica gelled pile—sufficient silicate was used to form a coherent gel on the surface of the pile
D	Alumina/silica gelled pile—this treatment was similar to (C), except that sodium aluminate was added to form a more insoluble gel
E	In-depth alumina/silica gelled pile—a mixed gel was formed beneath the surface of the pile to obtain protection from erosion and mechanical disruption of the gel

The pH and the dissolved iron content of the samples were determined. The wash volumes used were converted to inches of water so that the results could be examined in terms of rainfall. Pennsylvania, for example, receives an average of 40 inches of rain per year.

One comment on the iron analysis is pertinent. Most of the analyses are for samples passed through conventional laboratory filter paper. Late in the program, it was demonstrated that this procedure was inadequate for removing finely divided iron hydroxide. Calculations of the total soluble, chemically complexed ferric iron as a function of pH indicate negligible amounts (<1 ppm) at and above pH 6. Experimental results confirm this conclusion. Therefore, the total dissolved iron values include suspended solid iron hydroxide for those samples with high pH. This applies to Piles B and C for the first 80 inches of equivalent rainfall; Pile D: 240 inches; Pile E: over 280 inches.

Examination of the wash effluent analysis data for the control pile (A) shows that the effluent composition was comparable to that found in actual field studies. The pH was about 2.0 and the iron content was 1,000 to 5,000 ppm.

The neutralization treatment (B) is effective for about 120 inches of equivalent rainfall in causing a reduction in the acid mine drainage produced by the washing of coal mine refuse. After this period, the treatment's effectiveness is reduced, but still yields acid flows at a lower level than untreated refuse.

This reduction in effectiveness is undoubtedly associated with the washing out of the silica from the pile and as the silicate is washed out, the refuse is again exposed to the wash water. However, even after much of the silicate is washed out, the AMD production never approaches the level of the untreated pile (A).

For treatment C (surface gelled with silica), as with the neutralized pile, the pH is increased and iron content is depressed. The silica is also slowly being washed out of this pile, although the rate is somewhat lower, i.e., about 40 ppm for Pile C versus 60 ppm for Pile B. It would be anticipated that the gel will eventually lose its integrity and allow the production of AMD. However, for the first 80 inches of equivalent rainfall, the appearance of the surface of the gel did not change appreciably and the wash water continued to run off the surface of the pile, not penetrating to the interior. The iron content of the effluent water was slowly increasing, but still remained an order of magnitude lower than the untreated pile.

After 80 inches of equivalent rainfall, the pile was disturbed. Several cracks were made in the surface of the gel and the washings were continued. The pile was then violently disturbed by turning the surface over. Examination of the interior of the pile showed very little internal gel, none of it continuous. Washings were continued.

The detailed data show that breaking the gel on the surface of the pile allowed water to enter the interior, resulting in the formation of AMD. Comparison with Pile A, however, reveals that although the pH drops and the iron and sulfate increase sharply, all values are lower than the untreated pile.

Pile D was given a thin sealant layer of alumina/silica gel and was washed with water on a daily basis. It is evident that not only are the iron and sulfur compositions at reduced, stable levels, but the silica in the effluent is much lower than in the silica gel protected pile as expected. This result indicates that the pile will retain its stability for a longer period of time.

The continuous dropping of the wash water on certain spots on the surface of the pile caused local erosion faults in the continuous gel structure which were large enough to permit water to enter the interior of the pile and start to create AMD. These faults in the gel structure were due to the fact that the washing was not completely random, but tended to drop selectively in certain spots (this would not be the case under ordinary rainfall conditions). Since these faults were due to an experimental artifact and were not representative of the system in general, it was decided to seal the holes with gel and continue the test, making an effort to "rain" on the pile in a more random manner. The pile was therefore repaired, and the analysis data showed that the protection against AMD formation was resumed immediately.

The best results were obtained from the pile treated in depth with the alumina/silica gel. The decreased solubility of the gel enhanced the long term stability of the gel, while the deeper layer of gel increased its resistance to erosion and cracking. This fact can be confirmed upon considering that Pile E had been violently disturbed on day 1 of the treatment period, yet the wash effluent showed the lowest concentrations of pollutants. Clearly, this is a very promising treatment technique.

In summary, examination of the sulfur and iron removal rates for the various treatments shows that they are all markedly superior to the untreated condition. Each treatment has its own advantages and disadvantages which will determine its use in specific situations. The neutralization technique appears to have the shortest lifetime, yet it is undoubtedly the cheapest since it would use the least amount of treatment material. Its lifetime can be increased through the use of aluminate in the neutralization procedure to reduce the solubility of the sealing precipitate.

The surface treatment with silicate gel alone is not as durable as that with the alumina/silica gel because of higher gel solubility, but it is cheaper than the mixture. The surface treatment with the alumina/silica gel is less expensive than the in-depth treatment, but the in-depth treatment is certainly the most effective. The proper treatment method must be chosen by optimization of both cost and effectiveness criteria.

Gel Weatherability Tests

A series of tests were run to determine the effect of temperature extremes on the stability and effectiveness of the gel treatment. The tests were performed at two levels: small laboratory heating and freezing tests and larger outdoor freezing tests. The indoor tests were carried out by forming sealant layers of gel on 1-lb piles set up in small flat containers.

Effect of High Temperature on Alumina/Silica Gel: There were two purposes in performing high temperature tests on alumina/silica gel in refuse piles. One was to see how the gel withstood simulated summer conditions and the other was to see if the gel technique could be used on hot piles where the inside of the refuse pile was burning.

The "summer" test consisted of placing 1 kg of refuse in a shallow dish and forming the alumina/silica gel on the small pile. The treated pile was then placed under an infrared lamp which was adjusted so that the ambient temperature was between 85° and 90°F. During the one-week period of the test, the gel slowly dried out and took on the appearance of dried plaster.

At the end of the week, the pile was sprayed with water to determine the stability of the gel and its ability to prevent AMD generation. The washing was continued for 17 days (the results are shown in Table 13.2). It was apparent that the gel was slowly washing away, and at the end of the test period there was little evidence of gel structure. However, when the pile dried out, the particles of refuse were covered with a very thin layer of white material which was apparently silica or alumina or a mixture of both. The results in Table 13.2 show that although the sealant layer of gel per se was gone, the residual silica or alumina continued to prevent AMD generation. It should be noted that in these tests 50 liters of water are equivalent to about 110 inches of rainfall.

A second test was run to simulate the treatment of a hot pile. Here, a 1-kg pile of refuse was treated with alumina/silica gel and placed in an oven at 130°F, which is approximately the surface temperature of a hot pile. After two days of baking, a 15-day series of water washings was started. The effluent analysis is given in Table 13.3. The visual results were very similar to the lower temperature test: the gel hardened during baking and was eroded during washing. Despite the fact that the continuity of the sealant layer was interrupted, no AMD was produced during the equivalent of 120 inches of rainfall.

The temporarily high pH of the effluent is undoubtedly due to the washout of the gel-forming materials as silicate and aluminate. Here again, the refuse retained a white layer when it was dried at the end of the equivalent of 110 inches of rainfall. This indicates that there is still residual silica adhering to the refuse which should seal the rock and thus prevent or at least minimize AMD generation.

TABLE 13.2: SIMULATED SUMMER TEST

Day	Cumulative Volume of Wash,* ℓ	pH
1	0.5	12
4	2	8.3
7	5	7.5
9	10	7.2
11	15	7.2
14	20	7.0
15	30	7.0
16	40	6.8
17	50	6.5

* 0.46 ℓ of water is equivalent to 1 in. of rain.

TABLE 13.3: SIMULATED HOT PILE

Day	Cumulative Volume of Wash,* ℓ	pH
1	5	10.8
2	10	8.0
3	15	7.2
4	20	7.0
7	25	7.0
8	30	7.0
9	35	6.8
10	40	6.8
11	45	6.6
14	50	6.6
15	55	6.5

* 0.46 ℓ of water is equivalent to 1 in. of rain.

Source: PB 198 427

Antifreeze Selection: Several experiments were carried out to develop an antifreeze system that would prevent the interstitial water in the gel from freezing. The tests consisted of gelling small laboratory piles weighing 500 grams each, set up on a watch glass, and keeping them in the freezer at a temperature of -8°C for various periods of time.

Sodium Chloride — Experimentation using NaCl as an antifreeze, in which the concentrations of the solution range from 5 to 10%, showed that the higher concentrations of NaCl form a water soluble gel with the silicate solution alone. When 500-g laboratory piles were gelled by adding 10% NaCl to the alumina/silica gel (in which the concentration of SiO_2 ranged from 5 to 10%), a gel was formed that resisted freezing at -8°C. However, when the pile was sprayed with water, the NaCl was washed out. The chemical analysis showed that 93% of the added NaCl was removed during washing.

Calcium Chloride — This salt was ineffective as an antifreeze additive, since it formed a gel with the silicate solution which broke on freezing. At the same time, it reacted with sodium aluminate and formed a white precipitate of $Ca(OH)_2$.

Ammonium Chloride — These experiments showed that a water-soluble gel was formed by the addition of NH_4Cl to the silicate solution with the release of NH_3 gas. Again, chloride ion was detected in the wash effluent. Also, NH_4Cl reacts with the sodium aluminate to yield the gelatinous precipitate of aluminum hydroxide.

Ethyl Alcohol — This material formed a strong gel with the concentrated solution of silicate, but the gel was found to be soluble in water. When ethyl alcohol was mixed with sodium aluminate (5%) and then added to a waterglass solution containing 4% SiO_2, it was found that the resultant gel has some solubility in water. The solubility of the gels formed using ethyl alcohol prevents its use as an antifreeze additive.

Prestone Ethylene Glycol — This material was also ineffective, since it formed a soluble gel.

Polyethylene Glycol 400 — This material was also ineffective, since it formed a soluble gel.

Propylene Carbonate — This material formed a gel which was found to decompose when sprayed by acid mine water. Carbon dioxide gas was evolved, indicating that the carbonate was decomposed by the acidic water, thus destroying the gel.

Polyethylene Glycol 20,000 — This material formed freeze-resistant gels when added to the silicate and aluminate mixtures. The concentrations of polyethylene glycol 20,000 used ranged from 5 to 10%. This was added to silicate solutions containing from 4 to 8% SiO_2 and sodium aluminate solutions containing 5 to 10% aluminate.

At low additive concentrations, the gel was quite weak in consistency. Although the higher concentrations gave stronger gels, it was found that this gel tends to separate from rock particles on spraying with water.

Glycerin — When glycerin was added (10% concentration) to the silicate and sodium aluminate, it produced a gel with characteristics similar to those produced from polyethylene glycol 20,000.

Effects of Low Temperature on Alumina/Silica Gel: Two outdoor tests were set up to evaluate the effect of low temperatures on the alumina/silica gel. Both were performed in 3-gal pails containing about 25 lb of the refuse. One sample was treated with the gel in a manner similar to Pile E of the large scale test group, while the other sample was treated with a gel containing 0.5% by weight of polyethylene glycol (20,000).

Both pails were placed outdoors in an exposed place and allowed to endure the full effect of a New England winter. During the 45 days of the test, the temperature ranged between 3° and 53°F. There was snow on eight days and rain on seven days. At the end of this six-week period, the pails were examined and it was found that virtually no gel was visible. Upon drying, a layer of white material, which was probably residual alumina and silica, could be seen on the surface of the refuse pieces.

The liquid that had collected in the bottom of each pail was drained out and analyzed with the results shown in Table 13.4. It is clear that although the gel was no longer sealing the refuse, the acid production of the material was very low. It should be noted that the pails did not have drain holes in the bottom so the water could not drain out, thus the refuse was soaked in water for much of the six weeks of the test.

In order to see if the refuse that had been used in this test was still protected from producing AMD under more normal drainage conditions, samples from each test were placed in a dish and washed with water. The washings were collected and analyzed as shown in Table 13.5. It would seem that the residual alumina and silica on the refuse surface maintained a protective layer, despite the absence of a coherent gel, and still prevented the production of AMD.

TABLE 13.4: ANALYSIS OF WATER FROM OUTDOOR TEST

Test	pH	Acidity, ppm $CaCO_3$	Soluble Silica, ppm
With antifreeze	5.9	30	50
Without antifreeze	5.8	50	80

TABLE 13.5: ANALYSIS OF EFFLUENT WATER FROM REFUSE FROM OUTDOOR TEST

Test	pH	Acidity, ppm $CaCO_3$	Soluble Silica, ppm
With antifreeze	6.2	10	15
Without antifreeze	6.1	15	20

Source: PB 198 427

Treatment of Acid Mine Drainage

Silicate Treatment of Artificial Acid Waters: The purpose of these experiments was to study the reaction of ferrous and ferric ions (present as sulfates) with silicate ions (present as waterglass) over a range of concentrations. The goal was to determine the effectiveness of waterglass in precipitating or complexing the iron.

The test procedure was to add 10 ml of ferrous (or ferric) sulfate solution of a chosen concentration to 10 cc of sodium silicate (waterglass) solution of a given concentration. The iron was kept in the range of 100 to 1,200 ppm in order to simulate actual acid waters. After the reagents were mixed, the solution was filtered to remove any resultant precipitate and the filtrate was analyzed for pH and iron concentration. The filtrate was then treated with 0.5 ml of concentrated (10%) sodium aluminate solution to test the presence of free iron ions in the presence of silicates. It had been determined previously that the aluminate would precipitate free iron, but would not react with iron in any complex form. The aluminate-treated solution was filtered to remove precipitates and the filtrate was again analyzed for pH and iron content.

Data obtained when ferric sulfate was treated with waterglass and sodium aluminate show that there are three different situations:

No precipitate from silicate addition and no precipitate from aluminate addition. In this case, it is clear that the addition of the silicate caused the complexing of the iron into soluble iron silicates. The addition of the sodium aluminate could not break the iron silicate complex, although the aluminate would ordinarily precipitate all the iron. The analytical procedure for determining iron was chosen so that both free iron ions and complexed iron could be detected.

No precipitate from silicate addition, but some precipitate from aluminate addition. It appears that in this case the addition of silicate caused some of the iron to complex, but not all of it. When the aluminate is added, the free iron ions are precipitated.

Silicate addition causes precipitation, and aluminate addition produces further precipitation. In this case, the silicate causes the precipitation of the iron as iron silicate. The iron silicate precipitates at pH values below about 7, but not above that. Therefore, if insufficient silicate is added to raise the pH from the acid level of the iron sulfate solution to a point above 7, the iron silicate will precipitate. If the pH is raised above 7 by waterglass addition, the iron is complexed and cannot then be precipitated.

The amount of waterglass needed to cause the iron to complex without precipitating can be determined by means of a graph where the iron and silicate concentrations in the mixed solution are plotted for the condition where all the iron is complexed. The minimum silicate required for any concentration of iron is determined by drawing a line through the origin and the data points representing the least amount of silicate that was used to complex a given amount of iron.

The ratio of silicate to iron along this line is a constant and has a value of six parts of waterglass (100% sodium silicate with a weight ratio of SiO_2/Na_2O of 3.22) to one part of iron. This is a rather high value and will undoubtedly prove to be very costly, but the work does establish the concept of being able to make the iron inert and not subject to precipitation. The advantage of this treatment would be that the acid mine water could be treated without the necessity of settling and collecting a hard-to-handle, low-solids precipitate.

A similar series of experiments was performed with ferrous sulfate solutions. When the data are plotted, it is not as obvious how the minimum line should be drawn, but a line can be drawn that will create a region that includes all other points. The minimum weight ratio is again about six parts silicate to one part iron; there is some evidence that suggests that the minimum line should be steeper than that drawn. This would give a lower ratio of silicate to iron.

Oxidation of Ferrous to Ferric in the Presence of Silicate: The previous experimentation raised the question of how stable the ferrous silicate complex was to oxidation of the iron. Since it is known that most of the iron content of acid waters is in the ferrous form, this might have some bearing on the treatment. An iron silicate complex solution was formed by mixing 100 ml of ferrous sulfate solution (2,122 ppm as ferrous) with 100 ml of sodium silicate solution (0.15 M). This mixture should be fully complexed.

Portions of this mixture were analyzed for ferrous at various time intervals. For each sample of the mixture that was analyzed, a portion of the original ferrous solution was also analyzed for ferrous as a control. The results are shown in Table 13.6 and plotted in Figure 13.1. The data show clearly that the presence of silicate accelerates the rate of ferrous oxidation. The fact that the control sample of ferrous sulfate did not appear to oxidize at all over a period of almost three days is somewhat surprising, although it is known that without aeration this oxidation takes place quite slowly.

Precipitation of Ferrous and Ferric Iron Using Sodium Aluminate: Since experimentation showed that sodium aluminate could be used to precipitate all the iron in acid water, a series of experiments was run to determine the minimum amount of aluminate needed to do this and to compare the results with lime treatment.

The results of all the tests indicated that there is no advantage to using sodium aluminate to precipitate iron from acid mine drainage.

Other Experiments

Optimization of Alumina/Silica Gel Concentration: In order to optimize the composition of the alumina/silica gel for in-depth gelation, a series of tests was run using several different combinations of solution concentrations. The criteria used for optimization were gelation time and the strength of the gel. The optimum solution was considered to be a mixture of 3% silica solution with 3% aluminate, or a final solution containing 1.5% of each.

Stability of Alumina/Silica Gel in Contact with AMD: In order to determine the stability of the alumina/silica gel when it was contacted with AMD, a series of static and dynamic tests was run. The static tests consisted of forming the gel either in the bottom of a beaker or jar and then pouring in an AMD formed by passing water over the coal mine refuse. The system was allowed to sit for several weeks. In the dynamic test, the gel was formed in a beaker and then transferred in pieces into a jar. AMD was put into the jar and periodically shaken vigorously.

The most obvious result was that the iron in the AMD precipitated quite rapidly in the form of iron hydroxide. This was probably due to the high pH of the gel

TABLE 13.6: OXIDATION OF FERROUS TO FERRIC IN PRESENCE OF SILICATE

Time Elapsed, hr	Ferrous Concentration With Silicate, ppm	Ferrous Concentration Without Silicate, ppm	Amount Ferrous Oxidized, %
0.33	770	1060	27.4
1	614	1060	42.1
3	470	1060	55.7
5	357	1060	66.3
70	90	1060	91.5

FIGURE 13.1: OXIDATION OF FERROUS TO FERRIC IN PRESENCE OF SILICATE

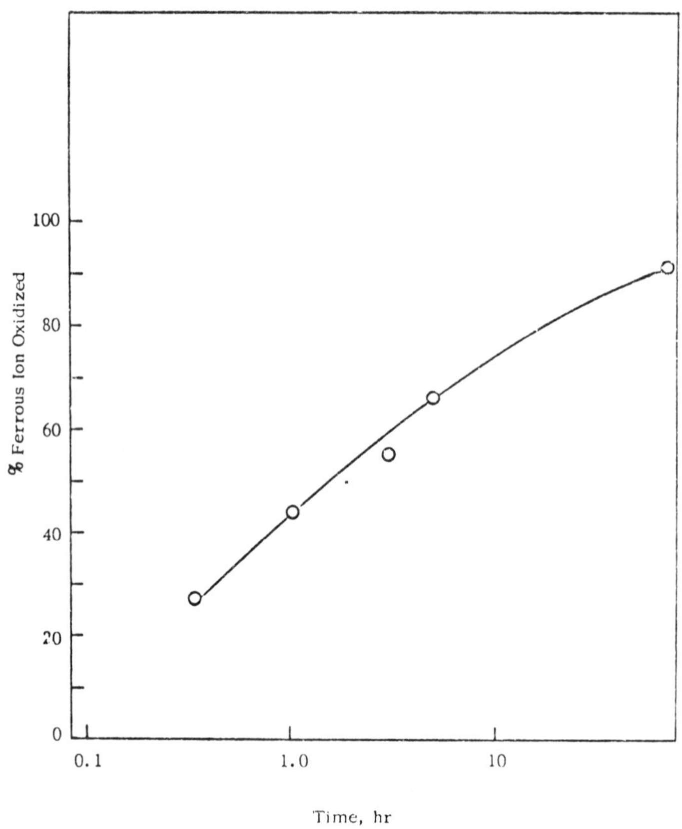

Source: PB 198 427

material. The supernatant solutions were periodically analyzed to see the effect on the AMD composition. Table 13.7 shows the results of such an analysis made after 10 days. It can be seen that the presence of the gel caused most of the acidic constituents of the AMD to be precipitated out of solution.

TABLE 13.7: ANALYSIS OF SUPERNATANT AMD IN GEL STABILITY TESTS

	Analysis of AMD	AMD + Silica/ Alumina (Shaken)	AMD + Silica/ Alumina (Without Shaking)
pH	2.3	8.5	4.5
Acidity on $CaCO_3$, ppm	3950	—	180
Total dissolved iron	1117	5	110
Ferrous	120	—	0
Sulfate	3260	760	1330
Soluble silica	0	200	80

Source: PB 198 427

The gel itself did not appear to be affected in any way, although it was hard to see if there was any change in the shaken test. The gel mass was covered with yellowish-brown iron hydroxide which might have masked any erosion.

Analysis of Bacteria Content of Wash Effluent: A sample of the wash effluent from the control pile (Pile A) was taken after about 100 inches of equivalent rainfall had passed through the pile. This sample was analyzed for bacteria content. The analysis showed a bacteria count of 2.4×10^4 cells/ml as compared to an average cell count of about 100 in the effluent from a coal mine. The cell concentration in effluents from existing refuse piles is probably somewhat higher than that of the coal mine effluent.

The very high cell count in the effluent from the control pile washing helps explain the fact that almost all the effluent solutions showed very high ferric and very low ferrous concentrations. It has been shown that bacteria in acid solutions promote the oxidation of ferrous ion with the rate of oxidation varying directly with the bacteria concentration. Thus the fact that the effluent from the laboratory piles shows abnormally high bacteria counts is consistent with the very high ferric concentrations.

The reasons for the abnormally high bacteria count are hard to determine in the absence of controlled experimentation, but the environmental conditions were certainly conducive to the propagation of biological organisms. The refuse was kept in a highly humid atmosphere at temperatures consistently above 60°F and the pyrite content was relatively high. These conditions undoubtedly encouraged the growth of the microorganisms and the resulting ferrous oxidation.

Conclusions

The following conclusions are based on the experimental work accomplished during this program:

Neutralization of coal mine refuse by treatment with waterglass (sodium silicate) is an effective technique for preventing or minimizing acid mine drainage (AMD) generation for a period in excess of 120 inches of equivalent rainfall.

Sealing the coal mine refuse pile surface to water percolation with silicate-based gels is effective in preventing or minimizing AMD generation over a long period, perhaps in excess of 400 inches of equivalent rainfall.

Alumina/silica gels are more effective in prolonged sealing of the refuse structure than are the silica gels because of the lower water solubility of the alumina/silica gel structures.

Surface gels are not stable to extremes in temperature. Both heating and freezing cause coagulation of the gel with the resultant loss in continuity. Such broken gels no longer seal the refuse structure, but the residual effects of the treatment minimize the generation of AMD over prolonged periods of time, similar to the effect of neutralizing the material with silicate.

In-depth gelation of refuse piles using alumina/silica gels should be effective in prolonged prevention or retardation of AMD generation regardless of the environmental temperature variations.

The treatment of AMD water with sodium silicate and sodium aluminate in a holding pond is no more effective than the conventional application of lime. There is no advantage in settling rates or precipitate volume.

CARBONATE BONDING

The project involves the investigation of a process that has several applications in combating water pollution and other problems created by coal refuse piles. The need for such a process arises from the fact that as rain water percolates down through the spoil bank, the reaction between pyrites, air and water generates acid-forming constituents which are carried in the water table or streams surrounding the locale. Thus, to minimize this serious pollution problem, an effective process for sealing off spoil banks or utilizing coal refuse is required.

One application of the process is the sealing of coal refuse piles with a layer of carbonate-bonded refuse to exclude air and moisture from the pile, and thus eliminate the formation of acid drainage. A second application of this process is the elimination of refuse piles by utilizing the refuse material as filler in the construction of carbonate bonded road pavement and in the manufacture of low-cost bricks and similar products.

In this process, lime hydrate is thoroughly blended with coal refuse and then treated with carbon dioxide to form strong calcium carbonate (calcite) crystals. The calcite crystals act as a cementing matrix that turns the coal refuse into a coherent and impervious structure. The carbonate bonding process relies on the simple reaction that occurs between carbon dioxide and lime hydrate. In the reaction:

$$Ca(OH)_2 + CO_2 \longrightarrow CaCO_3 + H_2O$$

carbon dioxide reacts with lime hydrate to produce calcium carbonate (calcite) and water. The reaction is well known and occurs naturally upon the exposure of lime hydrate to the atmosphere. However, the rate of conversion of lime hydrate to calcium carbonate is relatively slow because of the small percentage of carbon dioxide contained in the air. It is of interest to note that although the reaction in air is slow, the reaction proceeds rapidly in concentrated carbon dioxide. Because of its high strength, the calcite crystal affords an opportunity for exploiting lime hydrate as a cementing material to bind coal refuse into a coherent structure.

Most of the experimentation prior to this work has been conducted with various types of soils and not with coal refuse materials. Therefore, this bench-scale experimental study was undertaken to determine the operating parameters for generating a strong carbonate bonded coal refuse and to determine the feasibility of the carbonate bonding process for the elimination of acid mine water formation from coal refuse piles. In particular, the effects of lime hydrate content, moisture content, partial pressure of carbon dioxide in carbonating gas, carbonation time, and the external surface area of coal refuse on the compressive strength of the coal refuse were determined. In addition, the air and water permeability and the freeze-thaw resistance were determined for carbonate bonded coal refuse.

Before proceeding with a discussion of the experimental work, it would be well to discuss some of the theoretical aspects of the carbonate bonding reaction and the results from prior experimentation with other fill materials. The relationship between the compressive strength of carbonate bonded materials and process operating variables is as yet not fully understood. Prior experimentation on soils, clays and sands indicates that compressive strength increases as:

(1) The particle size of the fill material and lime hydrate decreases,
(2) An optimum water content is selected for each specific fill material,
(3) The external surface area of the fill material increases,
(4) The carbon dioxide partial pressure in the carbonating gas is balanced so that the rate of heat generation within the composite equals the rate of heat loss to the surrounding gas phase,
(5) The time of carbonation increases, and
(6) The lime hydrate concentration increases.

Decreasing the particle size (along with a wider particle size distribution) of the fill material decreases the porosity of the composite which is to be carbonate bonded. Decreasing the porosity yields a more dense composite and enhances the formation of a continuous calcite matrix. As the compressive strength of the composite is mainly a function of the continuity of the calcite crystal structure, a minimum porosity condition increases the strength of the carbonate bonded material.

As the surface area of the lime hydrate increases, its rate of solution in water is increased. This causes the carbon dioxide to react with a saturated lime hydrate solution and results in a rapid precipitation of a fine calcite crystal matrix. The rapid formation of a multitude of crystals enhances additional crystalline growth so that bridging and growth can occur between crystals with a resulting increase in strength.

In general, the greater the external specific surface area of the fill material, the greater will be the rugosity or roughness of the surface of the particles. This condition provides numerous bridging channels between particles for the growth of calcite crystals and increases the strength of the carbonate bonded material.

The effect of the partial pressure of carbon dioxide in the carbonating gas on the carbonation reaction is directly related to the gas phase mass transfer of carbon dioxide to the liquid film. Prior experimentation has shown that the rate of the carbonate bonding reaction is directly proportional to the partial pressure of carbon dioxide. Since the carbonate bonding reaction is highly exothermic, the heat released during this reaction can be controlled by varying the partial pressure of carbon dioxide in the gas phase.

Depending upon the water content, and the thermal characteristics and packing of the fill material, it is possible to crack the resulting carbonate bonded material by causing the reaction to occur too rapidly. Due to a high carbonation rate sufficient heat is generated to produce steam within the interior of the composite. If the packing is too dense, rapid steam evolution from the carbonate bonded material is inhibited and can result in sufficient internal pressure to crack or rupture the composite. Consequently, the partial pressure of carbon dioxide must be optimized so that internal pressure cracks are not generated and the strength characteristics of the carbonate bonded material can be maintained.

Materials and Apparatus

Raw Materials Used: Four types of coal refuse were investigated, a relatively fresh (unburnt) bituminous, a highly oxidized (burnt) bituminous, a relatively fresh unoxidized (unburnt) anthracite, and a highly oxidized (burnt) anthracite. Each of the samples was initially screened to remove lumps and yield a −10 mesh sample. A portion of the −10 mesh sample was subjected to grinding to obtain a −20 mesh and −60 mesh size fraction for use in subsequent experimentation.

TABLE 13.8: TYPICAL SCREEN ANALYSIS

Fineness

Passing	100	U.S. Mesh	100.00%
Passing	200	U.S. Mesh	99.78%
Passing	300	U.S. Mesh	99.64%
Passing	325	U.S. Mesh	99.50%

60% finer than 20 microns
20% finer than 14 microns
8% finer than 10 microns

Source: PB 198 230

The lime hydrate used in these experiments is a dolomitic hydrate. A typical screen analysis for this superfine lime hydrate is presented in Table 13.8. This type of lime hydrate was selected because prior experimentation has shown it to be suitable for carbonate bonding.

Carbonate Bonding Technique and Apparatus: The preparation of coal refuse for carbonate bonding consisted of mixing in sufficient water to bring the water content to a specified value. This mixture was allowed to stand at least 24 hours to permit an equilibrium of the water throughout the mixture. Lime hydrate was then thoroughly blended into the coal refuse. To permit a rapid scanning of parameters and to minimize the quantities of materials and size of the carbonate bonding apparatus, the lime hydrate coal refuse mixture was placed in a small die and compacted to yield a 1⅛ inch diameter by 1 inch high briquette. A compaction load of 6,000 psi was used to form the briquettes of these experiments.

In the commercial application of the carbonate bonding process lower compaction loads would be used in the order of 1,000 psi. It was decided to use a higher compaction load for experimental purposes so that the briquettes would not have loose edges and would be of exact dimensions. The briquetting pressure was then removed and the soft, easily crushed briquette was placed in a reactor and contacted with a carbon dioxide rich gas for a specified time. The briquette was removed and the compressive strength determined. In this manner, the compressive strength of a carbonate bonded coal refuse was determined for a particular coal refuse and particle size, lime hydrate and water content, carbon dioxide concentration and carbonate bonding time.

The carbonate bonding reactor is shown schematically in Figure 13.2. Tanks of carbon dioxide and nitrogen were used as the source of these gas constituents. Experiments were run with carbonate bonding gas containing 20, 60 and 100% carbon dioxide concentrations. The carbon dioxide and nitrogen were individually metered, combined in a mixing chamber and admitted to the carbonate bonding reactor.

The reactor consisted of a Plexiglas tube, 2 inches i.d. by 24 inches long. The formed briquettes were arranged on a metal grate which slid into the reactor tube. Six reactor tubes were used to permit segregation of the various coal refuse briquettes. For all experimentation, a gas flow of 0.55 scfm was used which resulted in a superficial gas velocity of 0.43 fps in each individual reactor tube.

The procedure used to carbonate bond the coal refuse briquettes consisted of placing the briquettes in the reactor tubes. The flow rates of nitrogen and carbon dioxide were then adjusted to the desired flows. During this initial adjustment the flow of the resulting gas mixture was made to bypass the reactor tubes. Once the desired gas mixture was obtained, a stop watch was started and the bypass closed causing the gas mixture to enter the reactor tubes. When the specified carbonate bonding reaction time was reached the carbon dioxide was cut off and the nitrogen allowed to flow through the reactor tubes for several additional minutes to flush out all remaining reaction gas. The carbonate bonded coal refuse briquettes thus formed were removed and either tested for their compressive strength or subjected to other tests.

FIGURE 13.2: CARBONATE BONDING REACTOR SYSTEM

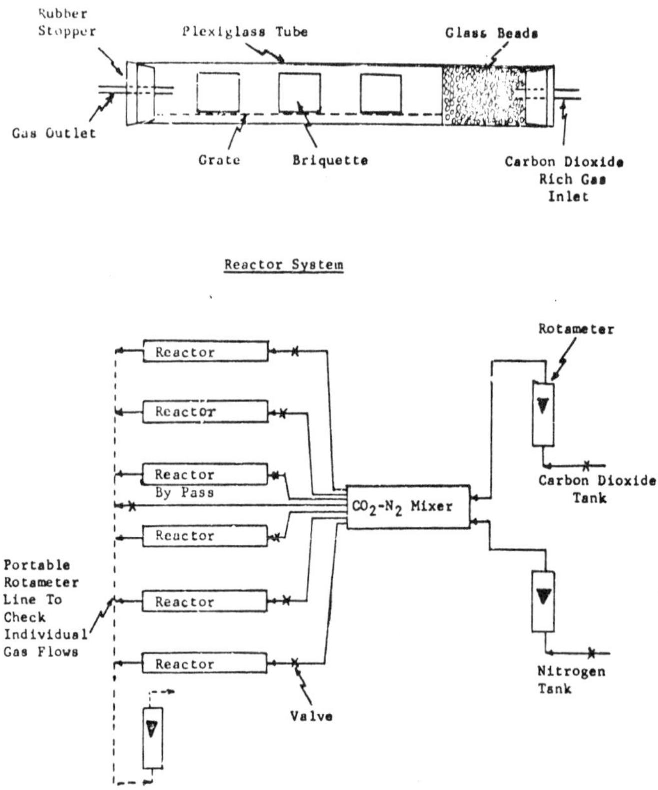

Source: PB 198 230

Experimental Laboratory Study

For convenience in presentation, the experimental laboratory study was divided into two categories—those experiments which pertain to the effect of various carbonate bonding parameters on the compressive strength of carbonated coal refuse and those experiments which were used to determine the physico-chemical properties of the resulting carbonate bonded coal refuse. In the first category, the effects of coal refuse type and particle size, lime hydrate and water content, carbonate bonding reaction time, and carbon dioxide concentration, on the compressive strength of the carbonate bonded coal refuse were ascertained.

In addition, the effect of curing time on compressive strength will be discussed. The curing time is defined as the total time elapsed after carbonate bonding is officially completed (removal from reactor). During this study it was observed that the compressive strength increased significantly for some time after carbonation. Included in the second category of experiments were air and water permeability and freeze-thaw testing of carbonate bonded coal refuse.

Effect of Carbonate Bonding Parameters on Compressive Strength: The structural integrity of carbonate bonded coal refuse is a prime consideration of any of its possible applications. Since the process is being considered for the sealing of coal refuse piles, the sealing layer must have sufficient strength to withstand vehicular and pedestrian traffic. Consequently, in this short feasibility study it was expedient to select one strength property as representative of the structural integrity of the resulting carbonate bonding material.

For this study compressive strength was selected because of its applicability toward the intended use of the carbonate bonding process and because of its ease of determination. The determination of the compressive strength of a solid material is normally accomplished under conditions in which the rate of application of load is precisely controlled. In view of the limited objectives of this study and the numerous uncontrollable parameters associated with carbonate bonding, this degree of precision was deemed unnecessary. Consequently, a simple technique was devised to determine compressive strength.

The briquette was placed in a press and the press plates closed until a small load resulted on the briquette. The press operator then applied a slow but approximately reproducible rate of loading on the briquette until failure occurred. The load at failure was recorded and used to calculate the compressive strength of the briquette.

Effect of Lime Hydrate Content on Compressive Strength: To determine the effect of lime hydrate content on the compressive strength of the resulting carbonate bonded briquettes, a series of carbonate bonded coal refuse briquettes was prepared in which the lime hydrate content was varied between 3 and 12% by weight and the water content varied in the range of 3 to 12% by weight. From prior experimentation with other fill materials (soils, ores, etc.) it has been found that optimum carbonate bonding strength normally occurs in these ranges of lime hydrate and water.

The percent of water was based on the original dry weight of the coal refuse (and thus corresponds to the moisture that will be measured in exposed refuse piles) whereas the percent of lime hydrate was based on the weight of wet coal refuse. The compressive strength was determined immediately after carbonate bonding was completed (zero curing time). Results clearly indicate that the compressive strength increased as the percent of lime hydrate increased for each of the refuse materials studied.

Effect of Curing Time on Compressive Strength: The maximum compressive strength obtained for each coal refuse was about 1,500 psi at 12% lime hydrate, and maximum strengths occurred at 12, 12, 9 and 12% water for unburnt and burnt bituminous and unburnt and burnt anthracite coal refuse respectively. The resulting compressive strengths were rather low judging from prior experimental results using other fill materials for carbonate bonding which yielded compressive strengths in excess of 2,000 psi. Consequently, additional experiments were conducted to determine if there is a curing time effect on compressive strength.

To show the effect of curing time on compressive strength, carbonate bonded briquettes were prepared in the usual manner except that the compressive strength was determined as a function of the time elapsed after carbonate bonding was completed. This was done for each coal refuse at 12% lime hydrate and the optimum percent of water. For these experiments the –10 mesh particle size fraction, a 100% carbon dioxide gas mixture and a 4-hour carbonate bonding time were used. Compressive strengths were determined at various times after carbonate bonding was completed for a total time of 600 hours. The compressive strength increases significantly during the first 150 hours of curing time and then levels off to a relatively constant value.

Specifically, for the unburnt bituminous coal refuse, the compressive strength increased from about 1,500 psi to 4,400 psi after a curing time of approximately 150 hours. In order to make these findings applicable to a wide variety of coal refuses, Figure 13.3 was prepared and is a plot of the reduced compressive strength, defined by $(S_t - S_0)/(S_{00} - S_0)$, as a function of the curing time in hours: where,

S_t = compressive strength at any time

S_0 = compressive strength at time zero (immediately after carbonate bonding)

S_{00} = final compressive strength at times beyond 150 hours

FIGURE 13.3: REDUCED COMPRESSIVE STRENGTH vs CURING TIME

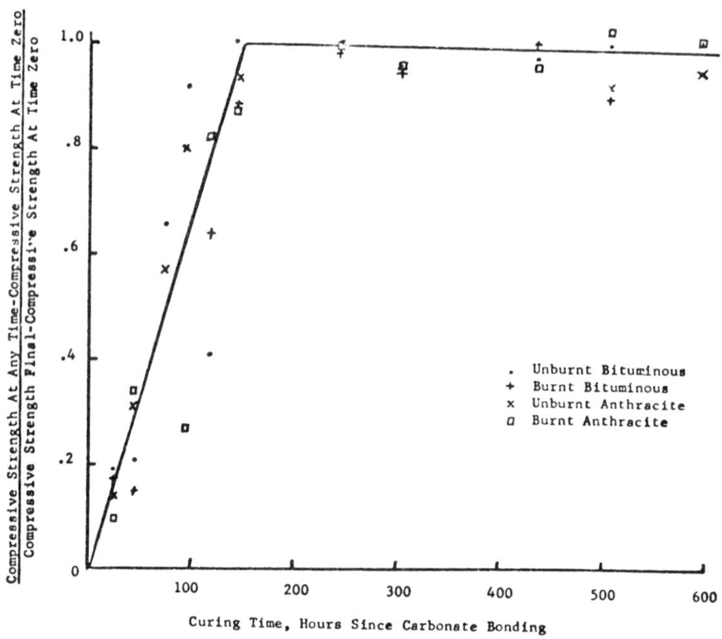

Source: PB 198 230

This figure indicates that a curing time of approximately 150 hours is required to develop the ultimate compressive strength. It is believed that a recrystallization of the calcium carbonate crystal matrix is occurring to yield a stronger structure. Based on this finding, compressive strength determination for all subsequent experimental investigations was performed on briquettes which were cured at least 150 hours.

Effect of Water Content on Compressive Strength: The effect of coal refuse particle size and water content on carbonate bonded compressive strength was determined. In these experiments the coal refuse contained 12% lime hydrate which was reacted with pure CO_2 for eight hours. Considering only the results for the -10 mesh material, the maximum compressive strength results at the higher water content. Specifically, the optimum water content is 12% for all coal refuses except the burnt bituminous which increased to the maximum percent of water investigated (15%).

Based on these limited results, it appears that coal refuse will be amenable to carbonate bonding at 12 to 15% water. The results show a gradual variation in compressive strength with changes in water content. These results are encouraging since it appears that quite usable compressive strengths are obtained without resorting to unrealistic water content controls which would be a decided disadvantage to the large scale application of this process.

It is also advantageous that the observed optimum water contents occur at relatively high values, because in the commercial application of the process it is more practical to add water rather than dry the coal refuse. Of course, lime rather than lime hydrate could be used if lower water content levels were required. Lime would reduce the free moisture by reacting with water to form lime hydrate.

The unburnt bituminous coal refuse exhibited a compressive strength-water content relationship which was unexpected. The compressive strength did not exhibit its usual functional dependence (i.e., increase as particle size decreases), but the optimum moisture varied in a somewhat random manner. Specifically, the optimum water content for maximum compressive strength was 12, 9 and 12% for the -10, -20 and -60 mesh particle size fractions respectively.

Effect of Particle Size on Compressive Strength: Again excluding the unexplainable results for unburnt bituminous, results show that the compressive strength increases as the particle size of the coal refuse decreases. This is expected since a decrease in particle size (along with a wider size distribution) normally results in a more dense composite thus enhancing the formation of a strong continuous calcite crystal matrix. Additionally, the higher surface area permits more calcite-to-refuse contacts for cementing and holding the materials together.

The smallest economically feasible particle size coal refuse should be used in the commercial application of the process to obtain maximum strength (provided the composite material is not so dense as to retard the transport of carbon dioxide into the material, and to prevent the venting of steam formed by the highly exothermic nature of the carbonate bonding reaction). If steam is prevented from rapidly exiting the carbonate bonded material, the internal pressure will build up and cause cracking with a resulting loss in strength. The rate of the carbonate bonding reaction can be controlled by varying the partial pressure of the carbon dioxide in the gas.

Effect of Carbon Dioxide Concentration and Carbonate Bonding Reaction Time on Compressive Strength: To determine the effect of carbon dioxide concentration and carbonate bonding reaction time on compressive strength, three separate experiments were conducted using a 20, 60 and 100% carbon dioxide gaseous mixture. Briquettes (12% lime hydrate, 12% H_2O) were removed from the reaction tubes after 2, 5 and 8 hours for each of the above gaseous mixtures.

The results of these experiments for the –10 mesh coal refuse show that, as is generally the rule, the compressive strength increases as the carbon dioxide concentration increases for a given carbonate bonding reaction time. This occurs because the extent of carbonation increases with carbon dioxide concentration. Both the 60% and 100% carbon dioxide gaseous mixtures yield higher compressive strengths than the 20% gaseous mixture. However, for long carbonation times (less than or equal to 8 hours) compressive strengths tend to converge to a common value. This is to be expected since the ultimate strength is a strong function of the degree of carbonation.

For burnt bituminous and unburnt anthracite the 60% CO_2 gaseous mixture yields a higher compressive strength than the 100% CO_2 gaseous mixture. This is probably caused by the carbonate bonding reaction occurring too rapidly at the higher concentration. Steam generation was too rapid causing a multitude of hair line fractures which have reduced the compressive strength.

The effect of carbonate bonding reaction time is as expected—the compressive strength increased in each case as the time increased. More important is the observation that for an 8-hour carbonate bonding time the compressive strength is not significantly different at the various carbon dioxide concentrations. For example, the compressive strength of unburnt bituminous after 8 hours of carbonate bonding varies between 4,200 and 4,350 psi for carbon dioxide concentrations of 20, 60 and 100%.

Thus, compressive strength (obtained under optimum conditions) is a strong function of the degree, or extent, of the completion of the carbonate bonding reaction. This result is encouraging because in the commercial application of the process it is economically more feasible to utilize a 20% carbon dioxide gaseous mixture derived from the combustion of a carbonaceous fuel rather than expensive pure CO_2.

Additional Properties of Carbonate Bonded Coal Refuse: In addition to strength, the resulting carbonate bonded coal refuse must possess other requisite properties to produce suitable construction materials. The carbonate bonded product must be able to prevent or minimize the permeation of water and air (oxygen) to inhibit acid mine water formation and must also possess a resistance to weathering. If carbonate bonded coal refuse cannot maintain its structural integrity during repeated freezing and thawing, then it cannot be an effective construction material.

The properties that were investigated and presented in this section are: the air and water permeability of coal refuse and the compressive strength after repeated freeze-thaw cycles. A search of the ASTM standards did not yield an applicable standard determination for air and water permeability; therefore, special equipment and procedures were established for these determinations which were based on the commonly accepted theory of permeability. Where applicable, cement briquettes were used as a basis for comparison. These cement briquettes were prepared using 1 part (by weight) Portland Type One cement, 2.75 parts Ottawa Sand and 0.5 part water. The cement and sand were dry mixed, the water added slowly, and the wet mixture blended thoroughly. The cement briquettes were allowed to cure 28 days.

Air Permeability of Carbonate Bonded Coal Refuse: The air permeability of carbonate bonded coal refuse briquettes was determined using the apparatus developed. The apparatus consisted of a 1¼ inch i.d. by 24 inch high Plexiglas chamber on the bottom of which was cemented a coal refuse briquette. The top of the chamber was sealed with a Plexiglas plate containing a pressure gauge. The air permeability was defined as the volumetric flow rate of air (ft^3/hr) per cross-sectional flow area (ft^2) per unit pressure driving force ($lb/ft^2/ft$).

The chamber was pressurized with gas and the change in chamber pressure with time was measured to determine the permeability of the coal refuse briquette. Initial chamber pressures of 30, 15 and 5 psig were used. For an experimental run the chamber pressure was allowed to decrease to a pressure of 1 psig before termination. An average air permeability was then determined using the results of all the experiments.

In this experiment the air permeability for -10, -20 and -60 mesh unburnt bituminous coal refuse briquettes was determined. For comparative purposes, experiments were also run on a cement briquette of the same dimensions as the coal refuse briquettes. The results are presented in Table 13.9 where the air permeability, the ratio of the permeability of the coal refuse briquettes to the cement briquette, and the average particle diameter of the coal refuse are presented.

TABLE 13.9: AIR PERMEABILITY OF UNBURNT BITUMINOUS COAL REFUSE

Sample	Air Permeability	Permeability of Coal Refuse/Permeability of Cement	Average Particle Diameter, cm
Unburnt bituminous -10 mesh	2.7×10^{-4}	9	0.046
Unburnt bituminous -20 mesh	6.0×10^{-5}	2	0.013
Unburnt bituminous -60 mesh	2.4×10^{-5}	0.8	0.008
Cement	3.0×10^{-5}	-	-

Source: PB 198 230

It is seen for the -10 mesh briquettes that the permeability ratio is 9 times that of the cement briquette whereas the ratio is 2 and 0.8 for the -20 and -60 mesh coal refuse. This infers that the smaller particle size fractions (-20 mesh or lower) carbonate bonded coal refuse compare quite well with cement and would provide an adequate seal against air permeation. In order to make the air permeability results generally applicable to all types of coal refuses, Figure 13.4 was prepared. Figure 13.4 is a plot of the air permeability for unburnt bituminous coal refuse vs the average particle diameter and can be used to estimate the permeability of a coal refuse provided the particle size distribution is known.

FIGURE 13.4: AIR PERMEABILITY vs AVERAGE PARTICLE DIAMETER

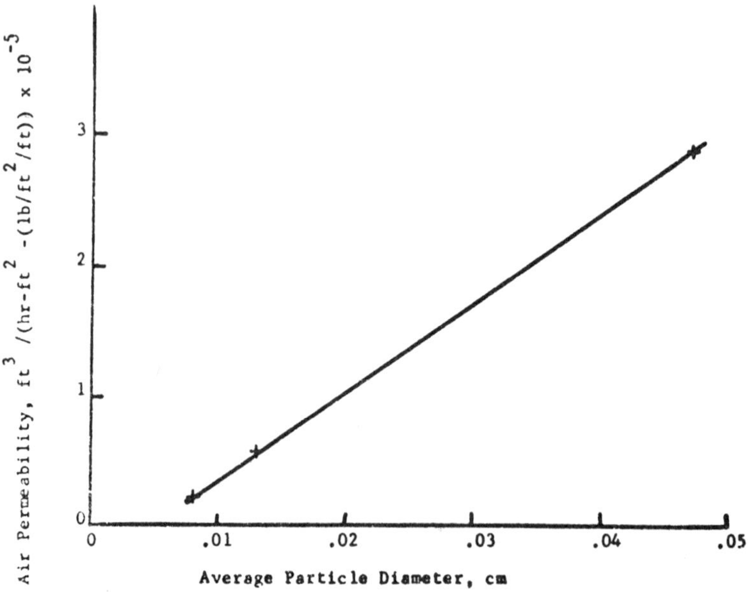

Source: PB 198 230

Water Permeability of Carbonate Bonded Coal Refuse: The water permeability of coal refuse was determined in much the same manner as air permeability. The apparatus for the water permeability determination consisted of a 1¼ inch i.d. by 24 inch high Plexiglas primary chamber sealed on top with a Plexiglas plate containing a valve and on the bottom with a coal refuse briquette. For this test, an additional chamber is attached to the primary chamber and located beneath the briquette to collect the water permeating through the briquette and to prevent unrealistic permeation rates due to convective air flows about the exposed surface of the briquette.

In addition, water permeating through carbonate bonded coal refuse will probably enter into surroundings which will normally be saturated with water vapor. The lower chamber, therefore, provides a water vapor saturated atmosphere to better simulate actual conditions and provide more meaningful results. The lower chamber contains a small hole on its side to vent evaporated water and prevent a build up of pressure.

The water permeability test was conducted by initially filling the primary chamber with water to a level of 20 inches and, on a daily basis, measuring the amount of water lost due to permeation. After the quantity of water loss was determined, this amount was added back into the primary chamber to maintain a constant pressure driving force.

This test was run for 14 days and an average rate of water permeation determined. For all practical purposes this test was run on a steady state basis in which the water pressure driving force was constant; consequently, the water permeability of coal refuse can be directly calculated since the flow area of the briquette, the thickness of the briquette, the constant water pressure driving force, and the average rate of water permeation are known.

The water permeability of -10, -20 and -60 mesh unburnt bituminous coal refuse and a cement briquette is shown in Table 13.10.

TABLE 13.10: WATER PERMEABILITY OF COAL REFUSE

Sample	Permeability*	Permeability Ratio**
Unburnt bituminous, -10 mesh	5.5×10^{-9}	4.0
Unburnt bituminous, -20 mesh	2.5×10^{-9}	1.8
Unburnt bituminous, -60 mesh	2.5×10^{-9}	1.8
Cement	1.3×10^{-9}	

* Water permeability in lb/[hr-ft^2-(lb/ft^2/ft)]
** Permeability of coal refuse/permeability of cement

Source: PB 198 230

Water permeability is the mass flow rate of water per cross-sectional flow area per unit pressure driving force. Also shown in Table 13.10 is the water permeability ratio defined as water permeability of coal refuse divided by the water permeability of cement. It is interesting to note that carbonate bonded coal refuse compares quite well with cement in prevention of water permeation. This is particularly so at the lower particle size fractions of -20 and -60 mesh where the permeability ratio is 1.8. Results indicate that carbonate bonded coal refuse would provide an adequate seal against water permeation and minimize acid mine water formation.

Freeze-Thaw Test on Carbonate Bonded Coal Refuse: Freeze-thaw tests were performed on a series of -10, -20 and -60 mesh unburnt bituminous coal refuse briquettes. The freeze-thaw test was based on a modification of ASTM C290-67. The objective of this test was to determine compressive strength of coal refuse briquettes as a function of the number of freeze-thaw cycles to which they had been subjected. The briquette compressive strength as a function of the number of freeze-thaw cycles for -10, -20 and -60 mesh unburnt bituminous coal refuse was determined. The compressive strength in the 20 to 35 freeze-thaw cycle range tends to stabilize at a relatively constant value of 700 to 900 psi for each of the particle size fractions.

It must be noted that subjecting a composite material of small dimensions (1⅛ inch diameter by 1 inch high briquettes) to 35 freeze-thaw cycles is an extreme test. The ASTM standard test method contemplates the use of specimens not less than 3 inches nor more than 5 inches in width and depth and not less than 14 inches nor more than 16 inches in length. The normal erosion of surface edges which occurs during freeze-thaw testing does not seriously effect large specimens, whereas for small briquettes the effect on strength is quite dramatic as evidenced by the results of this study. An examination of the coal refuse briquettes after 35 cycles revealed an erosion of about ¼ inch of material almost completely around all edges. This reduced the area on which the compressive strength test load was applied causing the briquette to fail at lower loads and indicate a lower compressive strength based on the original one square inch area.

For comparison, cement briquettes were also subjected to freeze-thaw testing. The cement briquettes had an initial compressive strength of 2,000 psi which reduced to 1,200 psi after 35 freeze-thaw cycles. An examination of the cement briquettes after 35 cycles revealed an erosion of about ⅛ inch of material almost completely around all edges.

Even under extremely unfavorable test conditions, the carbonate bonded coal refuse yielded compressive strengths in the order of 700-900 psi. Consequently, it should adequately withstand the effects of severe changes in temperature.

Operating Parameters for Carbonate Bonding Process: The economics of coal refuse carbonate bonding depend primarily on four controllable factors: the particle size of the coal refuse, quantity of lime hydrate required, carbon dioxide concentration and time of carbonation.

In this study, particle size effects were evaluated over a rather narrow range limited by the size of the briquette die employed. However, based on the results obtained, it is reasonable to expect that the carbonate bonding process for the conversion of coal refuse to paving materials will accommodate fill material containing a top size of one inch. To maintain air and water permeability conditions as developed in this report, it will only be necessary to specify a particle size distribution so that minimum porosity (i.e., maximum density) will be achieved. Data are available in literature that will permit an estimate of the proper size consist to use in commercial applications.

Depending upon strength criteria established for commercial applications, this study has shown that a maximum of 12% lime hydrate will yield compressive strengths approaching 4,400 psi for coal refuse. Should lower compressive strength be acceptable, the amount of lime hydrate can be decreased accordingly.

It has been found that adequate carbonate bonded coal refuse strengths have been obtained when a 20% carbon dioxide gaseous mixture is used for reaction times of 8 hours. Assuming that faster set times are not required, a simple inexpensive carbon dioxide gas generator can be used to yield carbon dioxide concentrations of about 15 to 20%. Faster set times will require the use of carbon dioxide generators producing a higher purity gas.

When using coal refuse material, the data indicate that a moisture content of about 9 to 15% will be required to obtain optimum strength. This moisture requirement can readily be realized at the site.

Applications of Carbonate Bonded Coal Refuse

The applications of carbonate bonded coal refuse are numerous. The carbonate bonded process utilizing coal refuse could be employed to (1) construct secondary roads utilizing the formed carbonate coal refuse as is, (2) provide a strong inexpensive base material for major road construction, (3) provide a first class road when used in conjunction with a thin layer of tar or similar material, (4) provide a strong inexpensive sealant layer on existing coal refuse piles, and (5) make bricks. These applications, except (4) above, have the advantage that coal refuse can be worked off thus eliminating unsightly coal refuse piles. Also the current production of coal refuse could be utilized in the process to prevent the formation of new or the enlargement of old coal refuse piles.

Although no work has been completed on the field testing of the carbonate bonding process employing coal refuse or spoil bank material, the bench scale study and earlier experimentation (on a 4-foot square patch of carbonate bonded soil) show that the following proposed commercial concept is feasible for both sealing coal refuse piles and constructing roads.

The coal refuse or spoil bank material is tilled to a predetermined depth, depending on the particular application. During the tilling operation, lime hydrate and water (if necessary) in the proper proportions, are introduced into the tilled material. The tilling action not only loosens the fill, but is also effective for blending and making a homogeneous mixture of lime and fill material.

Once the tilling and blending step has been completed, perforated plastic pipe is placed on four-foot centers underneath the surface of the blended material. The perforated plastic pipe is expendable and remains in place under the carbonate bonded material. The plastic pipe is connected to a portable carbon dioxide generator which can be a burner for coal refuse, coal, oil or any other carbonaceous fuel which generates products of combustion containing carbon dioxide. Once the plastic pipe is in place and connected to the generator, the blended material is then tamped down by means of a roller. The flue gases are pumped under pressure (approximately 5 lbs/in^2) into the perforated plastic pipe.

At this time the portable carbon dioxide generator is started and gaseous carbon dioxide along with the inert components of the flue gas are pumped into the perforated pipe. As the carbon dioxide permeates upward through the lime hydrate-fill material mixture, the carbonate bonding reaction starts and hardens the blend. The conversion of the lime hydrate to the carbonate form occurs with the visible evolution of steam. When the steam has stopped emanating from the pavement, the reaction is completed and the spoil bank or coal refuse material will be surface hardened with a concrete-like covering.

Cost analysis, which is included in PB 198 230, indicates that substantial economies can be realized if the carbonate bonding process is utilized as a refuse pile sealant or road construction technique. This should induce coal producers to cover or utilize their coal refuse, thereby minimizing the formation of acid water.

BIBLIOGRAPHY

The following reports used in the preparation of this book are available from

National Technical Information Service
U.S. Department of Commerce
5285 Port Royal Road
Springfield, Virginia 22151

PB 198 230 — LaRosa, P.J., Karnavas, J.A. and Pelczarski, E.A., *Carbonate Bonding of Coal Refuse,* 1971.

PB 198 258 — Andreuzzi, F.C., *A Method for Extinguishing and Removing Burning Coal Refuse Banks,* 1970.

PB 198 427 — Tyco Laboratories, Inc., *Silicate Treatment for Acid Mine Drainage Prevention, Silicate and Alumina/Silica Gel Treatment of Coal Refuse for the Prevention of Acid Mine Drainage,* 1971.

PB 208 820 — Jasinski, R. and Gaines, L., *Electrochemical Treatment of Acid Mine Waters,* 1972.

PB 222 936 — Pearson, F.H. and Nesbitt, J.B., *Combined Treatment of Municipal Wastewater and Acid Mine Drainage,* 1972.

PB 226 725 — Staff, Bureau of Mines, *Methods and Costs of Coal Refuse Disposal and Reclamation,* 1973.

PB 227 734 — Intorre, B.J., Kaup, E.G., Hartman, J.L., Feiler, H.D. and Szostak, R.M., *Evaluation of Ion Exchange Processes for Treatment of Mine Drainage Waters,* 1974.

PB 228 868 — Ciolkosz, E.J., Kardos, L.T., Beers, W.F., *Soil as a Medium for the Renovation of Acid Mine Drainage,* 1973.

PB 229 649 — Unz, R.F. and Lieberman, M.T., *The Microbiology of Acid Mine Water Treatment in Packed Bed Columns,* 1973.

PB 230 756 — Gupta, M.K., *Development of a High Product Water Recovery System for the Treatment of Acid Mine Drainage by Reverse Osmosis,* 1974.

PB 232 764 — Franco, N.B. and Balouskus, R.A., *Electrochemical Removal of Heavy Metals from Acid Mine Drainage,* 1974.

PB 241 402 — Warner, D.L., *Rationale and Methodology for Monitoring Groundwater Polluted by Mining Activities,* 1974.

PB 245 130 — A Report to the National Science Foundation by the Study Committer to Assess the Feasibility of Returning Underground Coal Mine Wastes to the Mined-Out Areas, Environmental Studies Board/Board on Energy Studies, National Academy of Sciences/National Academy of Engineering, *Underground Disposal of Coal Mine Wastes,* 1975.

The following reports used in the preparation of this book are available from

Superintendent of Documents
U.S. Government Printing Office
Washington, D.C. 20402

14010 DDH — Truax-Traer Coal Company, *Control of Mine Drainage from Coal Mine Mineral Wastes, Phase I, Hydrology and Related Experiments,* 1971.

EPA-430/9-73-011 — U.S. Environmental Protection Agency, *Processes, Procedures and Methods to Control Pollution from Mining Activities,* 1973.

EPA-670/2-73-081 — Penrose, R.G., Jr. and Holubec, I., *Laboratory Study of Self-Sealing Limestone Plugs for Mine Openings,* 1973.

EPA-670/2-73-092 — Stoddard, C.K., *Abatement of Mine Drainage Pollution by Underground Precipitation,* 1973.

EPA-670/2-73-093 — Lovell, H.L., *An Appraisal of Neutralization Processes to Treat Coal Mine Drainage,* 1973.

EPA-670/2-73-100 — Wilmoth, R.C., *Applications of Reverse Osmosis to Acid Mine Drainage Treatment,* 1973.

The following U.S. patents used in the preparation of this book are available from

Commisioner of Patents and Trademarks
Washington, D.C. 20231

U.S. Patent 3,516,931	Birch, J.J.; June 23, 1970; assigned to Barnes & Tucker Company
U.S. Patent 3,617,559	Cywin, A.; November 2, 1971; assigned to U.S. Secretary of the Interior
U.S. Patent 3,617,560	Deul, M. and Mihok, E.A.; November 2, 1971; assigned to U.S. Secretary of the Interior
U.S. Patent 3,617,562	Cywin, A. and Mihok, E.A.; November 2, 1971; assigned to U.S. Secretary of the Interior
U.S. Patent 3,725,259	DePree, D.O.; April 3, 1973; assigned to Aerojet-General Corporation
U.S. Patent 3,795,609	Hill, R.D., Wilmoth, R.C. and Scott, R.B.; March 5, 1974; assigned to Administrator of the Environmental Protection Agency
U.S. Patent 3,823,081	Treharne, R.W. and Wright, D.E.; July 9, 1974; assigned to Kettering Scientific Research, Inc.

STRIP MINING OF COAL
Environmental Solutions 1976

by William S. Doyle

Pollution Technology Review No. 27

Late in 1974 the Office of Research & Development of the U.S. Environmental Protection Agency made the statistically well founded projection that for the remainder of the 20th century surface-mined coal will have to account for over 50% of our nation's production of this fuel.

Strip mining can be done responsibly without permanent damage to land and water. Technology exists for effective reclamation of mined lands, and such reclamation is being performed in some areas. Authorities emphasize the importance of planned premining (fully discussed in this book) and point out that reclamation is less costly and more effective when integrated with the mining operation.

This book, based on 19 government reports issued from 1967 through 1974, describes surface mining of coal, land use and methods, land reclamation technology plus sediment and erosion control. Acid mine drainage, its sources, prevention and correction, as well as the mechanism of reclaiming acid strip mine lakes are discussed. Specific studies on revegetation, use of spoil amendments etc. are included. A whole chapter is devoted to West Germany's approach to the problem of strip-mined lands. The final two chapters are devoted to costs, economics, and financing.

A partial and condensed table of contents follows here. Chapter headings are given, followed by examples of important subtitles.

ISBN 0-8155-0611-2 **352 pages**

OIL FROM COAL 1975

by Francis W. Richardson

Chemical Technology Review No. 53

The extensive coal deposits within the continental U.S. and other countries represent a vast, largely untapped energy reserve. The technology for converting coal into gas and liquid fuels continues to proliferate with increased emphasis being placed on liquefaction. The research efforts of the last decade are now being expanded to include pilot and large scale studies for obtaining liquid hydrocarbons via coal technology.

For liquefaction the energy conversion factor is higher than for gasification, and the products are more easily stored and transported from production site to point of use. Earlier research tried to make a crude oil that could be processed into gasoline by conventional refining techniques. While this remains a primary objective, increasing attention is being given to making a low ash, low sulfur oil, that can be burned as fuel, thus freeing natural crude oil and gas for processing into gasoline.

This book presents the operating details of many coal liquefaction processes which are now undergoing extensive field operations here and abroad. Over 170 processes, as described in the U.S. patent literature, are covered.

A partial and very condensed table of contents follows here. Numbers in parentheses indicate the number of processes per topic or chapter. Chapter headings and some of the more important subtitles are given here.

ISBN 0-8155-0588-4 **387 pages**

COAL CONVERSION TECHNOLOGY 1976

by I. Howard-Smith and G. J. Werner

Chemical Process Technology Review No. 66

The industrialized countries of the world are showing renewed and increasing activity in the area of synthetic fuels production from coal. Existing commercial coal conversion processes, such as the Fischer-Tropsch synthesis, the destructive distillation of coal, methanation etc. resulting in liquid and gaseous fuels, are again becoming economically competitive with petroleum.

The many processes and techniques of coal conversion have as a basic concept the transmutation of coal into forms acceptable to our transportation and heating equipments. To accomplish this, high-sulfur coals must be desulfurized, high-ash coals must be demineralized, and, most important of all, solid coal must be depolymerized into liquid and gaseous products that can be ignited and burned with facility.

Over 100 processes, in various stages of development, are available to carry out these procedures and are discussed fully in this book.

Originally prepared as an "in-house" report for the Millmerran Coal Pty. Ltd. of Brisbane, Australia, it is intended to provide readily accessible and concise data on all major activities in coal conversion technology. A special feature are the flow charts which clearly illustrate all the principal processes. A partial and very condensed table of contents follows here. Chapter headings are given, and some of the more important subtitles are also included.

GENERAL PROCESSES (36)

Bergius
Catalytic Coal Liquefaction
Clean Coke Process
COALCON Hydrocarbonization
COALPLEX
COED
COG (Coal-Oil-Gas Refinery)
COGAS
CONSOL Synthetic Fuel
COSTEAM
Extractive Coking Process
EXXON Liquefaction
Fischer-Tropsch Synthesis
GARRETT Coal Pyrolysis
H-Coal
Intermediate Hydrogenation
LEAS Liquefaction
LURGI-RUHRGAS
Methanol Synthesis
Molten Salt Catalysis
Project Lignite—SRL
Projects Seacoal & Seacoke
Solvent Extraction: CRE & UOP
SRC (Solvent Refined Coal) Processes
COMBUSTION ENGINEERING
MITSUI
PAMCO
SYNTHOIL
TOSCOAL
Underground Liquefaction
UNION CARBIDE

HIGH BTU COAL GASIFICATION (28)

Agglomerating Ash
AIR PRODUCTS Recycle Process
ATGAS/PATGAS
BIANCHI
BI-GAS
CHEVRON Gasification
CO_2 Acceptor
Electric Arc Gasification
Electrofluidic Gasification
GEGAS
HYDRANE
HYGAS
KOPPERS-TOTZEK
SYNTHANE
WINKLER

LOW & MEDIUM BTU GASIFICATION (48)

AVCO Arc Coal
BABCOCK & WILCOX
CONSOL Fixed Bed
KELLOGG Fixed Bed
ICI Moving Burden
In Situ Gasification
KERPELY PRODUCER
MARISCHKA
Molten Salt Process
OTTO RUMMEL Slag Bath
PANINDCO
PINTSCH HILLEBRAND
RUHRGAS VORTEX
THYSSEN GALOCSY
UGI Blue Water Gas

ISBN 0-8155-0614-7 **133 pages**

OIL FROM SHALE AND TAR SANDS 1975

by Edward M. Perrini

Chemical Technology Review No. 51

This book contains 213 process descriptions related to the retorting and refining of shale oil and the separation of oil from tar sands. Current events continue to point in favor of these crude materials as future sources of petroleum energy.

Oil shale is found in diverse regions of the world, but the richest and most extensive deposit discovered so far is in the Colorado-Wyoming-Utah triangle, where most of the land belongs to the U.S. Government.

Oil recovery from shale rock is admittedly difficult. With the *in situ* methods oil is obtained from shale under ground, by injecting hot gases and collecting the oil through holes in the rock. By alternate methods shale rock is mined and mixed with small preheated balls. Heat from the balls retorts the shale, and the oil vapors are condensed in a condensing chamber. While not in use in the U.S., various techniques using nuclear reactions have been devised.

Vast resources of petroleum oils are also found in tar sands. One of the largest of these reserves is the Athabasca deposit in Canada. Oil is obtained by mining the sands with huge shovels and applying aqueous separation techniques.

This book is a survey of the U.S. patent literature. It contains large, descriptive excerpts. A partial and condensed table of contents follows here. Numbers in parentheses indicate a plurality of processes per topic. Chapter headings and some of the more important subtitles are given.

ISBN 0-8155-0583-3 **307 pages**

UTILIZATION OF WASTE HEAT FROM POWER PLANTS 1974

by David Rimberg

Pollution Technology Review No. 14

Energy Technology Review No. 3

Present-day steam-driven turbo-electric power plants in the United States discharge as waste heat an amount of energy roughly equivalent to twice their total electricity-generating capacity. This energy is most difficult to utilize, because it is degraded in temperature. The effluent water is warm, but it is far from being near the boiling point. It constitutes a necessary, but unwanted, by-product of the energy conversion process for generating electricity.

The growing quantities of waste heat discharged, and the increasing ecological anxieties about undesirable growths, energy utilization and thermal discharge problems, have stimulated an examination of methods for productively using energy presently wasted on the environment.

Part I of this book discusses the reasons for present-day ineffective utilization, but it also shows ways and means to promote more efficient energy usage.

Part II assesses the cause, magnitude, and possible effects of heat discharges into water from steam electric power plants and related condenser cooling systems. Also contained in this section is a discussion of the thermodynamics of the electric power generation cycle detailing the reasons for the "inefficiencies" in by-product heat generation.

Ultimately all the accessible waste energy appears as low temperature heat, and the term "waste heat utilization" refers to the performance of useful functions with this heat before it is discharged into the environment. Within the framework of this book, the minimum temperature of the effluent water considered for subsequent use is about 38°C or 100°F. These uses are discussed in **Part III:** Food production in agriculture, hydroponics and aquaculture (fish farming) and the use of heat in wastewater treatment.

In the past the dissipation of waste heat was accomplished by wet and dry cooling towers and cooling ponds, lakes and streams. These methods are now being challenged by some sectors of society, and industry is being forced to consider their environmental impact. In this regard **Part IV** discusses the research needs necessary to equate the complicated interactions of the physical, engineering, biological, and social aspects of this waste heat problem.

This Pollution Technology Review is based on studies conducted by industrial and engineering firms or university research teams under the auspices of various governmental agencies, e.g. The National Water Commission, Oak Ridge National Laboratory Federal Water Pollution Control Administration, Office of Water Resources Research, Environmental Protection Agency, National Science Foundation, and the Department of Commerce.

A partial and condensed table of contents follows here:

ISBN 0-8155-0555-8 **175 pages**

POLLUTION CONTROL IN THE PETROLEUM INDUSTRY 1973

by H. R. Jones

Pollution Technology Review No. 4

The petroleum industry can logically be divided into three major divisions: (1) Production, (2) refining, (3) subdividing and distributing. Production includes the operations involved in locating and drilling oil fields, removing oil from the ground, pretreatment at the well site, and transporting the crude to the refinery. The extent of pretreatment depends on the type of crude oil, and usually comprises removal of gas and brine. Refining includes the operations necessary to convert the crude into saleable products. Subdividing and distribution involves handling the finished petroleum products.

All of these activities are or can be sources of emissions into the air and water. This book shows you practical technology for avoiding or reducing these harmful emissions, and how to clean up after unavoidable damage has been done. There is also a separate chapter on the costs and economics of pollution control.

Chapter headings are given here, followed by examples of important subtitles.

ISBN 0-8155-0494-2 322 pages

SOLAR ENERGY FOR HEATING AND COOLING OF BUILDINGS 1975

by Arthur R. Patton

Energy Technology Review No. 7

Solar energy can be used for indirect heating purposes in many ways. The information in this book has been limited to so-called low temperature solar thermal processes. Designs requiring photocells or other thermoelectric generators and lenses or reflecting mirrors plus tracking equipment have been excluded.

Low temperatures are the easiest to obtain, and the necessary collectors are fairly simple in construction. A black surface is used to absorb the sun's rays, this surface is usually covered with glass and the collector is insulated on the back and sides against heat loss. Water or some other heat transfer fluid is passed through the collector and can reach temperatures from 60°C (140°F) to about 95°C (203°F). The thermal energy is then stored in a heat storage system (perhaps based on the latent heat of fusion of selected salts). Coupled to the heat storage system are heating loops to furnish heat by convection and to operate an air conditioning system. In most temperate zones an auxiliary heater, operated with conventional fuels, must also be connected.

Large scale applications designed for schools and similar building are beginning to appear or are in the planning stage. This book describes in detail several large scale feasibility studies with designs suitable for institutions and industrial plants.

Descriptions are based on studies conducted by industrial or engineering firms or university research teams under the auspices of various government agencies. A partial and condensed table of contents follows here.

ISBN 0-8155-0579-5 **328 pages**

WIND POWER 1975

by Daniel M. Simmons

Energy Technology Review No. 6

Wind is a free, clean, inexhaustible source of energy. In previous centuries, when dependence on time and deadlines for production and shipments were not as important as they are now, wind served mankind well, despite becalmed and stormy intervals. Wind moved sailing ships of all sizes and drove windmills to grind seeds and other materials, saw wood and to pump water. In later years it was used to produce some electrical current of the DC type to charge storage batteries.

There are three phases to harnessing the power of the wind. The first is the selection of a suitably windy site. This necessitates studies of wind behavior in possible locales, including ocean sites. Secondly a suitably designed windmill and generator to convert the mechanical to electrical energy or to convert seawater to hydrogen and oxygen. The third and rather fundamental difficulty is the available choice of storage batteries or other storage means, so that energy produced at peak times can be used when the wind is not blowing.

The various phases for developing wind power are considered in the first three chapters of this book. The next three describe wind power developments in the U.S., Canada, U.S.S.R., Germany, Denmark, France, U.K., Sweden and other countries in Asia and Africa. The last chapter describes commercially available apparatus.

This Energy Technology Review is based primarily on international studies conducted by industrial and engineering firms or university research teams under the auspices of various governments and governmental agencies.

A partial and condensed table of contents follows here.

ISBN 0-8155-0575-2 **300 pages**

OCEAN FLOOR MINING 1975

by John S. Pearson

Ocean Technology Review No. 2

This book represents a sincere attempt to gather together and to collate the available information concerning the state of the art of deep sea mining. It is hoped that it may prove helpful to those interested in tapping these valuable ocean resources.

Although manganese nodules now occupy center stage on the undersea minerals scene, there are other promising prospects. Natural barium sulfate, for instance, which is practically insoluble in water, is needed in ever increasing quantities as an additive to drilling muds for oil wells. Promising deposits of this substance have been found off Alaska. Tin-bearing sands from the ocean floor are likely to prove commercially exploitable.

Ocean floor mining offers many advantages which are not possible with traditional land mining. As a whole new concept it lends itself readily to automation. The minerals are available without removing the so-called "overburden," without the use of explosives, and with very little drilling. Transportation by boat, so very desirable for mined ores, is already available without the now prohibitively expensive building of canals connecting to inland waterways.

A partial and condensed table of contents follows. Chapter headings and some of the more important subtitles are given here.

ISBN 0-8155-0569-8 **205 pages**